Clearance of Flight Control Laws: Optimal Control Approach

Clearance of Flight Control Laws: Optimal Control Approach

Joseph Z. Ben-Asher
Professor, Faculty of Aerospace Engineering, Technion Israel Institute of Technology

Johannes Diepolder
Research Associate, Institute of Flight System Dynamics, Technical University of Munich

Volume 269
Progress in Astronautics and Aeronautics

Seetha Raghavan, Editor-in-Chief
Embry-Riddle Aeronautical University
Daytona Beach, Florida

AIAA
American Institute of Aeronautics and Astronautics, Inc.

American Institute of Aeronautics and Astronautics, Inc.
12700 Sunrise Valley Drive, Suite 200, Reston, VA 20191-5807

1 2 3 4 5

ISBN 978-1-62410-734-4
Ebook ISBN 978-1-62410-735-1

TABLE OF CONTENTS

PREFACE

The majority of the content covered in this book is based on the methods and examples from [1]. Our primary motivation behind the publication of this material in the form of a book is to facilitate the adoption of these methods by domain experts facing practical clearance tasks.

This book is intended for practitioners. In particular, it is designed for flight dynamics and control engineers without a strong theoretical background in the fields of optimization and optimal control. Moreover, in addition to the rewritten/restructured content from [1], this book features an additional part ("Tools") specifically focusing on software solutions for the application of the presented methods. These tools include the Counter Optimization Library, COLIBRY, and the FSD Optimal Control Toolbox, FALCON.m. Both tools have been developed and published by the Institute of Flight System Dynamics at the Technical University of Munich*.

In industry practice, rigorous testing of technical systems is considered one of the most important development tasks. To provide an example from software engineering, it is a common conception that in enterprise application development (as a rule of thumb), one to three lines of test code are required for each line of production code [2]. Most technical systems nowadays feature a considerable portion of software and follow much the same development paradigms. Flight control systems are no exception here, and the criticality paired with the high complexity poses a major challenge regarding rigorous and exhaustive testing. One of the substantial challenges in this particular case is contributed to the dynamic nature of the system under test. Ultimately, the goal of aircraft control law clearance is to ensure the safety of the flight system, i.e., to show that it cannot be driven to unsafe states considering all possible parameter combinations as well as external inputs. Typical parameters in flight systems can be related to mass/mass distribution, aerodynamic properties, or reference flight conditions, to name a few. External inputs are, in the majority of cases, pilot commands or wind gusts. It should be clarified that the notions of "parameters" and "inputs" in this book describe two fundamentally different types of influence. Parameters are meant to model influences that remain constant over time. However, inputs (pilot commands or wind gusts) are allowed to be general functions of time. It is noted that, given the current state-of-the-art, inputs represent a considerably more challenging type of influence for testing dynamic systems.

In practice, the prevailing approach for performing parameterized tests is often based on the realization of a considerable number of simulations for different parameter combinations. Because of the large number and types of parameters, such an approach is very rarely comprehensive in covering all feasible cases. To stress the analogy with program testing once more, Dijkstra addresses this

*www.fsd.ed.tum.de/software

very problem in [3] within the section "On the Reliability of Mechanisms" from a general point of view. He illustrates the challenges of establishing correctness by means of a simple example (testing a "multiplier"):

But how is the correctness established in a convincing manner? As long as the multiplier is considered as a black box, the only thing we can do is "testing by sampling", i.e. offering to the multiplier a feasible amount of factor pairs and checking the result.... Whole classes of in some sense "critical" multiplications may remain untested and in view of the reliability justly desired, our quality control is still most unsatisfactory. Therefore it is not done that way. The straightforward conclusion is the following: a convincing demonstration of correctness being impossible as long as the mechanism is regarded as a black box, our only hope lies in not regarding the mechanism as a black box. I shall call this "taking the structure of the mechanism into account". [3, p. 4] The shortcomings of the testing strategies based on analyzing "as many samples as possible" assuming that the set of test samples provides reasonable coverage is aptly summarized by Dijkstra [3, p. 7] in the corollary: *Program testing can be used to show the presence of bugs, but never to show their absence.* An alternative approach that is not solely based on exhausting the search space using "testing by sampling" can be seen in the application of optimization methods for the worst-case search. The core idea here is not to merely check a large number of samples but to leverage the efficiency as well as effectiveness of optimization algorithms and search for the right test cases (i.e., test cases violating the criterion). The corresponding optimization problem in this context should express the goal to bring the system as close as possible to (or beyond) a certain threshold regarding the criterion under investigation. If an optimization method is capable of identifying a global solution over its admissible search space, this directly implies that there cannot exist a worse solution (in the sense of the criterion) than the one identified as a solution to the optimization problem. Most importantly, if such a solution does not violate the criterion, the system is cleared over the whole admissible search space.

This idea has been introduced in the last decades to the flight control law clearance task [4, 5]. In particular, for worst-case input analysis, the field has seen the successful application of various parameter optimization methods [6–18]. Under these approaches to worst-case input analysis, the dynamic nature of the system is essentially encapsulated/hidden in the objective function, meaning that the mapping from the parameters and the parameterized inputs to the criterion is considered directly without exposing the dynamic model to the optimization method. This allows the application of general parameter optimization algorithms for solving the respective clearance problems computationally. Almost exclusively, the optimization algorithms employed for these tasks are so-called global methods implying that there is a higher chance of identifying a global solution compared to "local" methods such as gradient-based approaches using Newton-type iterations. These global methods are usually based on some form of sampling strategy in order to improve the cost and have shown to be very effective for a large class of optimization problems. This is the case even if the objective function is noisy, which poses challenges for many algorithms outside this class. Moreover, these methods regard the objective function as a black box that reduces the complexity of the testing task as the intricacies of the system under test do not need to be treated directly by the optimization method.

Following this development, [19] approaches the aircraft worst-case input problem from an optimal control perspective. The essential novelty under this approach is the transition from finite-dimensional to infinite-dimensional optimization problems. The primary motivation for studying the problem within the infinite-dimensional problem class is that the problem of finding inputs (controls) that are worst-case (optimal) in the sense of the criterion is naturally expressed and studied using optimal control theory. In optimal control theory, the dynamic system is an essential part of the problem formulation. As such, this approach may be viewed as more of a "white box" approach in which we peek inside the black box and explicitly take the dynamic system into account by treating the clearance problem as a dynamic optimization problem.

It should be mentioned that the exposure of the dynamic system does not come without a cost. In fact, the analysis from an optimal control point of view can, in many cases, become quite challenging, and often the encapsulation of the complexity must be the way to go in practice. However, we are convinced that the optimal control-based methods suggested in this book can be seen as a valuable extension to the existing optimization-based clearance methods. In particular, a transfer of recent advances in this field from research to industry practice is thought to be beneficial for enhancing the safety level of flight control systems. After all, among other benefits, optimal control analysis enables us, for certain clearance formulations, to reveal the structure of the optimal solution based on the characteristics of the dynamic system, which can give additional insights into aircraft clearance problems or be used as sanity checks for numerical solutions. Last, but not least, the optimal control point of view not only allows us to stand on the shoulders of giants regarding the theoretical foundation, such as Pontryagin's Maximum Principle, but also establishes a link to the vast number of computational methods that have been specifically developed for this particular problem class in the last decades.

The research conducted at the Institute of Flight System Dynamics (FSD) at the Technical University of Munich (TUM) is foundational to the content in this book. We extend our thanks to the head of the institute, Prof. Dr.-Ing. Florian Holzapfel, for the insights, support, and resources he provided over the years. His ideas regarding application of the optimal control-based clearance methods described in this book constitute an essential part of this work. This also includes his role as supervisor of the dissertation [1] that forms the basis for large parts of this book. In this context, we also want to thank Prof. Dr.-Ing. habil. Christian Breitsamter from the Chair of Aerodynamics and Fluid Mechanics at TUM for his role as the chair of the examination committee. Last, but not least, special thanks go to all our colleagues and friends at the FSD office. The positive atmosphere and numerous discussions significantly contributed to this book. This also includes the former and active contributors/maintainers of the FSD toolboxes FALCON.m/COLIBRY and, in particular, David Braun for his assistance with the COLIBRY chapter.

Johannes Diepolder would like to thank his family, Luise and Wolfgang Diepolder, Sophie Stettner, and Elisabeth Diepolder, for their steady support. Moreover, he extends thanks to his coauthor for the guidance during the time as a doctoral candidate and thereafter; the kind encouragement and support throughout the periods at the TUM and Technion are deeply appreciated.

Joseph Z. Ben-Asher wishes to thank his wife Eti Ben-Ari for her continuous support throughout the sabbatical periods in Munich dedicated mostly to this research.

Joseph Z. Ben-Asher and Johannes Diepolder
July 2024

HOW TO READ THIS BOOK

The guiding intention behind the presentation of the material in this book is to provide domain experts with the required background for an appropriate application of the methods and interpretation of numerical results. The content is divided into three main parts:

I **Fundamentals**

II **Methods**

III **Tools**

These parts build on each other in the sense that the clearance methods depend on the underlying theoretical fundamentals, and clearance tools provide implementations of the clearance methods at a suitable level of abstraction.

In the first part, we introduce the building blocks of the direct discretization approach for optimal control problems: optimization (Chapter 1) and simulation (Chapter 2). Furthermore, Chapter 3 discusses optimal control problems in continuous time. The second part, which focuses on clearance analysis methods, contains the core topics of this book. The following types of analysis are covered for linear and linear-parametric systems (Chapter 4 and Chapter 5) as well as nonlinear-parametric systems (Chapter 6):

- *Nominal Analysis*: Are there worst-case inputs violating a clearance criterion?
- *Sensitivity Analysis*: What is the influence of parameters on (optimal) worst-case solutions?
- *Combined Analysis*: Is there a worst-case combination of parameters and inputs violating a clearance criterion?

Additionally, Chapter 7 covers set-based approaches regarding the joint consideration of multiple criteria for linear and nonlinear systems. The third part illustrates the use of software for the application of clearance methods in practice. First, we discuss the COLIBRY framework in Chapter 8, which can be used to implement the presented methods for linear and nonlinear systems. Additionally, the FALCON.m toolbox described in Chapter 9 is useful for modeling general nonlinear-parametric optimal control-based clearance problems using direct methods.

It should be mentioned that the progressive disclosure of complexity of the content in this book is not only a guiding principle for the sequence of the topics as a whole but as well for the detail of presentation within each topic.

Hence, we explicitly structure the material into three categories:

- **Essential**: This content is deemed vital to apply the presented methods.
- **Optional**: This content provides deeper insights into each of the topics. Despite its importance, it should be possible for the reader to follow the

essential content without going through this optional content at first read. Examples of this category are the majority of proofs and other flavors of the same topic (e.g., alternative methods/approaches).

- **Supplemental**: This includes all other content.

The "essential" category represents the default category within the main text body of the book. Within the linear format prescribed by a book, the "optional" category is realized as follows: Albeit important, this content is marked with a side-note "May be skipped at first read," including an embedded arrow icon.[†] Finally, all content regarding the last category ("supplemental") is collected in the appendix A–I.

Additionally, we provide a synopsis regarding the main takeaways at the beginning of sections in a "Precap" box. In this box, we summarize the most important information/essential content. These summaries are not necessarily self-contained and may as such reference other content within the book. In particular, only the top layer of content, "what," is presented; all layers below, "why" and "how," are left for the reader to explore in the respective sections.
The following mathematical notation is used throughout this book:

Special Symbols:

- $\mathbf{e}_i$ denotes a vector for which the ith component is equal to one, and all other components are equal to zero.
- $\mathbf{1}$ denotes a vector or matrix of appropriate size with all entries equal to one.
- $\mathbf{0}$ denotes a vector or matrix of appropriate size with all entries equal to zero.
- $\boldsymbol{\infty}$ denotes a vector or matrix of appropriate size with all entries equal to infinity.

Scalars, Vectors, and Matrices:

- *Scalar*: v (lowercase symbol) denotes a scalar $v \in \mathbb{R}$.
- *Vector*: $\mathbf{v}$ (bold, lowercase symbol) denotes a column vector $\mathbf{v} \in \mathbb{R}^n$ with n components v_i, $i = 1, \ldots, n$.
- *Matrix*: $\mathbf{M}$ (bold, uppercase symbol) denotes a matrix $\mathbf{M} \in \mathbb{R}^{n\times m}$ with n rows and m columns and n times m components $m_{i,j}$, $i = 1, \ldots, n$, $j = 1, \ldots, m$.

Derivatives:

- The derivative $\partial \mathbf{v}/\partial \mathbf{w}$ of a vector $\mathbf{v} \in \mathbb{R}^n$ with respect to a vector $\mathbf{w} \in \mathbb{R}^m$ yields a matrix with n rows and m columns.

[†] Clicking on this icon in a Portable Document Format (PDF) lets the reader jump over such sections.

- In the special case of the derivative $\partial v/\partial \mathbf{w}$ of a scalar $v \in \mathbb{R}$ with respect to the vector $\mathbf{w} \in \mathbb{R}^m$, the derivative yields a row vector with m components.
- In the special case of the derivative $\partial \mathbf{v}/\partial w$ of a vector $\mathbf{v} \in \mathbb{R}^n$ with respect to the scalar $w \in \mathbb{R}$, the derivative yields a column vector with n components.

Complex Numbers:

- $\Im[x]$ denotes the imaginary part of x.
- $\Re[x]$ denotes the real part of x.

Nullary Sums and Products:

- A sum in which the number of terms is zero is defined as the additive identity. For example, the sum $\sum_{i=1}^{j} x_i,\ x_i \in \mathbb{R}$ is defined as 0 for $j < i$.
- A product in which the number of factors is zero is defined as the multiplicative identity. For example, the product $\prod_{i=1}^{j} x_i,\ x_i \in \mathbb{R}$ is defined as 1 for $j < i$.

PART I

Fundamentals

Most of the state-of-the-art approaches to optimization-based flight control law clearance concentrate on "black-box" optimization methods. These approaches are based on the application of general parameter optimization methods to minimize/maximize a simulation output of the aircraft closed-loop system under test. In the context of optimal control, we identify this black-box approach as the "single shooting" method, a particular form of direct optimal control methods (see center in Fig. 1). This embedding into the world of optimal control allows us to apply the theoretical background and modeling techniques as well as leverage the numerous numerical methods available in this field.

From the available direct optimal control methods, single shooting is straightforward to implement using "off-the-shelf" simulation and optimization methods. "Unfortunately, the shooting method also suffers from one major disadvantage. In particular, a small change in the initial condition can produce a very large change in the final conditions" ([20], p. 95). This property can render a numerical solution challenging.

Besides single-shooting, there exist other direct optimal control approaches that have proven highly effective in practice. Especially, segmentation approaches applied in multiple-shooting or collocation methods appear to be beneficial from a numerical perspective. However, segmented collocation methods lead to larger problems with a higher number of optimization variables and constraints. Nonetheless, these problems are loosely coupled, which is a favorable property for certain types of numerical optimization methods.

Unfortunately, the majority of direct optimal control methods represent "leaky abstractions" when it comes to problems of realistic size in flight control law clearance applications. This implies that although optimal control frameworks have improved to provide user interfaces at a fairly high level of abstraction, it is still required, to a certain extent, to understand aspects regarding the underlying methods in order to successfully apply these methods in practice.

In particular, direct optimal control methods require two fundamental building blocks: optimization (Chapter 1) and simulation (Chapter 2). In this context, "simulation" refers to the type of discretization method used to convert the clearance optimal control problem in continuous time to a discretized version.

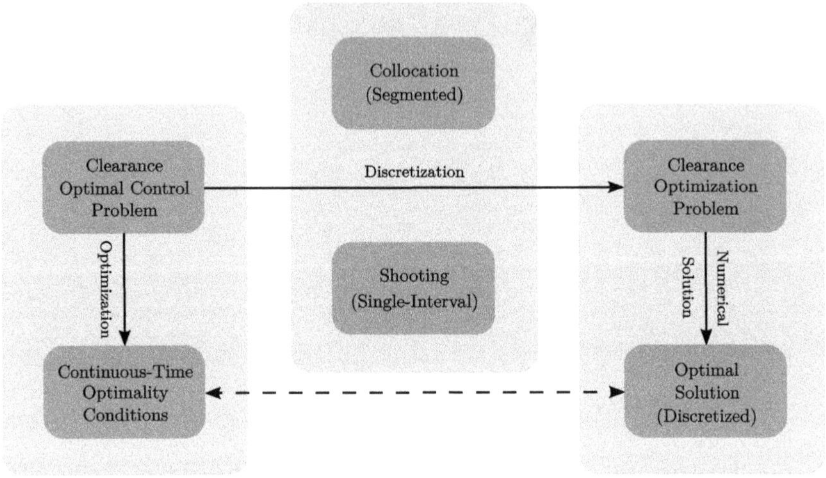

Fig. 1 Flight control law clearance based on optimal control methods.

Depending on the particular type of problem, a numerical solution of the discretized problem can then be obtained by a suitable optimization algorithm (see right side in Fig. 1). The optimization algorithms used for segmented collocation methods are in the majority of cases gradient-based. This implies that they rely on the computation of partial derivatives in the step computation and for checking the optimality of solution candidates. The efficient and accurate computation of derivatives as well as the determination of the associated sparsity patterns is of high importance for this type of method.

Finally, we note that, besides the direct approach, it is also possible to formulate the optimality conditions ("optimize") directly for the clearance optimal control problem in continuous time (see left side in Fig. 1). The fundamental concepts regarding this approach are outlined in Chapter 3. In particular, the optimality conditions for optimal control problems in continuous time discussed in this chapter allow us to make statements about the type of optimal control functions that can be expected for certain classes of clearance problems. Moreover, the optimality conditions of optimal control problems in continuous time and optimization problems resulting from the direct approach can be related/matched to provide sanity checks regarding the numerical solution.

CHAPTER 1

Optimization

1.1 OPTIMIZATION PROBLEMS

Clearance Optimization Problem

Find a decision vector $\mathbf{z} \in \mathbb{R}^n$ bounded by $\underline{\mathbf{z}}, \overline{\mathbf{z}} \in \mathbb{R}^n$ that minimizes the objective function $J : \mathbb{R}^n \to \mathbb{R}$ subject to the equality constraints $\mathbf{a}(\mathbf{z}) = \mathbf{0}$, $\mathbf{a} : \mathbb{R}^n \to \mathbb{R}^{n_a}$:

$$\begin{aligned} &\underset{\underline{\mathbf{z}} \leqslant \mathbf{z} \leqslant \overline{\mathbf{z}}}{\text{minimize}} \quad J(\mathbf{z}) \\ &\text{subject to} \quad \mathbf{a}(\mathbf{z}) = \mathbf{0} \end{aligned} \tag{1.1}$$

Optimization represents the first fundamental building block forming the underlying basis for the optimal control approach to flight control law testing. Here, and in what follows, *optimization* refers to the minimization of a cost function using an appropriate choice of the decision vector subject to equality and inequality constraints.

In general, there exists a large variety of optimization problem formulations. These formulations are often equivalent; however, they are stated in a different form depending on the most natural representation for a specific application. The formulation of the clearance optimization problem (1.1) involves a scalar cost function, a decision vector, general equality constraints, as well as bound constraints, and it represents a prototype for typical problem formulations in flight control law testing. Particularly from a methodical point of view, this form is sufficiently general and close to formulations arising from segmented collocation as well as shooting methods. It is noted that, if required, a formulation including general inequality constraints can also be modeled by Eq. (1.1) using suitable standard transformations.

Let us first discuss the interpretation of the cost, the decision vector, as well as the constraints in the context of flight control law clearance problems from a high-level perspective. From this perspective, we may state that, in general terms, the objective of an optimization problem expresses "the goal we seek to achieve," the decision vector represents the "means at our disposal," and constraints set the "limits to what is considered admissible."

For the application of extremal finding methods to flight control law testing, our goal is to transfer the aircraft closed-loop system under test from a reference condition to an unsafe state. Whether or not a state is regarded as unsafe is determined by the clearance criterion expressed through a state-dependent metric with limits defining the acceptable values. Let us suppose that a lower limit $\underline{J}$ of such a metric is given, and we are interested in finding if it is possible to achieve values beyond this limit. In this case, the cost of the optimization problem $J(\mathbf{z})$ to be *minimized* can be used to model the criterion as the optimization method attempts to locate a point toward or beyond $\underline{J}$. Even though the optimization process could in fact be stopped once the lower limit $\underline{J}$ has been reached, it is usually preferred in practice to continue minimizing the cost and seeking the lowest possible value representing the "worst-case solution." In other cases where an upper limit $\overline{J}$ is given for the criterion, for example, for the maximum angle of attack criterion, it may seem natural to reformulate the clearance problem as a maximization problem. Note, however, that in such a case it is always possible to simply negate the cost function; after all, a minimization of the negative cost is equivalent to its maximization. Therefore, we note that Eq. (1.1) is also sufficiently general for these types of clearance problems.

The means at our disposal to minimize the cost are represented by the elements in the decision vector $\mathbf{z}$. These elements can model, for example, static parameters related to the aerodynamic coefficients, weight and weight distribution or input parameterizations of pilot inputs, as well as other disturbances, such as wind. Clearly, some bounds need to be prescribed for these elements to ensure meaningful solutions from a physical perspective. Such bounds can be imposed by defining suitable vectors $\underline{\mathbf{z}}$ and $\overline{\mathbf{z}}$, effectively limiting the admissible search space explored during the course of optimization.

The equality constraints $\mathbf{a}(\mathbf{z}) = \mathbf{0}$ are not necessarily introduced for all types of clearance problems. One case in which these equality constraints are required is the application of numerical optimal control methods in segmented form. This form can be advantageous from a numerical point of view as it leads to an exposure of the discretized states as well as the integration method to the optimization algorithm, ultimately increasing the overall problem sparsity. In particular for segmented collocation methods, the states at discretization time points are introduced as decision variables. Additional equality constraints, which represent the collocation conditions, enforce that the dynamics of the aircraft closed-loop model are satisfied along the trajectory. The fact that the states are part of the decision vector also allows us to directly impose limits on these variables. This can, for example, be used in optimal control formulations to introduce physical limits such as actuator rate and position bounds.

The particular forms of clearance optimization problems stemming from numerical optimal control methods, applied to the flight control law clearance task, are discussed extensively throughout Part II.

1.2 TRANSFORMED PROBLEMS

Transformed Clearance Optimization Problems

The clearance optimization problem (1.1) as well as its linearized form

$$\begin{aligned} \underset{\underline{\mathbf{z}}\leqslant\mathbf{z}\leqslant\overline{\mathbf{z}}}{\text{minimize}} \quad & \mathbf{J}^T\mathbf{z} \\ \text{subject to} \quad & \mathbf{A}\mathbf{z} = \mathbf{b} \end{aligned} \tag{1.2}$$

with cost coefficient vector $\mathbf{J} \in \mathbb{R}^{n}$, equality constraint matrix $\mathbf{A} \in \mathbb{R}^{n_a \times n}$, and corresponding vector $\mathbf{b} \in \mathbb{R}^{n_a}$, can be transformed without loss of generality to the compact form

$$\begin{aligned} \underset{\mathbf{s}\geqslant\mathbf{0}}{\text{minimize}} \quad & J(\mathbf{s}) \\ \text{subject to} \quad & \mathbf{c}(\mathbf{s}) = \mathbf{0} \end{aligned} \tag{1.3}$$

with nonnegative decision variables $\mathbf{s} \in \mathbb{R}^{n_s}$ and the augmented equality constraint function $\mathbf{c} : \mathbb{R}^{n_s} \to \mathbb{R}^{n_c}$ as well as its linearized form

$$\begin{aligned} \underset{\mathbf{s}\geqslant\mathbf{0}}{\text{minimize}} \quad & \mathbf{J}^T\mathbf{s} \\ \text{subject to} \quad & \mathbf{C}\mathbf{s} = \mathbf{d} \end{aligned} \tag{1.4}$$

with an appropriate vector $\mathbf{d} \in \mathbb{R}^{n_c}$ and the constraint Jacobian $\mathbf{C} \in \mathbb{R}^{n_c \times n_s}$.

Before proceeding to the discussion of the optimality conditions and numerical methods for solving the clearance optimization problem, let us discuss the transformed problem (1.3). This form constitutes a more compact representation compared to problem (1.1) and is used together with its linearized form (1.4) for theoretical developments in this book.

May be skipped (↴) at first read.

Consider a minimization of the cost $J\left(\mathbf{z}^+ - \mathbf{z}^-\right)$ using the decision vector

$$\begin{bmatrix} \mathbf{z}^+ \\ \mathbf{z}^- \\ \overline{\mathbf{s}} \\ \underline{\mathbf{s}} \end{bmatrix} \geqslant \mathbf{0}$$

subject to the equality constraints

$$\begin{bmatrix} \mathbf{a}\left(\mathbf{z}^{+}-\mathbf{z}^{-}\right) \\ \mathbf{z}^{+}-\mathbf{z}^{-}-\overline{\mathbf{z}}+\overline{\mathbf{s}} \\ \underline{\mathbf{z}}-\mathbf{z}^{+}+\mathbf{z}^{-}+\underline{\mathbf{s}} \end{bmatrix} = \mathbf{0} \tag{1.5}$$

This minimization problem is of the form of Eq. (1.3). Note that with the interpretation of $\mathbf{z}^{+}$ and $\mathbf{z}^{-}$ as the nonnegative and nonpositive parts of the decision vector $\mathbf{z}$ in Eq. (1.1), i.e.,

$$\mathbf{z} = \overbrace{\mathbf{z}^{+}}^{\geqslant \mathbf{0}} - \underbrace{\mathbf{z}^{-}}_{\geqslant \mathbf{0}}$$

the first constraint vector entry in Eq. (1.5) is equivalent to

$$\mathbf{a}\left(\mathbf{z}\right) = \mathbf{0}$$

and the following vector entries

$$\mathbf{z}-\overline{\mathbf{z}}+\overline{\mathbf{s}} = \mathbf{0}$$
$$\underline{\mathbf{z}}-\mathbf{z}+\underline{\mathbf{s}} = \mathbf{0}$$

together with $\underline{\mathbf{s}}, \overline{\mathbf{s}} \geqslant \mathbf{0}$ are equivalent to

$$\mathbf{z}-\overline{\mathbf{z}} \leqslant \mathbf{0}$$
$$\underline{\mathbf{z}}-\mathbf{z} \leqslant \mathbf{0}$$

Clearly, with $\overline{\mathbf{s}}$ being nonnegative, the only way for $\mathbf{z}-\overline{\mathbf{z}}+\overline{\mathbf{s}}$ to be zero is if the term $\mathbf{z}-\overline{\mathbf{z}}$ is nonpositive (similar for the lower bound).

Finally, we note that this type of modeling also allows us to formally introduce general inequality constraints of the form

$$\mathbf{g}\left(\mathbf{z}\right) \leqslant \mathbf{0}$$

into the clearance optimization problem (1.1). In this case, the inequality constraints can be modeled by additional equality constraints

$$\mathbf{g}\left(\mathbf{z}\right)+\mathbf{s}_0 = \mathbf{0}$$

as well as additional decision variables $\mathbf{s}_0 \geqslant \mathbf{0}$.

It should be stressed that the transformed problem (1.3) still represents the same cost and constraints as the original problem. Merely, the inequality constraints from problem (1.1) are implicitly imposed by the additional equality constraints and inequalities ensuring nonnegativity of the decision variables.

An important special case for optimization problems occurs if the cost and constraints are linear-affine. Such a problem can directly be obtained through the linearization of all components of the nonlinear problem around some reference point. In clearance applications, this reference point is typically represented by some trim condition around which we seek to test the closed-loop system.

The linearized cost and constraints in Eq. (1.2) around a suitable reference decision vector $\tilde{\mathbf{z}}$ are expressed as

$$\left.\frac{\partial J}{\partial \mathbf{z}}\right|_{\mathbf{z}=\tilde{\mathbf{z}}}(\mathbf{z}-\tilde{\mathbf{z}}) = \overbrace{\left.\frac{\partial J}{\partial \mathbf{z}}\right|_{\mathbf{z}=\tilde{\mathbf{z}}}}^{\mathbf{J}^T}\mathbf{z} - \overbrace{\left.\frac{\partial J}{\partial \mathbf{z}}\right|_{\mathbf{z}=\tilde{\mathbf{z}}}\tilde{\mathbf{z}}}^{\text{const}}$$

$$\mathbf{a}|_{\mathbf{z}=\tilde{\mathbf{z}}} + \left.\frac{\partial \mathbf{a}}{\partial \mathbf{z}}\right|_{\mathbf{z}=\tilde{\mathbf{z}}}(\mathbf{z}-\tilde{\mathbf{z}}) = \underbrace{\left.\frac{\partial \mathbf{a}}{\partial \mathbf{z}}\right|_{\mathbf{z}=\tilde{\mathbf{z}}}}_{\mathbf{A}}\mathbf{z} + \underbrace{\mathbf{a}|_{\mathbf{z}=\tilde{\mathbf{z}}} + \left.\frac{\partial \mathbf{a}}{\partial \mathbf{z}}\right|_{\mathbf{z}=\tilde{\mathbf{z}}}\tilde{\mathbf{z}}}_{-\mathbf{b}}$$

Note that the minimizer of an optimization problem is invariant regarding a constant shift of the cost. Thus, we can drop the constant factor in the linearized cost for the problem formulation.

Similarly, the constraints of the transformed problem (1.3) can be obtained from a linearization around a reference point $\tilde{\mathbf{s}}$:

$$\mathbf{c}|_{\mathbf{s}=\tilde{\mathbf{s}}} + \left.\frac{\partial \mathbf{c}}{\partial \mathbf{s}}\right|_{\mathbf{s}=\tilde{\mathbf{s}}}(\mathbf{s}-\tilde{\mathbf{s}}) = \underbrace{\left.\frac{\partial \mathbf{c}}{\partial \mathbf{s}}\right|_{\mathbf{s}=\tilde{\mathbf{s}}}}_{\mathbf{C}}\mathbf{s} + \underbrace{\mathbf{c}|_{\mathbf{s}=\tilde{\mathbf{s}}} + \left.\frac{\partial \mathbf{c}}{\partial \mathbf{s}}\right|_{\mathbf{s}=\tilde{\mathbf{s}}}\tilde{\mathbf{s}}}_{-\mathbf{d}}$$

This linearized form (1.4) of the transformed problem represents the so-called standard form for linear optimization problems (also called, for historical reasons, "linear programming problems").

1.3 OPTIMALITY CONDITIONS

First-Order Necessary and Second-Order Sufficient Conditions

Based on the Lagrangian for the transformed problem (1.3),

$$\mathcal{L}(\mathbf{s}, \boldsymbol{\lambda}, \boldsymbol{\mu}) = J(\mathbf{s}) + \boldsymbol{\lambda}^T\mathbf{c}(\mathbf{s}) - \boldsymbol{\mu}^T\mathbf{s}$$

with multipliers $\boldsymbol{\lambda} \in \mathbb{R}^{n_c}$ and $\boldsymbol{\mu} \in \mathbb{R}^{n_s}$, the local first-order necessary conditions of optimality at a candidate solution require

$$\frac{\partial \mathcal{L}}{\partial \mathbf{s}} = \mathbf{0}, \qquad \frac{\partial \mathcal{L}}{\partial \boldsymbol{\lambda}} = \mathbf{0}, \qquad \text{diag}(\mathbf{s})\boldsymbol{\mu} = \mathbf{0}, \qquad \boldsymbol{\mu}, \mathbf{s} \geqslant \mathbf{0}$$

If an optimization problem exhibits a convex cost and linear-affine constraints, this condition represents a sufficient condition for a global minimum. In the general case, a second-order sufficient condition requires in addition that, assuming strict complementarity, the projected Hessian

$$\mathbf{N}^T \frac{\partial^2 \mathcal{L}}{\partial \mathbf{z}^2} \mathbf{N}, \qquad \mathbf{N} = \text{Null}\left(\hat{\mathbf{C}}\right)$$

is positive definite. Here, $\hat{\mathbf{C}}$ represents the Jacobian of all active inequality and equality constraints at an optimal point. For all optimality conditions, it is assumed that $\hat{\mathbf{C}}$ has full row rank.

The following discussions are concerned with the statement of optimality conditions of the transformed problem (1.3) as well as optimization methods for solving these conditions numerically. For this purpose, let us first define the Lagrangian of Eq. (1.3) by adjoining the function $\mathbf{c}$ of the equality constraints $\mathbf{c}(\mathbf{s}) = \mathbf{0}$ with multiplier vector $\boldsymbol{\lambda} \in \mathbb{R}^{n_c}$ and the inequality constraints $-\mathbf{s} \leqslant \mathbf{0}$ with multiplier vector $\boldsymbol{\mu} \in \mathbb{R}^{n_s}$ to the cost with multiplier $l_0 \in \mathbb{R}$:

$$\mathcal{L}\left(\mathbf{s}, l_0, \boldsymbol{\lambda}, \boldsymbol{\mu}\right) = l_0 J(\mathbf{s}) + \boldsymbol{\lambda}^T \mathbf{c}(\mathbf{s}) - \boldsymbol{\mu}^T \mathbf{s}$$

For all necessary and sufficient optimality conditions at a candidate solution $\mathbf{s}, \boldsymbol{\mu}, \boldsymbol{\lambda} = \hat{\mathbf{s}}, \hat{\boldsymbol{\mu}}, \hat{\boldsymbol{\lambda}}$, it is assumed in the following that the gradients of the equality constraints as well as the inequality constraints that are "active" (satisfied with equality) are linearly independent. This allows us to normalize the multipliers such that $l_0 = 1$ and guarantees the multipliers' uniqueness (see, e.g., [21], Corollary 2.3.39). At such a candidate solution, the *Karush-Kuhn-Tucker* (KKT) conditions based on the stationarity of the Lagrangian together with the complementarity conditions constitute first-order necessary conditions of optimality.

Local Optimality Condition: The Lagrangian is required to be stationary with respect to the decision variable vector $\mathbf{s}$:

$$\frac{\partial \mathcal{L}}{\partial \mathbf{s}} = \mathbf{0} \tag{1.6}$$

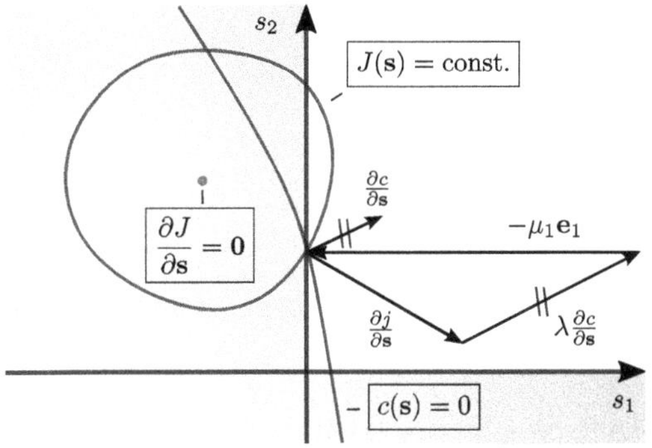

Fig. 1.1 Optimality condition in two dimensions featuring an equality constraint $c(s) = 0$ (multiplier λ) and an active inequality constraint $s_1 \geqslant 0$ (multiplier μ_1).

which implies

$$\frac{\partial J}{\partial \mathbf{s}} + \sum_{k=1}^{n_c} \lambda_k \frac{\partial c_k}{\partial \mathbf{s}} - \sum_{i=1}^{n_s} \mu_i \mathbf{e}_i^T = \mathbf{0} \tag{1.7}$$

This condition states that at an optimal solution, the negative gradient of the cost must be expressed as a linear combination of the constraint gradients. Intuitively, this linear combination may be compared to an equilibrium of forces. Such an equilibrium is established by the multipliers scaling the "constraint forces" (constraint gradients) and the "cost force" (negative cost gradient) "pushing" in the direction of steepest descent of the cost (see Fig. 1.1).

Feasibility Condition (Equality Constraints): Feasibility regarding the equality constraints is expressed by requiring the Lagrangian to be stationary with respect to the multiplier vector $\boldsymbol{\lambda}$:

$$\frac{\partial \mathcal{L}}{\partial \boldsymbol{\lambda}} = \mathbf{0} \tag{1.8}$$

which implies

$$\mathbf{c}(\mathbf{z}) = \mathbf{0}$$

Complementarity Condition (Inequality Constraints): For the inequality constraints, the complementarity conditions require for $i = 1, \ldots, n_s$:

$$s_i, \mu_i \geqslant 0, \; s_i \mu_i = 0 \tag{1.9}$$

Let us take a closer look at the complementarity condition and note that, as illustrated in Fig. 1.2, a condition of the form

$$s, \mu \geqslant 0, \; s\mu = 0$$

mathematically describes a "corner."

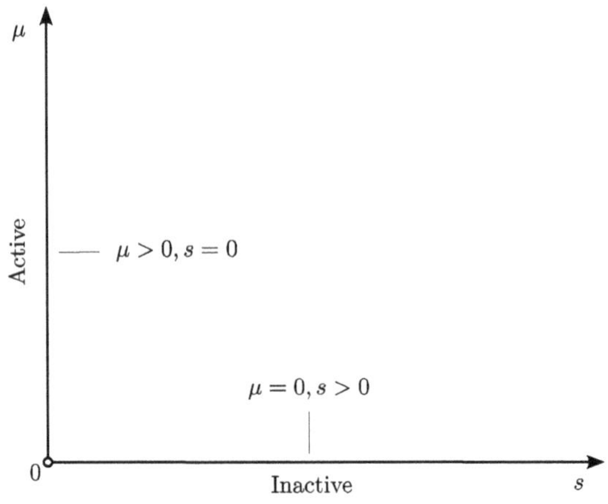

Fig. 1.2 Complementarity condition.

The ordinate in this illustration visualizes the *active* branch with $s = 0$ and $\mu > 0$. The abscissa depicts the *inactive* branch with $s > 0$ and $\mu = 0$. Moreover, the origin $\mu, s = 0$ describes a so-called *weakly active* point. During an iterative solution of the optimization problem, constraints may switch from the active branch to the inactive branch from one iteration to the next and vice versa. One strategy for dealing with this fact is to keep track of the set of active constraints that need to be respected for a certain iteration and update this set in each iteration. These methods are commonly referred to as "active-set methods." Note that this strategy effectively permits the problem's structure to evolve, as the constraints used to compute the next iterate can vary from one iteration to the next.

Another strategy is to not explicitly distinguish between active and inactive inequality constraints but to consider all constraints as "active to a varying degree." This strategy can be realized by smoothing the corner (see Fig. 1.3) with some nonnegative τ:

$$s, \mu \geqslant 0, \; s\mu = \tau \geqslant 0$$

Clearly, for $\tau = 0$ the original corner with the active and inactive branch is obtained. For higher values $\tau > 0$, both s and μ are *strictly positive*:

$$s, \mu > 0, \; s\mu = \tau > 0 \tag{1.10}$$

Regarding the complementarity condition, this implies that the active and inactive branches can only be approached in the respective limits.

Loosely speaking, under this strategy the constraints are all "more or less active," and the "degree of activeness" is determined by the position on the smooth curve for a particular $\tau > 0$. The problem can then be solved for decreasing values of τ. This then continuously transforms the problem toward the original problem

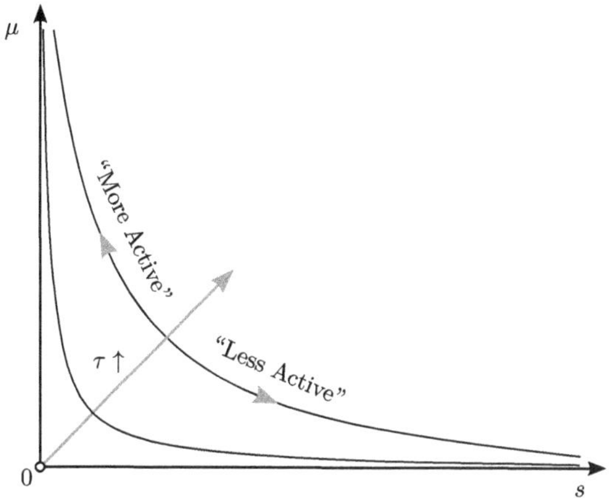

Fig. 1.3 Smoothed complementarity condition.

with $\tau = 0$ through a sequence of intermediate problems. Observe that the essential structure of the problem parameterized in τ does not change as all constraints are considered in each iteration.

While respecting $\mathbf{s}, \boldsymbol{\mu} > \mathbf{0}$, the residuum of the optimality conditions for some $\tau > 0$ to be solved is thus

$$\begin{bmatrix} \mathbf{J} + \mathbf{C}^T\boldsymbol{\lambda} - \boldsymbol{\mu} \\ \mathbf{c} \\ \mathbf{S}\boldsymbol{\mu} - \tau\mathbf{1} \end{bmatrix} = \mathbf{0} \tag{1.11}$$

Here, the first vector entry is the stationarity condition (1.7) in vectorial form with $\mathbf{C} = \frac{\partial \mathbf{c}}{\partial \mathbf{s}}$, the Jacobian of the contstraints, and $\mathbf{J}^T = \frac{\partial J}{\partial \mathbf{s}}$, the gradient of the cost. The last vector entry represents the smoothed version of the complementarity condition (1.9) with $\mathbf{S} = diag(\mathbf{s})$.

In each iteration of the solution process, the steps for the decision variables and multipliers need to be determined. A common strategy is to decompose the step construction into the computation of a search direction followed by determination of a suitable step size.

The search direction can be determined using Newton-type root-finding methods, that is, from a solution to the linear system

$$\begin{bmatrix} \mathbf{B} & \mathbf{C}^T & -\mathbf{I} \\ \mathbf{C} & \mathbf{0} & \mathbf{0} \\ \mathbf{M} & \mathbf{0} & \mathbf{S} \end{bmatrix} \begin{bmatrix} \mathbf{d}_\mathbf{s} \\ \mathbf{d}_\lambda \\ \mathbf{d}_\mu \end{bmatrix} = - \begin{bmatrix} \mathbf{J} + \mathbf{C}^T\boldsymbol{\lambda} - \boldsymbol{\mu} \\ \mathbf{c} \\ \mathbf{S}\boldsymbol{\mu} - \tau\mathbf{1} \end{bmatrix} \tag{1.12}$$

with $\mathbf{M} = \text{diag}(\boldsymbol{\mu})$ and $\mathbf{B}$ representing the Lagrangian Hessian or a suitable approximation thereof (e.g., based on the L-BFGS algorithm [22]). Note that the matrix on the left side of Eq. (1.12) is obtained from the derivatives of the residuum (1.11) with respect to the decision vector $\mathbf{s}$, the Lagrange multipliers of the equality constraints $\boldsymbol{\lambda}$, and the inequality constraints $\boldsymbol{\mu}$.

Once the search directions are computed, a practical implementation for determining the step size for Newton directions is to perform a backtracking line search. This procedure starts with a step size $\alpha > 0$ and reduces this step size $\alpha \leftarrow \rho\alpha, \rho \in (0, 1)$ until the step is deemed suitable (e.g., based on an Armijo backtracking line search satisfying the Wolfe conditions) [23].

Engineering Note 1.3.1: Solver Output and Options in IPOPT

IPOPT [24] is a popular NLP (nonlinear programming) solver employed by a large number of optimization and optimal control frameworks. Understanding its standard output and how it is affected by certain options can be valuable for monitoring and interpreting the progress or troubleshooting when applying the clearance methods presented in this book. The columns provided in the standard output of IPOPT for each iteration are as follows [25]:

iter	objective	inf_pr	inf_du	lg(mu)	||d||	lg(rg)	alpha_du	alpha_pr	ls

In clearance applications, the second column (label `objective`) shows the value of the criterion in the current iteration. As a general rule of thumb, it is typically beneficial for the performance of the numerical solution to scale this value such that it ranges between -0.5 and 0.5 throughout the solution process. A multiplicative scaling factor can be applied either internally in the user-supplied cost function or by setting the IPOPT option `obj_scaling_factor`. Note that if such scaling is applied through this IPOPT option, the effect is not reflected in the solver output as IPOPT displays the unscaled objective value. Note further that IPOPT minimizes by default, which is in alignment with our clearance problem formulation (1.1). As outlined in Sec. 1.1, a negative scaling factor effectively yields a maximization problem.

The next two columns labeled `inf_pr` and `inf_du` in the solver output display feasibility [Eq. (1.8)] and optimality [Eq. (1.6)] for the current iterate. These quantities in the output represent the largest absolute values of the first two elements in the residuum [Eq. (1.11)], specifically the infinity norm of $\frac{\partial \mathcal{L}}{\partial \mathbf{s}}$ $(= \mathbf{J} + \mathbf{C}^T\boldsymbol{\lambda} - \boldsymbol{\mu})$ for `inf_du` and the infinity norm of $\frac{\partial \mathcal{L}}{\partial \boldsymbol{\lambda}}$ $(= \mathbf{c})$ for `inf_pr`. If, for example, a segmented collocation method is employed for testing the aircraft closed-loop system, the collocation conditions are imposed as equality constraints. In this case, the reported feasibility in IPOPT provides a measure for the maximum violation of the dynamics along the trajectory in the current iteration. As the optimization progresses, both feasibility and optimality are eventually driven to a positive value close to zero, depending on the termination tolerance. This tolerance can be set in the option `tol`. (The default setting of 10^{-8} is suitable for most applications.)

The largest absolute entry in $\mathbf{d}_s$ from Eq. (1.12) for the current iteration is shown in the column labeled `||d||`. Therefore, this output reports the largest step that could be taken for an element of the decision vector (containing, e.g., static parameters and the pilot inputs in our case) in the attempt to bring the aircraft closed-loop system closer to failure. Recall, however, that ultimately the step is determined by a backtracking procedure that determines a suitable step size for a particular search direction. Here, IPOPT allows the step size for the duals λ to be different, which in the experience of the developers increases the efficiency of the algorithm [24]. Both step sizes, the primal step size `alpha_pr` and the dual step size `alpha_du`, are reported in the solver output. Values close to one are often desirable for these step sizes as they imply the acceptance of a full step. If a full step is not acceptable, the backtracking procedure reduces the primal step size during line search, and the solver output shows values greater than one in the last column labeled `ls`.

The column labeled `lg(rg)` provides the logarithm (base ten) of a regularization factor that is introduced for the solution of the Newton system [see Eq. (1.12)]. The regularization factor times an identity matrix of appropriate size is added to the Hessian of the Lagrangian in the upper left block of the system [Eq. (1.12)]. This value is only applicable if the default setting `exact` is set for the option `hessian_approximation`. Using the `exact` setting is advisable as it often outperforms the Quasi-Newton update (see setting `limited-memory`), especially close to a minimizer where locally quadratic convergence can be observed. However, if no second derivatives are available or expensive to compute for a particular clearance application (e.g., if the Hessian is computed from a finite difference approximation over an expensive simulation), then switching to `limited-memory` can be a viable option.

Finally, the logarithm (base ten) of the current barrier parameter τ [see Eq. (1.10)] is given in the column labeled `lg(mu)`. There are two strategies `monotone` and `adaptive` that can be selected in the option `mu_strategy`. As opposed to the `monotone` strategy, the `adaptive` strategy does not require that the barrier parameter is driven to zero monotonically with an increasing number of iterations.

Computational experience from the solution of general optimal control problems, and in particular clearance optimal control problems, suggests that the `adaptive` strategy often outperforms the `monotone` strategy set by default.

For optimization problems with a convex cost and linear-affine constraints, the KKT conditions are not only necessary but also sufficient conditions ([21], Theorem 2.3.41). In this case, if an optimization problem returns a feasible point satisfying the KKT conditions, this solution constitutes a global minimum.

In the general case, however, additional conditions are required to state sufficient conditions of optimality. Under the assumption that second derivatives of the cost and constraint functions exist and strict complementarity holds, meaning there are no weakly active constraints, a local second-order sufficient condition for optimality can be stated in compact form.

This condition requires positive definiteness of the projected Hessian:

$$\mathbf{N}^T \frac{\partial^2 \mathcal{L}}{\partial \mathbf{z}^2} \mathbf{N} \tag{1.13}$$

where the columns of the matrix $\mathbf{N}$ span the nullspace of the Jacobian of all active inequality and equality constraints [23]. A necessary condition for local optimality is that the projected Hessian is semidefinite.

Before concluding this section, let us observe that the statement of the optimality conditions as well as the solution of these conditions involve the partial derivatives of the cost and constraint functions [see Eqs. (1.11)–(1.13)]. *Efficient and accurate procedures for computing these derivatives are of paramount importance regarding the optimal control approaches presented in this book.*

To a certain extent, the preprocessing, evaluation, and derivative computation of the involved functions is automatized by optimization and optimal control frameworks. For example, the optimal control framework FALCON.m used in this book automatizes this process by means of symbolic expressions. However, this particular tool-chain requires all functions to be representable by such expressions, a requirement rarely met for realistic aircraft closed-loop systems. More often than not, complex models include table data or other nondifferentiable functions, thus requiring either smooth reformulations or a manual implementation of the derivatives (often relying on the use of numerical differentiation methods). In FALCON.m, such an implementation can be realized through a decomposition of the model into symbolic and nonsymbolic subsystems. For nonsymbolic subsystems, however, the correctness of the derivatives as well as the sparsity patterns (i.e., the patterns of nonzero entries in the Jacobian and Hessian matrices) lies in the responsibility of the user. In this context, a basic understanding of commonly used differentiation methods in optimization and optimal control applications is essential to allow the use of real-world models for the clearance methods presented in this book. Hence, the following section is devoted to differentiation methods for optimization.

It should be mentioned that the selection of topics in this section does not represent a comprehensive guide for differentiation methods in general. Here, the reader is referred to the respective literature as well as the user guides and software manuals of optimization and optimal control software. However, the guiding intention behind this particular selection of topics should be understood as a discussion of *highlights and caveats* regarding general techniques applicable to optimal control-based flight control law clearance.

1.4 DIFFERENTIATION METHODS FOR OPTIMIZATION

1.4.1 SUBSYSTEM APPROACH

Derivative Propagation (Forward and Backward)

The derivative propagation along with the evaluation of subsystems $\mathbf{v}_i = \mathbf{f}_i(\cdot)$, each with vector-valued inputs $\mathbf{v}_j$, $j \in \mathcal{F}_i$, is defined in a forward sense $i = 1, \ldots, m$ as

$$\mathbf{s}_i = \sum_{j \in \mathcal{F}_i} \frac{\partial \mathbf{f}_i}{\partial \mathbf{v}_j} \mathbf{s}_j$$

With $\mathcal{B}_i = \{j : i \in \mathcal{F}_j\}$, the propagation in a backward sense $i = m - 1, \ldots, 0$ is defined as

$$\boldsymbol{\nu}_i^T = \sum_{j \in \mathcal{B}_i} \boldsymbol{\nu}_j^T \frac{\partial \mathbf{f}_j}{\partial \mathbf{v}_i}$$

All subsystems are assumed to be in sorted execution order with $\max(\mathcal{F}_i) < i$.

The differentiation methods presented in the following are discussed from a general subsystem modeling perspective based on the common form and procedures used in automatic differentiation frameworks [26] (see also [27–29]). In this context, let us consider the evaluation of a vector-valued function $\mathbf{f} : \mathbb{R}^n \rightarrow \mathbb{R}^{n_J}$ with input $\mathbf{z} \in \mathbb{R}^n$ and output $\mathbf{y} \in \mathbb{R}^{n_J}$:

$$\mathbf{y} = \mathbf{f}(\mathbf{z})$$

In clearance applications, such a function represents either the cost function modeling the clearance criterion, general constraints, or other functions related to the closed-loop system under test. In this setting, it is reasonable to assume that the procedure for evaluating $\mathbf{f}$ allows for a representation in which the atomic units are modeled by vector-valued functions $\mathbf{v}_i = \mathbf{f}_i(\cdot)$, $i = 1, \ldots, m$. In the following, we refer to these functions as *subsystems*. Each of these subsystems is evaluated with an arbitrary number of vector-valued inputs.

We require that the initial assignment $\mathbf{v}_0 = \mathbf{z}$, final assignment $\mathbf{v}_m = \mathbf{y}$, and all connections between these subsystems, meaning all input–output relationships, be represented as a directed, acyclic graph. This can be ensured by a sorted execution order in which all indices of the input vectors of a subsystem $\mathbf{f}_i$ are less than i.

Engineering Note 1.4.1: Subsystem Modeling Perspective

We prefer the general form of a subsystem-based approach as domain experts in flight control law development and testing are assumed to be familiar with environments (e.g., Simulink®) based on similar concepts. Moreover, the framework established from this point of view proves useful in several other contexts throughout this book, including the discussion on sparsity and numerical differentiation, the illustration of the FALCON.m model builder [29], as well as the derivation of first- and second-order sensitivities of clearance criteria. The close connection between sensitivity analysis for simulation methods and the so-called forward and backward (adjoint) procedure is well established and allows for an accessible derivation and representation in the context of clearance applications.

As a side note, some readers may have recognized that the assumed execution order of the graph representation is conceptually the same execution order in Simulink® which can be added as overlays by toggling "Information Overlays/Execution Order." For example, in the case of a model with three atomic subsystems, the following execution order is automatically annotated in Simulink® to the model with red numbers:

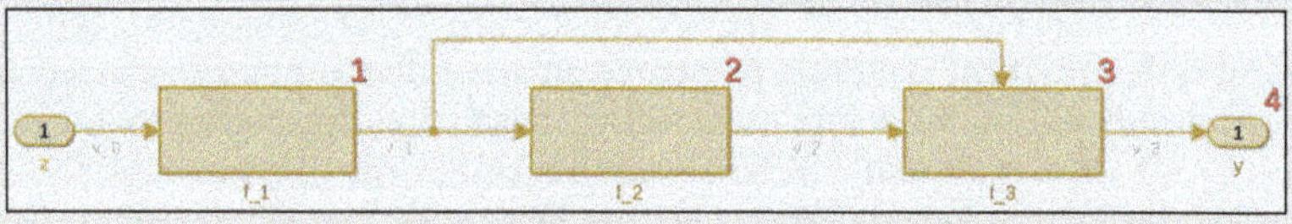

This modular Simulink® block structure is, for instance, used in the optimal control applications developed in Fisch [30] to derive the Jacobian and Hessian for simulation models from a subsystem modeling perspective.

An exemplary graph representation of the assumed general form with $m = 3$ subsystems $\mathbf{f}_1, \ldots, \mathbf{f}_3$ is illustrated in Fig. 1.4. In this illustration, two types of index sets are given:

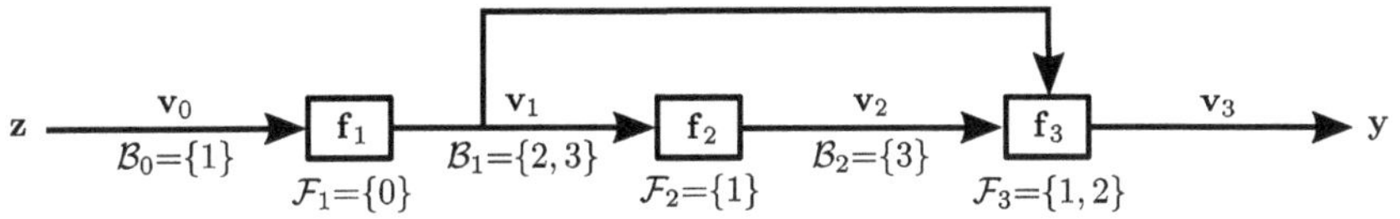

Fig. 1.4 Graph representation with three subsystems.

- The index sets $\mathcal{F}_i$, $i = 1, \ldots, m$ are defined for each subsystem $\mathbf{f}_i$ and collect all indices j of vectors $\mathbf{v}_j$ required for the evaluation of $\mathbf{f}_i$.
- The index sets $\mathcal{B}_i = \{j : i \in \mathcal{F}_j\}$, $i = 0, \ldots, m-1$ collect all indices of subsystems depending on $\mathbf{v}_i$.

Based on these notions, the procedures for the evaluation and derivative computation can be expressed in compact forms.

1.4.1.1 EVALUATION PROCEDURE

The procedure for the computation of $\mathbf{y} = \mathbf{f}(\mathbf{z})$ can be expressed by the sequential evaluation of

$$\mathbf{v}_0 = \mathbf{z}, \qquad \mathbf{v}_i = \mathbf{f}_i(\cdot),\ i = 1, \ldots, m, \qquad \mathbf{y} = \mathbf{v}_m \tag{1.14}$$

1.4.1.2 FORWARD DERIVATIVE PROCEDURE

The forward derivative procedure can be used to compute the derivatives of $\mathbf{f}$ with respect to an arbitrary $z \in \mathbf{z}$. For this procedure, additional vectors $\mathbf{s}_i$, $i = 0, \ldots, m$ corresponding to each $\mathbf{v}_i$ are introduced. The forward derivative procedure is given by the sequential evaluation of

$$\mathbf{s}_0 = \frac{\partial \mathbf{z}}{\partial z}, \qquad \mathbf{s}_i = \sum_{j \in \mathcal{F}_i} \frac{\partial \mathbf{f}_i}{\partial \mathbf{v}_j} \mathbf{s}_j,\ i = 1, \ldots, m, \qquad \frac{\partial \mathbf{f}}{\partial z} = \mathbf{s}_m \tag{1.15}$$

This method represents an application of the chain rule to the steps of Eq. (1.14) in a forward sense. As the forward derivative method proceeds in the same direction as Eq. (1.14), all pairs $(\mathbf{s}_i, \mathbf{v}_i)$, $i = 1, \ldots, m$ can be computed concurrently.

1.4.1.3 BACKWARD DERIVATIVE PROCEDURE

The backward procedure can be used to compute the derivatives of an arbitrary $y = f(\mathbf{z}) \in \mathbf{f}(\mathbf{z})$ with respect to $\mathbf{z}$. For this procedure, additional vectors $\boldsymbol{\nu}_i$, $i = 0, \ldots, m$ corresponding to each $\mathbf{v}_i$ are introduced. The backward derivative procedure is given by the sequential evaluation of

$$\boldsymbol{\nu}_m^T = \frac{\partial y}{\partial \mathbf{y}}, \qquad \boldsymbol{\nu}_i^T = \sum_{j \in \mathcal{B}_i} \boldsymbol{\nu}_j^T \frac{\partial \mathbf{f}_j}{\partial \mathbf{v}_i},\ i = m-1, \ldots, 0, \qquad \frac{\partial f}{\partial \mathbf{z}} = \boldsymbol{\nu}_0^T \tag{1.16}$$

As the backward derivative method proceeds in the opposite direction compared to the evaluation, we must first execute Eq. (1.14) and store the intermediate results $\mathbf{v}_i$, $i = 1, \ldots, m$. Then, in a second step, the backward procedure [Eq. (1.16)] can be executed.

Observe that, on the one hand, the forward procedure yields the derivatives of all elements in $\mathbf{f}(\mathbf{z})$ with respect to some $z \in \mathbf{z}$ in a single forward pass. The backward procedure, on the other hand, yields the derivatives of some $f(\mathbf{z}) \in \mathbf{f}(\mathbf{z})$ with respect to all elements in $\mathbf{z}$ in one backward pass. Therefore, if the number of elements in $\mathbf{f}(\mathbf{z})$ is smaller compared to the number of elements in $\mathbf{z}$, the backward procedure is expected to be more efficient than the forward procedure, and vice versa.

The forward and backward procedures are, in fact, sufficient to obtain derivatives of higher order. Note that the forward method [Eq. (1.15)] together with the evaluation procedure [Eq. (1.14)] can be interpreted as the evaluation of a general vector-valued function. The same holds for the backward method [Eq. (1.16)] together with the evaluation procedure [Eq. (1.14)]. Therefore, a recursive application of the derivative procedures can be employed to obtain higher derivatives.

In practice, differentiation frameworks allow the subsystems to be of varying "granularity," that is, the supported complexity of subsystems varies depending on the particular framework. On the one side of the spectrum, subsystems are only allowed to represent basic operations such as addition, multiplication, elementary functions (sin, cos,...), and so forth. Some elements of a program (e.g., functions such as the absolute value operator, conditionals, or external functions in compiled form) typically require special attention in these frameworks. On the other side of the spectrum, frameworks allow, or even encourage, subsystems of higher complexity. Let us provide some examples of such frameworks in the context of numerical optimal control methods.

ADOLC [27] is a well-established and general automatic differentiation framework for C++. This framework is based on overloaded operators and the use of an "active" data type `adouble` instead of the standard `double`. ADOLC is, for example, extensively used in the optimal control software PSOPT [31]. Another example of an automatic differentiation framework is the MATLAB® toolbox ADIGATOR [32] employed in the optimal control software GPOPS [33]. The Python toolbox CASADI [28] represents a generic framework including larger building blocks from automatic differentiation, optimization, and optimal control. Among other capabilities, it essentially allows the implementation of custom solvers from the ground up. An example of a MATLAB® framework that encourages the use of subsystems with a higher complexity is the model builder developed in Rieck [29]. This model builder constitutes an integral part of the optimal control toolbox FALCON.m [34] used in this book. It allows users to split functions into symbolic and nonsymbolic subsystems and uses the forward derivative procedure to compute derivatives. In this context, it should be mentioned that there exist typically two main reasons for not using a symbolic evaluation of the function **f** as a whole.

First, increased complexity of the involved expressions often results in a significantly larger build time spent on the symbolic derivative calculation, code generation, and compilation. Here, computer algebra systems are frequently brought to their limit, especially if symbolic simplifications and common subexpression elimination (CSE) (i.e., the automatic assignment of common subexpressions to temporary variables) are performed before or during the process of generating (compilable) code.

Engineering Note 1.4.2: CSE and Compilation

Because of the complexity of the aircraft closed-loop systems as models in optimal control frameworks, the symbolic expressions tend to become rather lengthy, which can significantly affect build time.

In FALCON.m, optimizations for the symbolic derivative computation including CSE can be switched off by setting the flag `OptimizeCode` to `false`. The performance hit from switching off the optimization of derivatives is often negligible if the symbolic expressions are translated to code, which is then processed by a compiler to machine code. In this case, symbolic simplifications and elimination of common subexpressions are often efforts spent in vain as compiler optimizations produce similar code anyway (see Appendix C for an example with different optimization levels and an output of the generated machine code).

Second, some functions (e.g., table data interpolation) do not allow for a direct evaluation with symbolic variables. In such cases, the user is required to implement a subsystem together with its derivatives either manually or based on some suitable approximation in a so-called derivative subsystem. If such a derivative subsystem is required, the manual implementation of the derivatives should mostly be preferred if the implementation effort allows for it. Otherwise, numerical differentiation and sparsity determination methods can be used in these subsystems.

In practice, the use of derivative subsystems is, in fact, often related to cases in which the differentiability requirements for gradient-based optimization methods are violated. Recall that Newton-type iterations as well as the necessary and sufficient conditions of optimality require the existence of first and second derivatives. A function that does not fulfill these requirements can typically also not directly be processed by a symbolic tool-chain. A practical strategy to treat such a problematic function is to introduce a "smoothed" version that preferably also allows for an evaluation with symbolic variables. Smoothed functions may, for example, be parameterized in some $s > 0$ such that the original problem is obtained in the limit $s \to 0^+$. The problem is then solved through a sequence of problems in which s is driven to some small positive number. Many of these smoothing techniques are based on smooth versions of basic functions.

Consider, for example, the step function

$$\text{step}(x) = \begin{cases} 0, & x < 0 \\ 1, & x \geqslant 1 \end{cases}$$

A smooth version of this function is given as a shifted and scaled version of the tangent hyperbolic (tanh) function with $s > 0$:

$$\text{step}(x) \approx \left[1 + \tanh\left(x/s\right)\right]/2$$

This smooth step function is useful in a variety of contexts, and it can be a valuable tool for the application of optimization and optimal control methods in flight control law clearance. A particularly important context is function blending, for example, for the purpose of interpolation. (See, e.g., the smooth interpolation function in the optimal control toolbox PSOPT [31] as well as Bittner [35] for a general description of blending in multiple dimensions.) To describe this blending approach, let us define the following two versions of the smooth step function with offset $o \in \mathbb{R}$ and smoothing factor $s \in \mathbb{R}^+$:

$$\begin{aligned} \lrcorner\!\ulcorner(x; o, s) &= \left\{1 + \tanh\left[\left(x - o\right)/s\right]\right\}/2, \\ \urcorner\!\llcorner(x; o, s) &= \left\{1 - \tanh\left[\left(x - o\right)/s\right]\right\}/2 \end{aligned} \tag{1.17}$$

These versions represent a smooth "step up" ($\lrcorner\!\ulcorner$) and a smooth "step down" ($\urcorner\!\llcorner$) function. Moreover, let us partition the space in each dimension, with index $i = 1, \ldots, d$ based on knot grids:

$$\mathcal{K}_i = \{k_1^i < k_2^i < \ldots < k_{m_i-1}^i < k_{m_i}^i\}$$

into consecutive intervals:

$$(-\infty, k_1^i],\ [k_1^i, k_2^i],\ \ldots,\ [k_{m_i-1}^i, k_{m_i}^i],\ [k_{m_i}^i, \infty)$$

Consider the vector function $\boldsymbol{\beta} : \mathbb{R} \to \mathbb{R}^{m+1}$ for some knot grid $\mathcal{K}$ of cardinality m in direction x:

$$\begin{aligned} \boldsymbol{\beta}^T(x; \mathcal{K}, s) = \big[&\urcorner\!\llcorner(x; k_1, s),\ \lrcorner\!\ulcorner(x; k_1, s)\urcorner\!\llcorner(x; k_2, s), \ldots, \lrcorner\!\ulcorner(x; k_{m-1}, s)\urcorner\!\llcorner(x; k_m, s), \\ &\lrcorner\!\ulcorner(x; k_m, s)\big] \end{aligned}$$

which collects a combination of consecutive univariate steps. Based on this definition, the Kronecker product ($\otimes$) then allows us to construct $n = \prod_{i=1}^{d}(m_i + 1)$ combined multivariate steps $\tilde{\beta}_1 : \mathbb{R}^d \to \mathbb{R}, \ldots, \tilde{\beta}_n : \mathbb{R}^d \to \mathbb{R}$ as the elements of

$$\tilde{\boldsymbol{\beta}}\,(x_1, \ldots, x_d; \mathcal{K}_1, \ldots, \mathcal{K}_d, s) = \boldsymbol{\beta}(x_1; \mathcal{K}_1, s) \otimes \ldots \otimes \boldsymbol{\beta}(x_d; \mathcal{K}_d, s)$$

Appendix A presents a specific two-dimensional example illustrating the construction.

Now, suppose that functions $f_1 : \mathbb{R}^d \to \mathbb{R}, \ldots, f_n : \mathbb{R}^d \to \mathbb{R}$ corresponding to each of the combined steps $\tilde{\beta}_1, \ldots, \tilde{\beta}_n$ are given. For these functions, we wish to construct a blended version $\tilde{f} : \mathbb{R}^d \to \mathbb{R}$ with

$$\tilde{f}(x_1, \ldots, x_d) \approx f_i\,(x_1, \ldots, x_d)$$

at all points where $\tilde{\beta}_i$ approaches one as s approaches zero. This blended function $\tilde{f} : \mathbb{R}^d \to \mathbb{R}$ can be constructed as follows:

$$\tilde{f}(x_1, \ldots, x_d; \mathcal{K}_1, \ldots, \mathcal{K}_d, s) = \sum_{i=1}^{n} f_i\,(x_1, \ldots, x_d)\,\tilde{\beta}_i\,(x_1, \ldots, x_d; \mathcal{K}_1, \ldots, \mathcal{K}_d, s)$$

Let us illustrate this blending approach with three functions representing a simplified model of the temperature T over height H for the troposphere ($H \leqslant 11\,\mathrm{km}$), lower stratosphere, and upper stratosphere ($H \geqslant 20\,\mathrm{km}$). The piecewise definition of this temperature profile is given as

$$T(H) = \begin{cases} T_t\,(H) = 288.15\mathrm{K} - 0.0065H, & H \leqslant 11\,\mathrm{km} \\ T_u\,(H) = 216.65\mathrm{K} + 0.001\,\big(H - 20\,\mathrm{km}\big), & H \geqslant 20\,\mathrm{km} \\ T_l\,(H) - 216.65\mathrm{K}, & \text{otherwise} \end{cases}$$

Based on tanh blending, we obtain

$$\begin{aligned} \overbrace{\tilde{T}\big(H; \{11\,\mathrm{km}, 20\,\mathrm{km}\}, s\big)}^{\tilde{f}} &= \overbrace{T_t\,(H)}^{f_1}\,\overbrace{\rceil\!\lfloor\big(H; 11\,\mathrm{km}, s\big)}^{\tilde{\beta}_1} \\ &\quad + \overbrace{T_l\,(H)}^{f_2}\,\overbrace{\rfloor\!\lceil\big(H; 11\,\mathrm{km}, s\big)\,\rceil\!\lfloor\big(H; 20\,\mathrm{km}, s\big)}^{\tilde{\beta}_2} \\ &\quad + \underbrace{T_u\,(H)}_{f_3}\,\underbrace{\rfloor\!\lceil\big(H; 20\,\mathrm{km}, s\big)}_{\tilde{\beta}_3} \end{aligned}$$

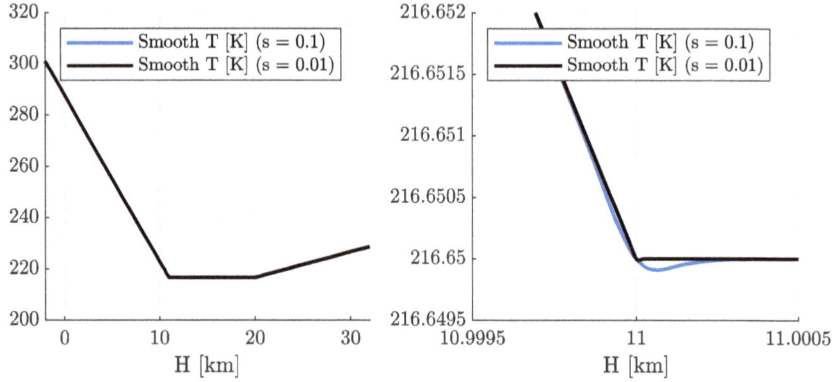

Fig. 1.5 Smoothed temperature curve *T* for different heights *H* (troposphere, lower and upper stratosphere).

The blended version for two different values of the smoothing parameter is shown in Fig. 1.5.

It should be mentioned that, despite its general applicability, whether the tanh blending approach should be the method of choice depends on the particular application. First, note that due to the global support of the tanh function, this type of blending becomes increasingly expensive as the number of contributing functions grows, because *all involved functions need to be executed regardless of the current evaluation point* (unless a pruning strategy is employed). Moreover, as can be seen in the detailed view of the example from Fig. 1.5, the smooth version exhibits a "bump" at the junction point. In particular, observe that the piecewise definition of the temperature over the three layers is convex, whereas the smooth version is not. Such a behavior can be problematic in certain situations, and smoothing based on other means, such as the square-root function, may be preferred in these cases.

For example, a smooth approximation of a limiter (a fairly common function in aircraft closed-loop models) based on square-root smoothing is given as [36]

$$\underline{\overline{x}}\,(x;s) = \frac{\overline{x}+\underline{x}}{2} + \frac{\sqrt{\left(x-\underline{x}\right)^2+s^2}}{2} - \frac{\sqrt{(x-\overline{x})^2+s^2}}{2} \tag{1.18}$$

See Fig. 1.6 for an illustration of this smooth limiter with limits $\underline{x} = -1$ and $\overline{x} = 1$ from below and above for two different values for the smoothing parameter s. Such a smooth limiter can be found from a smooth reformulation of the absolute value function.

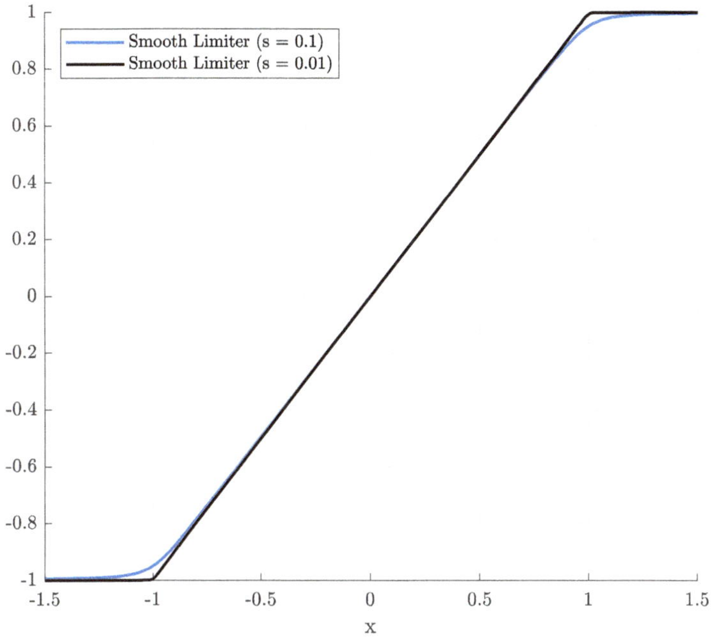

Fig. 1.6 Smooth limiters.

The absolute value function

$$|x| = \begin{cases} x, & x > 0 \\ 0, & x = 0 \\ -x, & x < 0 \end{cases}$$

May be skipped (↴ at first read.

is a common function used in control applications and, in particular, aircraft closed-loop systems. Obviously, the derivative of this function exhibits a discontinuity at $x = 0$.

Based on tanh blending, a smooth version of the absolute value function can be constructed as

$$|x| \approx -x \urcorner\!\!\llcorner(x; 0, s) + x \lrcorner\!\!\ulcorner(x; 0, s) = x \tanh(x/s)$$

For $s > 0$, the graph of this smooth version passes through the origin and, otherwise, lies strictly below the exact function. [Note that the approximation is symmetric, and $0 < \tanh(x/s) < 1$ for $0 < x < \infty$.]

Another smooth version of the absolute value function based on the square-root function is given as

$$|x| \approx \sqrt{x^2 + s^2}$$

For $s > 0$, the graph of this smooth version lies strictly above the exact function. (Note that $|x| = \sqrt{x^2}$, $x \neq 0$ and that the square root function is strictly monotonically increasing.)

Whether the tanh or the square root version should be used depends on the specific application.

Let us consider limiters, which can be expressed in terms of the absolute value function. For this purpose, we write the limiter with a lower limit $\underline{x}$:

$$\max\left(x, \underline{x}\right) = \begin{cases} x, & x \geqslant \underline{x} \\ \underline{x}, & x < \underline{x} \end{cases}$$

in the equivalent form

$$\max\left(x, \underline{x}\right) = \frac{x + \underline{x}}{2} + \frac{|x - \underline{x}|}{2} \tag{1.19}$$

Note that equivalence directly follows from

$$\boxed{x \geqslant \underline{x}}: \quad \frac{x + \underline{x}}{2} + \frac{x - \underline{x}}{2} = x$$

$$\boxed{x < \underline{x}}: \quad \frac{x + \underline{x}}{2} + \frac{\underline{x} - x}{2} = \underline{x}$$

Now, consider the application of both smooth versions for the absolute value function with $s > 0$ to Eq. (1.19). Note that the graph of the square root version lies strictly above the maximum value function:

$$\boxed{x \geqslant \underline{x}}: \quad \frac{x + \underline{x}}{2} + \frac{\sqrt{\left(x - \underline{x}\right)^2 + s^2}}{2} > x$$

$$\sqrt{\left(x - \underline{x}\right)^2 + s^2} > x - \underline{x}$$

$$\left(x - \underline{x}\right)^2 + s^2 > \left(x - \underline{x}\right)^2$$

$$\boxed{x < \underline{x}}: \frac{x+\underline{x}}{2} + \frac{\sqrt{(x-\underline{x})^2 + s^2}}{2} > \underline{x}$$

$$\sqrt{(x-\underline{x})^2 + s^2} > \underline{x} - x$$

$$(x-\underline{x})^2 + s^2 > (\underline{x} - x)^2$$

This typically represents a desirable property as it implies that the limiter is conservative regarding the lower limit, that is, the smooth limiter based on the square root function always yields values strictly greater than $\underline{x}$. For a smooth tanh version, it is, however, easy to find an example for which this is not the case (e.g., $x = 0$, $\underline{x} = 1$, $s = 1$).

The same line of reasoning presented for the maximum value function can be applied to a limiter with upper limit $\overline{x}$ represented by the minimum value function

$$\min(x, \overline{x}) = \begin{cases} x, & x \leqslant \overline{x} \\ \overline{x}, & x > \overline{x} \end{cases}$$

with the equivalent form

$$\min(x, \overline{x}) = \frac{x + \overline{x}}{2} - \frac{|x - \overline{x}|}{2}$$

Similar to the maximum value function, the smooth square root version of the minimum value function is conservative regarding the upper limit, whereas the tanh version yields values exceeding this limit. A limiter for x with lower limit $\underline{x}$ and upper limit $\overline{x}$ can be represented by a sequential application of two limiters, for example,

$$\underline{\overline{x}} = \min\left[\max(x, \underline{x}), \overline{x}\right] = \begin{cases} \overline{x}, & x > \overline{x} \\ x, & \underline{x} \leqslant x \leqslant \overline{x} \\ \underline{x}, & x < \underline{x} \end{cases}$$

which is equivalent to the expression

$$\underline{\overline{x}} = \frac{\overline{x} + \underline{x}}{2} + \frac{|x - \underline{x}|}{2} - \frac{|x - \overline{x}|}{2} \tag{1.20}$$

in terms of the absolute value function. Note that equivalence directly follows from

$$\boxed{x > \overline{x}}: \frac{\overline{x} + \underline{x}}{2} + \frac{x - \underline{x}}{2} - \frac{x - \overline{x}}{2} = \overline{x}$$

$$\boxed{\underline{x} \leqslant x \leqslant \overline{x}} : \frac{\overline{x}+\underline{x}}{2} + \frac{x-\underline{x}}{2} - \frac{\overline{x}-x}{2} = x$$

$$\boxed{x < \underline{x}} : \frac{\overline{x}+\underline{x}}{2} + \frac{\underline{x}-x}{2} - \frac{\overline{x}-x}{2} = \underline{x}$$

Based on the smooth square root version of the absolute value function, we directly obtain the smooth form [Eq. (1.18)] of the limiter [Eq. (1.20)].

It is worth mentioning that in the context of optimization there exists yet another natural possibility to realize the effect of a limiter through the introduction of inequality constraints. Consider, for example, the case that some system state x of the aircraft model has prescribed physical limits $\underline{x}$ from below and $\overline{x}$ from above. Introducing $\underline{x}$ and $\overline{x}$ as inequality bounds on the state variables, the optimal solution needs to respect these limits by an appropriate choice of the inputs to the system. Observe that under this type of modeling, the limit is not modeled as an inherent property inside the system but as an external constraint of the optimization problem. The condition if the limit is reached (constraint is active) or not (constraint is inactive) is thus directly resolved during the solution process through the complementarity condition. The particular method for solving the optimization problem ensures, in this case, that the bounds are respected without further modeling effort required by the user. This method constitutes the primary method used in the optimal control-based clearance methods presented in Part II to model physical limits.

1.4.2 SPARSITY PATTERNS

Sparsity Patterns

The entries of the sparsity pattern $\boxed{\mathbf{M}}$ associated with a matrix $\mathbf{M} \in \mathbb{R}^{n \times m}$ are defined as

$$\left(\boxed{\mathbf{M}}\right)_{i,j} = \begin{cases} 0, & (\mathbf{M})_{i,j} = 0 \\ 1, & \text{otherwise} \end{cases}$$

for all $i = 1, \ldots, n$ and $j = 1, \ldots, m$. Automatized methods for sparsity estimation of subsystems include symbolic computations, error propagation, conservative overapproximations, and random perturbations.

Another important aspect in the context of numerical optimization methods is the concept of sparse problem structures. These sparse structures reveal themselves in so-called *sparsity patterns* of the Jacobian and Hessian matrices and are heavily exploited in numerical algorithms.

In what follows, the term *sparsity pattern* refers to the pattern of zero and potential nonzero entries represented as a binary-valued matrix of the same shape as the associated matrix. An entry in such a pattern expresses if the associated matrix can exhibit a nonzero value at the corresponding location: A zero entry indicates that the value at this location is guaranteed to be zero. Conversely, a one entry indicates that the value at this location can be nonzero.

Engineering Note 1.4.3: Sparse Matrix Formats

If there are relatively few nonzero entries in the matrix, sparsity patterns (just as the associated matrices) are typically not represented as dense arrays but use sparse matrix formats. Among others, common sparse matrix formats are the coordinate format (COO), compressed sparse row (CSR) format, and compressed sparse column (CSC) format.

The COO format is often used for user interfaces (matrix construction and manipulation), whereas the CSR and CSC formats are primarily used as efficient data structures in algorithms. This is contributed to the more intuitive representation of the COO format compared to the other formats.

This COO format can be defined by triplets containing the coordinates (row and column indices) as well as the value at these coordinates. For example, the construction of a sparse matrix in MATLAB® is accomplished by providing three arrays for the row indices, column indices, and values to the sparse matrix constructor:

```
r = ...row indices
c = ...column indices
v = ...values
S = sparse(r,c,v);
```

Internally, such a COO representation is then converted by MATLAB® to a compressed format (CSC), which is favorable for certain algorithms involving sparse matrices.

In the context of optimization software tailored to large-scale problems, these sparse matrix formats are typically used to specify the sparsity patterns. For example, the IPOPT-MATLAB® interface [37] allows the user to provide the respective patterns, for example, for the constraint Jacobian based on sparse matrices:

```
function sparsityPattern = jacobianstructure()
  r = ...row indices
  c = ...column indices
  sparsityPattern = sparse(r,c,ones(size(r)));
```

```
end
funcs.jacobianstructure = @jacobianstructure;
ipopt(...call IPOPT with other inputs
      ,funcs);
```

Similarly, a sparse matrix representing the sparsity pattern of the lower triangular part of the Lagrangian Hessian can be provided.

Sparsity patterns represent an indicator for the coupling within an NLP. For example, in a problem with higher sparsity in the overall Jacobian matrix, there is a tendency for the cost and each of the constraint functions to be influenced by fewer decision variables. This setting is favorable as the underlying numerical algorithms employed in several NLP solvers are tailored to loosely coupled problems [24, 38, 39].

This fact also constitutes a key consideration in the design of numerical optimal control methods. For example, in the case of direct methods, which are the primary focus of this book, increased sparsity in the resulting NLP is achieved through a segmentation of the optimal control problem, for example, using segmented collocation methods (see Sec. 2.2.2). Such a segmentation embeds the problem in a higher-dimensional space through the artificial introduction of decision variables (providing additional degrees of freedom in the NLP) and equality constraints (effectively removing the additional degrees of freedom in the NLP). This strategy is applied extensively in the context of optimal control problems arising from flight control law clearance applications discussed in Part II.

Engineering Note 1.4.4: Sparsity Propagation

An important property for the determination of sparsity patterns is the fact that sparsity information can be propagated through matrix-vector multiplications. If we have a matrix $\mathbf{A}$ and a vector $\mathbf{b}$ with sparsity patterns $\boxed{\mathbf{A}}$ and $\boxed{\mathbf{b}}$, the multiplication

$$\mathbf{c} = \mathbf{A}\mathbf{b}$$

exhibits the sparsity pattern:

$$\boxed{\mathbf{c}} = \boxed{\boxed{\mathbf{A}}\boxed{\mathbf{b}}} \tag{1.21}$$

This property can be exploited to obtain sparsity patterns for the graph representation introduced in Sec. 1.4.1. As the derivative propagation in the forward as well as the backward procedure outlined in the last section only involves matrix-vector multiplications, sparsity information can be propagated in the same procedure by applying Eq. (1.21) to each step. As such, the problem of determining sparsity patterns can be reduced to determining the sparsity patterns of the subsystems (i.e., the vector-valued functions to be evaluated).

In optimal control applications, there exist several strategies to determine sparsity patterns of vector-valued functions.

One strategy is to let the user specify the pattern manually. On the one side, this option can be efficient from a computational perspective as the user has the possibility to provide the exact pattern. On the other side, practical evidence suggests that the human factor combined with the rather nonintuitive binary representation involved in this option can be problematic, rendering it rather error-prone. Observe that if a single zero/one entry in the sparsity pattern of some function is erroneous, it can jeopardize the correctness of the whole NLP.

A second strategy is to use random perturbations of the input values, which represent a method applicable in a very general context. Note, however, that for the correctness of the sparsity pattern, we require that a zero entry guarantees a zero value at the corresponding location in the associated matrix for all possible values in the input space. Clearly, using a sampling-based approach, this property is only guaranteed in some probabilistic sense. Moreover, care must be taken that the random samples are seeded consistently across multiple runs; otherwise, the solution of an optimization problem may work on Monday but fail the remaining days of the week. However, even in the case when working with a consistent seeding strategy, some change in the function may suddenly result in an erroneous sparsity pattern. For our clearance applications in which we also seek to perform automated tests (e.g., in build and test pipelines of version control systems), these properties tend to introduce "flakiness" in the solution of the optimization problems (i.e., for obscure reasons, the procedure sometimes works and sometimes does not).

Another option for sparsity detection is the "NaN-trick" that can be used to detect influences of input variables to output variables. This method is based on the idea to not evaluate the function purely with numbers but to propagate NaN "not-a-number" values through functions. Note that an entry in the Jacobian sparsity pattern indicates that there exists some influence of an input variable to an output variable. One way to reveal such an influence is to propagate an "error" through the function. If an input of some function is flagged as "erroneous," the outputs depending on this input should also be flagged as "erroneous" after an evaluation of the function. This flagging of a variable is accomplished by a special bit pattern that does not represent a floating-point number. As such, the NaN-trick essentially constitutes the exploitation of an error propagation mechanism to determine input–output dependencies. The propagation of NaN values often works, as evaluation environments, for the most part, quietly provide NaN values as outputs if NaN values are provided in the inputs to operators or functions. For example, in MATLAB®, the majority of the basic functions and operators (multiplication, addition, etc.) propagate such NaN values without throwing an evaluation error. The NaN-trick, however, has caveats that can result in erroneous sparsity patterns. First, observe that the sparsity pattern is local in the sense that all input values that are not flagged as NaN need to take on some specific value. Second, it only works out-of-the-box for Jacobian sparsity patterns. Note that we only have a single NaN-flag at our disposal carrying a single bit of information (whether there is a dependency or not). This property mainly affects efficiency and not correctness as for higher derivatives an overapproximation can be used. For example, a conservative approximation of the Hessian's sparsity pattern can be obtained from the sparsity pattern of a gradient vector:

$$\underbrace{\begin{bmatrix} 0 & 0 & 1 & 1 & 0 & 0 \end{bmatrix}}_{\text{Gradient Sparsity}} \rightarrow \underbrace{\begin{bmatrix} 0 & 0 & 0 & 0 & 0 & 0 \\ 0 & 0 & 0 & 0 & 0 & 0 \\ 0 & 0 & 1 & 1 & 0 & 0 \\ 0 & 0 & 1 & 1 & 0 & 0 \\ 0 & 0 & 0 & 0 & 0 & 0 \\ 0 & 0 & 0 & 0 & 0 & 0 \end{bmatrix}}_{\substack{\text{Hessian Sparsity} \\ \text{(Conservative)}}}$$

Last, but not least, the NaN-trick can lead to false-negative detections of influences, implying that some influences are silently missed.

Engineering Note 1.4.5: Caveats of the NaN-Trick for Detecting Influences

Regarding correctness, the following cases may occur in a general procedure detecting influences from inputs to outputs:

- Fail: The procedure fails and does not produce a result.
- Success: The procedure succeeds and produces a result.
  - Correct: An influence is detected that is actually present.
  - False Positive: An influence is detected that is actually not present.
  - False Negative: An influence is not detected that is actually present.

Another aspect is that a procedure may only provide local information, meaning that dependencies are not detected globally on the whole domain of the function, but the result is only valid for certain values of the inputs.

Clearly, it is desirable for a method not to fail for valid user implementations. The more concerning case, however, occurs if the procedure succeeds and produces a false-negative result. Consider, for example, the MATLAB® function given in Listing 1.1.

Listing 1.1 A function for sparsity estimation.

```
function y = f(x)
    %%%%%%%%%%%%%%%%%%%%%%%%%%%%%%%%%%%%%%%%%%%%%%%%%%%%%
    % A Function for Sparsity Estimation
    %%%%%%%%%%%%%%%%%%%%%%%%%%%%%%%%%%%%%%%%%%%%%%%%%%%%%
    % <INPUTS>
    % > x [2 x 1]: input.
    %
    % <OUTPUTS>
    % > y [1 x 1]: output.
    %
    %%%%%%%%%%%%%%%%%%%%%%%%%%%%%%%%%%%%%%%%%%%%%%%%%%%%%
    if (x(2) > 0)
        y = sin(x(1)) + sin(x(2));
    else
        y = 1;
    end
    %%%%%%%%%%%%%%%%%%%%%%%%%%%%%%%%%%%%%%%%%%%%%%%%%%%%%
end
```

The global Jacobian sparsity pattern of this continuously differentiable function is full as both elements in the input vector can influence the output.

Selected inputs are shown in Table 1.1.

TABLE 1.1 INPUTS WITH ANY (FINITE, NON-NAN) NUMBER *a*

f (x)	x
NaN	[NaN, 1]
1	[NaN, 0]
1	[a, NaN]

The first two input combinations from Table 1.1 reveal that the sparsity estimation procedure is in fact not always "global"; in the provided example, the estimation is only correct if the second entry in the input vector is strictly positive. An evaluation with the third input for any $a \in \mathbb{R}$ always yields a false-negative result. This is contributed to the behavior that the result of the expression `(NaN > 0)` in the conditional of the if-statement results in `false`. If no other measures are taken, these false-negative detections result in an erroneous sparsity pattern.

Typically, automatic differentiation frameworks also provide the option to obtain sparsity patterns. Unfortunately, there also exist caveats mainly related to nonsequential code execution. If the code contains branches, this can pose challenges for the implementation in such frameworks and requires special treatment (e.g., through the use of functions particularly designed for this purpose). For example, ADOL-C [27] provides a function `condassign` in two overloaded versions that can be used for the implementation of conditionals `if b > 0 then a = c else a = d` and `if b > 0 then a = c`.

Finally, a rather reliable way to determine sparsity patterns is based on symbolic expressions for the Jacobian and Hessian matrices. Clearly, assuming the correctness of the underlying symbolic framework, a "zero" expression guarantees that for all possible input values, this expression cannot return a nonzero value. Such an approach is followed in the model builder developed in Rieck [29]. Unfortunately, this strategy fails for certain operations. For example, in the case of the function provided in Listing 1.1, the evaluation with symbolic inputs would not produce a false-negative result but would fail with an evaluation error as conversion from a symbolic variable to a logical value is not defined. As a fallback in this case, the model builder allows users to implement a so-called derivative subsystem for modeling the problematic function (see Chapter 9).

1.4.3 NUMERICAL DIFFERENTIATION

Finite Difference Methods

Common finite difference methods for approximating partial derivatives with step-size $\delta_j \in \mathbb{R}^+$ are

$$\text{Forward}: \quad \frac{\partial \mathbf{f}}{\partial z_j} = \frac{\mathbf{f}(\mathbf{z} + \mathbf{e}_j\delta_j) - \mathbf{f}(\mathbf{z})}{\delta_j} + \mathcal{O}(\delta_j) \tag{1.22}$$

$$\text{Backward}: \quad \frac{\partial \mathbf{f}}{\partial z_j} = \frac{\mathbf{f}(\mathbf{z}) - \mathbf{f}(\mathbf{z} - \mathbf{e}_j\delta_j)}{\delta_j} + \mathcal{O}(\delta_j)$$

$$\text{Central}: \quad \frac{\partial \mathbf{f}}{\partial z_j} = \frac{\mathbf{f}\left(\mathbf{z} + \mathbf{e}_j \delta_j\right) - \mathbf{f}\left(\mathbf{z} - \mathbf{e}_j \delta_j\right)}{2\delta_j} + \mathcal{O}\left(\delta_j^2\right) \tag{1.23}$$

In addition to the indicated truncation errors, finite difference approximations suffer from round-off errors.

The formulas for the forward, backward, and central finite difference methods [Eqs. (1.22) and (1.23)] can be obtained based on Taylor expansions with different steps (and combinations thereof).

May be skipped at first read.

Assuming sufficient differentiability of the function $\mathbf{f}$, the expansion for a step of length $\delta_j \in \mathbb{R}^+$ in the direction $\mathbf{e}_j$ yields

$$\mathbf{f}\left(\mathbf{z} + \mathbf{e}_j \delta_j\right) = \mathbf{f}(\mathbf{z}) + \frac{\delta_j^1}{1!}\frac{\partial \mathbf{f}}{\partial z_j} + \frac{\delta_j^2}{2!}\frac{\partial^2 \mathbf{f}}{\partial z_j^2} + \frac{\delta_j^3}{3!}\frac{\partial^3 \mathbf{f}}{\partial z_j^3} + \frac{\delta_j^4}{4!}\frac{\partial^4 \mathbf{f}}{\partial z_j^4} + \ldots \tag{1.24}$$

Division by δ_j

$$\frac{\mathbf{f}\left(\mathbf{z} + \mathbf{e}_j \delta_j\right)}{\delta_j} = \frac{\mathbf{f}(\mathbf{z})}{\delta_j} + \frac{\partial \mathbf{f}}{\partial z_j} + \frac{\delta_j}{2!}\frac{\partial^2 \mathbf{f}}{\partial z_j^2} + \frac{\delta_j^2}{3!}\frac{\partial^3 \mathbf{f}}{\partial z_j^3} + \frac{\delta_j^3}{4!}\frac{\partial^4 \mathbf{f}}{\partial z_j^4} + \ldots \tag{1.25}$$

and truncation of all terms involving powers of δ_j greater than or equal to one, we arrive at

$$\frac{\partial \mathbf{f}}{\partial z_j} = \frac{\mathbf{f}\left(\mathbf{z} + \mathbf{e}_j \delta_j\right) - \mathbf{f}(\mathbf{z})}{\delta_j} + \mathcal{O}\left(\delta_j\right)$$

Similarly, for a step of length δ_j in the opposite direction, the expansion yields

$$\mathbf{f}\left(\mathbf{z} - \mathbf{e}_j \delta_j\right) = \mathbf{f}(\mathbf{z}) - \frac{\delta_j^1}{1!}\frac{\partial \mathbf{f}}{\partial z_j} + \frac{\delta_j^2}{2!}\frac{\partial^2 \mathbf{f}}{\partial z_j^2} - \frac{\delta_j^3}{3!}\frac{\partial^3 \mathbf{f}}{\partial z_j^3} + \frac{\delta_j^4}{4!}\frac{\partial^4 \mathbf{f}}{\partial z_j^4} + \ldots$$

Division by δ_j

$$\frac{\mathbf{f}\left(\mathbf{z} - \mathbf{e}_j \delta_j\right)}{\delta_j} = \frac{\mathbf{f}(\mathbf{z})}{\delta_j} - \frac{\partial \mathbf{f}}{\partial z_j} + \frac{\delta_j}{2!}\frac{\partial^2 \mathbf{f}}{\partial z_j^2} - \frac{\delta_j^2}{3!}\frac{\partial^3 \mathbf{f}}{\partial z_j^3} + \frac{\delta_j^3}{4!}\frac{\partial^4 \mathbf{f}}{\partial z_j^4} + \ldots \tag{1.26}$$

and truncation of all terms involving powers of δ_j greater than or equal to one, we arrive at

$$\frac{\partial \mathbf{f}}{\partial z_j} = \frac{\mathbf{f}(\mathbf{z}) - \mathbf{f}\left(\mathbf{z} - \mathbf{e}_j\delta_j\right)}{\delta_j} + \mathcal{O}\left(\delta_j\right)$$

Both a forward and backward step can be combined to improve the truncation error. Subtraction of Eq. (1.26) from Eq. (1.25) cancels all terms involving derivatives of even order (including order zero). As such, for the combined equation, it is sufficient to truncate terms involving powers of δ_j greater than or equal to two:

$$\frac{\partial \mathbf{f}}{\partial z_j} = \frac{\mathbf{f}\left(\mathbf{z} + \mathbf{e}_j\delta_j\right) - \mathbf{f}\left(\mathbf{z} - \mathbf{e}_j\delta_j\right)}{2\delta_j} + \mathcal{O}\left(\delta_j^2\right)$$

It is important to mention that the indicated truncation errors are not the only source of error in finite difference approximations. In particular, when using a floating point representation, the numerical error involving the subtraction in the numerator tends to increase for decreasing step-sizes. Note that this is contrary to the effect of the truncation error that tends to decrease for decreasing step sizes.

Engineering Note 1.4.6: Real-Valued Finite Difference Step Size

In practice, finite difference step sizes are typically adapted based on the magnitude of the input and the floating point precision. For example, in MATLAB®, the standard settings for the step size in optimization for the forward finite difference formula define the step size as $\sqrt{\epsilon}\max\left(|z_j|, 1\right)$ and the central step $\sqrt[3]{\epsilon}\max\left(|z_j|, 1\right)$, where $\epsilon \approx 2.2204e - 16$ is the value returned by the MATLAB® function `eps`.

The formulas introduced so far can also be used to obtain approximations to higher-order derivatives. Let us define

$$\mathbf{f}^{(1)}(\mathbf{z}) = \frac{\mathbf{f}\left(\mathbf{z} + \mathbf{e}_j\delta_j\right) - \mathbf{f}\left(\mathbf{z} - \mathbf{e}_j\delta_j\right)}{2\delta_j}$$

Then, the application of the central finite difference step δ_k in direction $\mathbf{e}_k$ yields an approximation for the mixed derivative $\frac{\partial^2 \mathbf{f}}{\partial z_j \partial z_k}$ as

$$\frac{\partial \mathbf{f}^{(1)}}{\partial z_k} \approx \frac{\mathbf{f}\left(\mathbf{z} + \mathbf{e}_j\delta_j + \mathbf{e}_k\delta_k\right) - \mathbf{f}\left(\mathbf{z} - \mathbf{e}_j\delta_j + \mathbf{e}_k\delta_k\right) - \mathbf{f}\left(\mathbf{z} + \mathbf{e}_j\delta_j - \mathbf{e}_k\delta_k\right) + \mathbf{f}\left(\mathbf{z} - \mathbf{e}_j\delta_j - \mathbf{e}_k\delta_k\right)}{4\delta_j\delta_k}$$

Note that for diagonal entries with $k = j$, this expression simplifies to

$$\frac{\partial \mathbf{f}^{(1)}}{\partial z_j} \approx \frac{\mathbf{f}\left(\mathbf{z} + 2\mathbf{e}_j\delta_j\right) - 2\mathbf{f}\left(\mathbf{z}\right) + \mathbf{f}\left(\mathbf{z} - 2\mathbf{e}_j\delta_j\right)}{4\delta_j^2}$$

Clearly, each recursive application of this numerical differentiation procedure for obtaining higher derivatives is expected to increase the error as the application of the finite difference formula represents an "approximation over an approximation."

In the context of finite difference methods, there exists yet another popular method for numerical differentiation, commonly referred to as the "complex-step method." Let us motivate this method by a scalar example using the function $f_1(x) = x^2$ to illustrate the mechanics involved in this method. The evaluation of f_1 at a perturbed value $x + i\bar{\delta}$ yields

$$f_1\left(x + i\bar{\delta}\right) = \left(x + i\bar{\delta}\right)^2 = x^2 + 2xi\bar{\delta} - \bar{\delta}^2$$

Taking the imaginary part on both sides and dividing by $\bar{\delta}$, that is,

$$\frac{\Im\left[x^2 + 2xi\bar{\delta} - \bar{\delta}^2\right]}{\bar{\delta}} = \frac{2x\bar{\delta}}{\bar{\delta}} = 2x$$

provides us with the exact derivative for the quadratic equation as the imaginary part carries the derivative scaled by the step length $\bar{\delta}$. Moreover, the real part

$$\Re\left[x^2 + 2xi\bar{\delta} - \bar{\delta}^2\right] = x^2 - \bar{\delta}^2$$

approximates x^2 for decreasing values of $\bar{\delta}$.

In general, the partial derivatives and values of a function based on a complex-valued step are approximated by

$$\frac{\partial \mathbf{f}}{\partial z_j} = \frac{\Im\left[\mathbf{f}\left(\mathbf{z} + \mathbf{e}_j i\bar{\delta}\right)\right]}{\bar{\delta}} + \mathcal{O}\left(\bar{\delta}^2\right) \tag{1.27}$$

and

$$\mathbf{f}\left(\mathbf{z}\right) = \Re\left[\mathbf{f}\left(\mathbf{z} + \mathbf{e}_j i\bar{\delta}\right)\right] + \mathcal{O}\left(\bar{\delta}^2\right) \tag{1.28}$$

The complex step method can be derived using the expansion introduced in Eq. (1.24) and substitution of $\delta_j = i\bar{\delta}$, $\bar{\delta} \in \mathbb{R}^+$:

$$\mathbf{f}\left(\mathbf{z} + \mathbf{e}_j i\bar{\delta}\right) = \mathbf{f}\left(\mathbf{z}\right) + \frac{\left(i\bar{\delta}\right)^1}{1!}\frac{\partial \mathbf{f}}{\partial z_j} + \frac{\left(i\bar{\delta}\right)^2}{2!}\frac{\partial^2 \mathbf{f}}{\partial z_j^2} + \frac{\left(i\bar{\delta}\right)^3}{3!}\frac{\partial^3 \mathbf{f}}{\partial z_j^3} + \frac{\left(i\bar{\delta}\right)^4}{4!}\frac{\partial^4 \mathbf{f}}{\partial z_j^4} + \dots$$

May be skipped (↴ at first read.

$$= \left[\mathbf{f}(\mathbf{z}) - \frac{\bar{\delta}^2}{2!} \frac{\partial^2 \mathbf{f}}{\partial z_j^2} + \frac{\bar{\delta}^4}{4!} \frac{\partial^4 \mathbf{f}}{\partial z_j^4} - \ldots \right]$$

$$+ i \left[\frac{\bar{\delta}}{1!} \frac{\partial \mathbf{f}}{\partial z_j} - \frac{\bar{\delta}^3}{3!} \frac{\partial^3 \mathbf{f}}{\partial z_j^3} + \ldots \right]$$

Taking the imaginary part on both sides,

$$\Im\left[\mathbf{f}\left(\mathbf{z} + \mathbf{e}_j i \bar{\delta}\right)\right] = \frac{\bar{\delta}}{1!} \frac{\partial \mathbf{f}}{\partial z_j} - \frac{\bar{\delta}^3}{3!} \frac{\partial^3 \mathbf{f}}{\partial z_j^3} + \ldots$$

rearranging,

$$\frac{\partial \mathbf{f}}{\partial z_j} = \frac{\Im\left[\mathbf{f}\left(\mathbf{z} + \mathbf{e}_j i \bar{\delta}\right)\right]}{\bar{\delta}} + \frac{\bar{\delta}^2}{3!} \frac{\partial^3 \mathbf{f}}{\partial z_j^3} + \ldots$$

and truncating all terms involving powers of $\bar{\delta}$ greater or equal to two, we directly arrive at the complex-step formula [Eq. (1.27)]. Similarly, taking the real part on both sides,

$$\Re\left[\mathbf{f}\left(\mathbf{z} + \mathbf{e}_j i \bar{\delta}\right)\right] = \mathbf{f}(\mathbf{z}) - \frac{\bar{\delta}^2}{2!} \frac{\partial^2 \mathbf{f}}{\partial z_j^2} + \frac{\bar{\delta}^4}{4!} \frac{\partial^4 \mathbf{f}}{\partial z_j^4} + \ldots$$

rearranging,

$$\mathbf{f}(\mathbf{z}) = \Re\left[\mathbf{f}\left(\mathbf{z} + \mathbf{e}_j i \bar{\delta}\right)\right] + \frac{\bar{\delta}^2}{2!} \frac{\partial^2 \mathbf{f}}{\partial z_j^2} - \frac{\bar{\delta}^4}{4!} \frac{\partial^4 \mathbf{f}}{\partial z_j^4} + \ldots$$

and truncating of all terms involving powers of $\bar{\delta}$ greater than or equal to two, we directly arrive at Eq. (1.28). Extensions of this method for higher derivatives exist (see, e.g., Lai and Crassidis [40]).

Note that the truncation error of the complex step approximation is the same as for the central finite difference method, but there is no subtraction in the numerator introducing round-off errors. Compared to the real-valued finite differences, this allows an unhindered step size reduction progressively reducing the truncation error.

Engineering Note 1.4.7: Complex Step Finite Difference Step Size

Clearly, the step sizes cannot be reduced arbitrarily due to the limited floating point representation of numbers. For example, in MATLAB®, the value returned by `eps(0)` is ≈ $4.9407e-324$. This number constitutes the smallest representable number above zero for the default floating point format with double precision in MATLAB®.

Even though the complex step method seems appealing as we expect a higher accuracy compared to real-valued finite differences, there exist several caveats that can render an application in practice cumbersome.

First and foremost, the method can only be applied to real-valued and analytic functions, and all elements in the algorithm for computing the function must be capable of complex-valued evaluations. In the case of aircraft closed-loop models, this requirement is often violated, and we must rely on reformulations to apply the method. In this context, it is interesting to note that, as shown in Martins et al. [41], the complex-step method can be interpreted as an inexact implementation of the forward derivative method introduced in Sec. 1.4.1 that improves toward smaller step-sizes $\bar{\delta}$. Here, the real part of the complex number approximates $\mathbf{v}_i$, and the imaginary part carries $\mathbf{s}_i\bar{\delta}$ (i.e., the derivatives scaled by the step-size $\bar{\delta}$). Therefore, it should not be surprising that functions posing difficulties for automatic differentiation frameworks are typically also problematic for the complex step method. For example, nonanalytic functions (e.g., the absolute value function), conditionals, and thus also table data interpolation schemes require special treatment for the most part. Moreover, the computational cost regarding memory usage as well as arithmetic operations for complex-valued evaluations is typically higher compared to the real-valued case.

Based on these drawbacks one may conclude that, even though there exist isolated use cases for this method, the outlined practical issues often outweigh the potential gain in accuracy compared to real-valued finite differences.

Finally, let us discuss an improvement to the real-valued finite difference method, referred to as "sparse finite differences." This method is commonly employed in practical optimization and optimal control frameworks (see, e.g., Betts [20], Becerra [31], and Patterson and Rao [33]) and combines the information contained in sparsity patterns introduced in Sec. 1.4.2 to improve the efficiency of finite difference approximations. Recall that the real-valued finite difference formulas presented in this section require function evaluations with perturbed input values. Each perturbation is typically associated with one function evaluation. However, using the information contained in sparsity patterns introduced in Sec. 1.4.2, it is possible to obtain the same output values based on multiple simultaneous perturbations in a single function evaluation.

May be skipped (↴) at first read.

Let us motivate the method based on an example. The columns of the Jacobian sparsity pattern for some function with five inputs and five outputs are given as

$$\begin{array}{ccccc} \mathbf{1} & \mathbf{2} & \mathbf{3} & \mathbf{4} & \mathbf{5} \\ 1 & 1 & 0 & 0 & 1 \\ 0 & 0 & 1 & 0 & 0 \\ 0 & 1 & 1 & 1 & 0 \\ 1 & 0 & 0 & 0 & 0 \\ 0 & 0 & 0 & 0 & 1 \end{array} \tag{1.29}$$

Two columns in the sparsity pattern are considered orthogonal if their inner product is zero. In this example,

- Column **1** is orthogonal to columns **3** and **4**.
- Column **2** is not orthogonal to any column.
- Column **3** is orthogonal to columns **1** and **5**.
- Column **4** is orthogonal to columns **1** and **5**.
- Column **5** is orthogonal to columns **3** and **4**.

Now consider a joint perturbation of the inputs corresponding to, for example, the first and third columns. According to the sparsity pattern, a perturbation of the first input can only lead to a change in the first and fourth outputs. If the third input was perturbed, only the second and third outputs may change. As such, there are no intersecting elements in the sets of outputs affected by these perturbations, and a single evaluation can be used to obtain the perturbed values.

The task to find suitable groups allowing for a joint perturbation in the case of general sparsity patterns can be abstracted to a so-called graph-coloring problem. From this abstract point of view, columns represent vertices in a graph, and edges are defined between two such vertices if the corresponding columns are not orthogonal. Then, the task of identifying the respective groups of orthogonal columns for a Jacobian matrix is equivalent to determining a solution to the *distance-1* coloring problem for this graph. This *distance-1* coloring problem consists of finding a vertex coloring such that no direct neighbors share the same color. The column indices in one perturbation group (i.e., the set of input indices eligible for a joint perturbation) are then represented in the graph by vertices with the same color.

In the case of the sparsity pattern of Eq. (1.29), a colored graph representation of the five columns would be given by

1	2	3	4	5
1	1	0	0	1
0	0	1	0	0
0	1	1	1	0
1	0	0	0	0
0	0	0	0	1

Graph Coloring
(Distance-1)

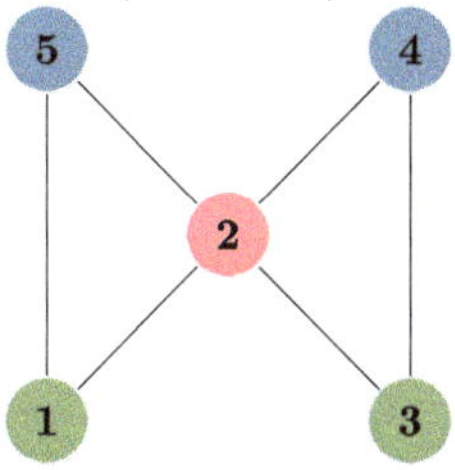

From an algorithmic point of view, the coloring problem is typically solved using greedy schemes with heuristics for the initial ordering (see, e.g., Gebremedhin et al. [42] for an extensive discussion on the topic).

CHAPTER 2

Simulation

2.1 INITIAL VALUE PROBLEMS

Simulation Problem

Find the solution $\hat{\mathbf{x}} : \mathcal{I} \to \mathbb{R}^{n_x}$ on $\mathcal{I} = [t_0, t_f]$, $t_f > t_0$ satisfying

$$\dot{\mathbf{x}}(t) = \mathbf{f}[\mathbf{x}(t), t]$$
$$\mathbf{x}(t_0) = \mathbf{x}_0$$

with $\mathbf{f} : \mathbb{R}^{n_x} \times \mathbb{R} \to \mathbb{R}^{n_x}$ and $\mathbf{x}_0 \in \mathbb{R}^{n_x}$ given.

The second fundamental building block for the optimal control-based clearance approaches presented in this book is devoted to the numerical solution of simulation problems. In the following, *simulation* is understood as the approximate solution of an initial value problem (IVP) on a time interval $[t_0, t_f]$ originating from $\mathbf{x}_0$ and subject to a first-order ordinary differential equation.

In clearance applications, the initial condition is often associated with a particular reference flight condition (e.g., at a trim point) for which we seek to test the control law. Moreover, the differential equation models the closed-loop aircraft system, including subsystems of the plant, servos, as well as the controller.

In practice, these dynamic equations are typically represented in computer-aided environments, such as Simulink®, which besides its modeling capabilities also allows users to directly perform numerical simulations. These approximate IVP solutions are in the majority of cases accomplished in a stepwise manner, that is, the solver starts at the provided initial condition and constructs an approximate solution, one step at a time, using a suitable step-size selection and integration method.

It should be mentioned that certain solvers allow for a variable step-size selection based on local error estimates (e.g., *ode45* in Simulink®). Methods of this type are not considered in the context of the optimal control methods in this book.

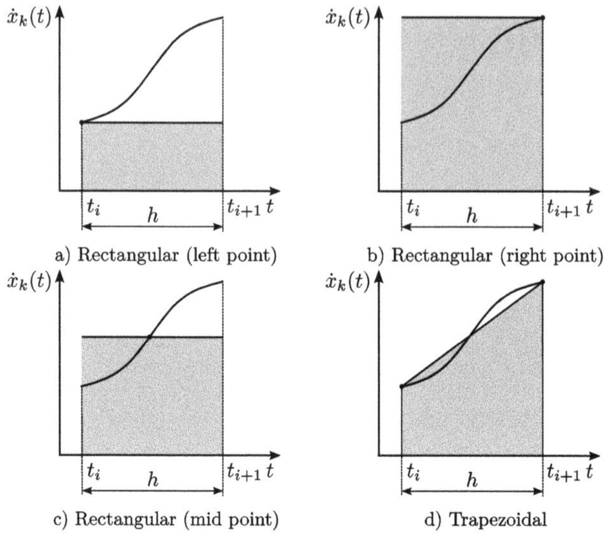

Fig. 2.1 Rectangular- and trapezoidal-shaped integral approximations.

The reason for not making use of such schemes is contributed to the numerical noise introduced by the step-size selection procedures, which is known to be problematic when combined with numerical optimization methods [20].

As such, we focus on methods for a fixed mesh with constant mesh length h. In Simulink®, this is equivalent to the specification of a fixed step-size solver type (*ode1*, *ode2*, ...) together with the step-size h for a given closed-loop model (i.e., an instance of the function **f** from the simulation problem).

Let us motivate the simulation methods presented in the following sections first from an intuitive perspective. To advance an approximate solution $\tilde{\mathbf{x}}(t_i)$ at some point t_i to an approximation $\tilde{\mathbf{x}}(t_{i+1})$ at the next point $t_{i+1} = t_i + h$, a simulation method needs to, in some form, approximate the following expression involving an integral term:

$$\tilde{x}_k(t_{i+1}) = \tilde{x}_k(t_i) + \int_{t_i}^{t_i+h} f_k[\mathbf{x}(t), t]\, dt, \; k = 1, \ldots, n_x$$

As a straightforward approximation to the integral term, solvers may use, for example, a rectangular- or trapezoidal-shaped approximation. Four possible choices are illustrated in Fig. 2.1 where the approximations are performed based on the value of the integrand at the left point, the right point, the midpoint, or both endpoints of the interval $[t_i, t_{i+1}]$.

The accuracy of the approximation clearly depends on the specific points where the integrand is evaluated and how these values are used to perform the integration. Consider, for example, the case in which the integrand is perfectly linear between t_i and t_{i+1}. In this case, the rectangular-shaped approximations from the left and right yield either over- or underestimates for the true value of the integral term. These two approaches would, in fact, only match exactly the true value of this term if the integrand was a constant function over the considered interval. Contrarily, the rectangular-shaped approximation using the midpoint and the trapezoidal-shaped approximation would both yield the exact solution for a linear function.

Note that so far we have assumed that the integrand can be evaluated at an arbitrary point to perform the approximation. However, in the provided examples, the value of the integrand is *explicitly* available only for the left-sided approximation. All other methods are in some form *implicitly* defined in terms of "future" values (i.e., values yet to be determined) that are not directly at the solver's disposal. This implies that for each step an implicit relation would need to be solved numerically. If it is not desirable to spend the effort for solving these implicit relations, it is necessary to use some strategy that provides us with an approximation of these "future" values. For example, in the case of the trapezoidal-shaped approximation, we could first take an auxiliary step using the left-sided approximation to obtain a prediction at t_{i+1}. This prediction could then be used with the value at the left point to (approximately) evaluate the trapezoidal-shaped area under the curve. This strategy is known as *Heun's method* (*ode2* in Simulink®). As it turns out, there exists a whole variety of strategies from a general class of methods (so-called Runge–Kutta methods) that allow us to perform approximations in each step using a particular sequence of intermediate *stages*. These methods form the basis for the majority of methods commonly employed in numerical simulation as well as optimal control frameworks and are the topic of this chapter.

2.2 RUNGE–KUTTA AND COLLOCATION METHODS

2.2.1 RUNGE–KUTTA METHODS

Runge-Kutta Methods

Runge–Kutta methods are employed to find an approximate solution $\tilde{\mathbf{x}} : \mathcal{G} \to \mathbb{R}^{n_x}$ to an initial value problem on a grid $\mathcal{G} = \{t_i = t_0 + ih : i = 0, \ldots, f\}$ with mesh length $h \in \mathbb{R}^+$ starting at $\tilde{\mathbf{x}}(t_0) = \mathbf{x}_0$. The stepwise construction of the approximate solution for each step $i \to i+1$ is defined by the step function $\boldsymbol{\psi} : \mathbb{R}^{n_x} \times \mathbb{R} \to \mathbb{R}^{n_x}$ that, given the current time-point t_i and state approximation $\tilde{\mathbf{x}}(t_i)$, returns an approximation at the next grid point:

$$\tilde{\mathbf{x}}(t_{i+1}) = \boldsymbol{\psi}\left[\tilde{\mathbf{x}}(t_i), t_i\right]$$

For Runge–Kutta methods, this step function is defined as

$$\boldsymbol{\psi}\left[\tilde{\mathbf{x}}(t_i), t_i\right] = \tilde{\mathbf{x}}(t_i) + h \sum_{q=1}^{s} b_q \mathbf{k}_q \left[\tilde{\mathbf{x}}(t_i), t_i\right] \tag{2.1}$$

and depends on the s intermediate-stage derivatives:

$$\mathbf{k}_j\left[\tilde{\mathbf{x}}(t_i), t_i\right] = \mathbf{f}\left(\tilde{\mathbf{x}}(t_i) + h \sum_{q=1}^{s} a_{j,q} \mathbf{k}_q \left[\tilde{\mathbf{x}}(t_i), t_i\right], t_i + c_j h\right), j = 1, \ldots, s \tag{2.2}$$

The coefficients $a_{j,q}, b_q, c_j, j, q = 1, \ldots, s$ are given for a particular Runge–Kutta method.

Runge–Kutta methods are so-called *one-step methods*. Observe that the step function [Eq. (2.1)] only depends on t_i as well as the state approximation at this point and returns an approximation of the state at the next point t_{i+1}. It is noted that there also exist so-called *multistep methods* that additionally use information corresponding to preceding $(t_{i-1}, t_{i-2}, \ldots)$ or following $(t_{i+2}, t_{i+3}, \ldots)$ steps. These methods are not considered in the context of this book. The weighted sum of intermediate-stage derivatives that, scaled by the step-length h, yields the state increment, is usually termed the *increment function*:

$$\boldsymbol{\phi}\left[\tilde{\mathbf{x}}(t_i), t_i\right] = \sum_{q=1}^{s} b_q \mathbf{k}_q \left[\tilde{\mathbf{x}}(t_i), t_i\right] \tag{2.3}$$

Using this definition, the step function can also be written as

$$\boldsymbol{\psi}\left[\tilde{\mathbf{x}}(t_i), t_i\right] = \tilde{\mathbf{x}}(t_i) + h\boldsymbol{\phi}\left[\tilde{\mathbf{x}}(t_i), t_i\right].$$

For notational purposes, we primarily use the step function throughout this chapter, as it conveniently encapsulates the complete step computation for general Runge–Kutta methods, as given in Eq. (2.1). Moreover, this view aligns nicely with the user interface typically provided in simulation software.

For illustration purposes, let us revisit the approaches introduced in the last section based on intuitive arguments and formally embed them in the framework of Runge–Kutta methods. In the case of the left-sided (forward Euler method), right-sided (backward Euler method), and centered (midpoint rule) approximation illustrated in the last section, a single derivative value is used to approximate the integral term to perform a step. Thus, the step function is defined for all three methods as

$$\boldsymbol{\psi}\left[\tilde{\mathbf{x}}(t_i), t_i\right] = \tilde{\mathbf{x}}(t_i) + h\mathbf{k}_1\left[\tilde{\mathbf{x}}(t_i), t_i\right]$$

with $b_1 = 1$. In addition, the intermediate-stage derivative $\mathbf{k}_1$ in the context of Runge–Kutta methods [see Eq. (2.2)] is defined by $a_{1,1} = 0$ and $c_1 = 0$ for the forward Euler method,

$$\mathbf{k}_1\left[\tilde{\mathbf{x}}(t_i), t_i\right] = \mathbf{f}\left[\tilde{\mathbf{x}}(t_i), t_i\right] \tag{2.4}$$

by $a_{1,1} = 1$, $c_1 = 1$ for the backward Euler method,

$$\mathbf{k}_1\left[\tilde{\mathbf{x}}(t_i), t_i\right] = \mathbf{f}\left\{\tilde{\mathbf{x}}(t_i) + h\mathbf{k}_1\left[\tilde{\mathbf{x}}(t_i), t_i\right], t_i + h\right\}, \tag{2.5}$$

and $a_{1,1} = 1/2$, $c_1 = 1/2$ for the midpoint rule,

$$\mathbf{k}_1\left[\tilde{\mathbf{x}}(t_i), t_i\right] = \mathbf{f}\left(\tilde{\mathbf{x}}(t_i) + h\frac{1}{2}\mathbf{k}_1\left[\tilde{\mathbf{x}}(t_i), t_i\right], t_i + \frac{1}{2}h\right) \tag{2.6}$$

Note that in the case of the forward Euler method, the stage derivative is explicitly given by Eq. (2.4). For the backward Euler method and midpoint rule, Eqs. (2.5) and (2.6) for the state derivatives are implicit as in both cases the respective stage derivative depends on itself. Based on the discussion in the last section, this can also be expected for the trapezoidal method. For this method, the step function with $b_1 = 1/2$ and $b_2 = 1/2$ is given by

$$\boldsymbol{\psi}\left[\tilde{\mathbf{x}}(t_i), t_i\right] = \tilde{\mathbf{x}}(t_i) + h\left(\frac{1}{2}\mathbf{k}_1\left[\tilde{\mathbf{x}}(t_i), t_i\right] + \frac{1}{2}\mathbf{k}_2\left[\tilde{\mathbf{x}}(t_i), t_i\right]\right)$$

with the first stage derivative defined by $c_1 = 0$ and $a_{1,1} = a_{1,2} = 0$:

$$\mathbf{k}_1\left[\tilde{\mathbf{x}}(t_i), t_i\right] = \mathbf{f}\left[\tilde{\mathbf{x}}(t_i), t_i\right]$$

as well as the second stage derivative defined by $c_2 = 1$ and $a_{2,1} = a_{2,2} = 1/2$:

$$\mathbf{k}_2\left[\tilde{\mathbf{x}}(t_i), t_i\right] = \mathbf{f}\left(\tilde{\mathbf{x}}(t_i) + h\left(\frac{1}{2}\mathbf{k}_1\left[\tilde{\mathbf{x}}(t_i), t_i\right] + \frac{1}{2}\mathbf{k}_2\left[\tilde{\mathbf{x}}(t_i), t_i\right]\right), t_i + h\right) \tag{2.7}$$

Note that Eq. (2.7) is implicit regarding the second stage $\mathbf{k}_2$. Employing the idea motivated in the last section to approximate the value for the derivative computation at $t_i + h$ by a forward Euler step, the second stage in Heun's method can also be explicitly represented by the following approximation:

$$\mathbf{k}_2\left[\tilde{\mathbf{x}}(t_i), t_i\right] = \mathbf{f}\left\{\tilde{\mathbf{x}}(t_i) + h\mathbf{k}_1\left[\tilde{\mathbf{x}}(t_i), t_i\right], t_i + h\right\}$$

The methods motivated so far are only a small subset of the integration schemes that can be represented as Runge–Kutta methods. For example, all standard simulation methods (*ode1*, …, *ode8*) in Simulink® may be classified using a particular construction sequence and can be expressed by coefficients $a_{j,q}, b_q, c_j, j, q = 1, \ldots, s$. Moreover, as discussed in Sec. 2.2.2, collocation methods, which form the underlying basis of many practical optimal control solvers, also fall under this category.

The coefficients $a_{j,q}, b_q, c_j, j, q = 1, \ldots, s$ of a Runge–Kutta method can be conveniently collected in a matrix $\mathbf{A} \in \mathbb{R}^{s \times s}$ as well as vectors $\mathbf{c} \in \mathbb{R}^s$ and $\mathbf{b} \in \mathbb{R}^{1 \times s}$ and represented in form of arrays (*Butcher tableaus*):

$$\begin{array}{c|c} \mathbf{c} & \mathbf{A} \\ \hline & \mathbf{b} \end{array}$$

Thus, the methods discussed so far can be compactly written in this array form as

Forward Euler, Backward Euler, Midpoint, Trapezoidal, Heun

$$\begin{array}{c|c} 0 & 0 \\ \hline & 1 \end{array} \qquad \begin{array}{c|c} 1 & 1 \\ \hline & 1 \end{array} \qquad \begin{array}{c|c} \frac{1}{2} & \frac{1}{2} \\ \hline & 1 \end{array} \qquad \begin{array}{c|cc} 0 & 0 & 0 \\ 1 & \frac{1}{2} & \frac{1}{2} \\ \hline & \frac{1}{2} & \frac{1}{2} \end{array} \qquad \begin{array}{c|cc} 0 & 0 & 0 \\ 1 & 1 & 0 \\ \hline & \frac{1}{2} & \frac{1}{2} \end{array}$$

Moreover, based on the coefficients given for a particular Runge–Kutta method, it is possible to directly check the order of a method. For a Runge–Kutta method to be of a certain order, so-called *order conditions* for the coefficients are required to hold. These conditions up to order three are summarized in Table 2.1. In addition, we require that the node condition $\sum_{q=1}^{s} a_{j,q} = c_j$ holds (i.e., the row sum of the coefficients in $\mathbf{A}$ is equal to the corresponding entries in $\mathbf{c}$).

To convince ourselves that the methods introduced based on intuitive arguments are Runge–Kutta methods of a certain order, the evaluations of the order

TABLE 2.1 RUNGE–KUTTA ORDER CONDITIONS UP TO ORDER THREE

	Conditions
Node	$\sum_{q=1}^{s} a_{j,q} = c_j$
Order 1	$\sum_{q=1}^{s} b_q = 1$
Order 2	$\sum_{q=1}^{s} b_q c_q = \frac{1}{2}$
Order 3	$\sum_{q=1}^{s} b_q c_q^2 = \frac{1}{3}$
	$\sum_{q=1}^{s} \sum_{j=1}^{s} b_j a_{j,q} c_q = \frac{1}{6}$

(see Butcher [43], Chapter 23, Section 234)

TABLE 2.2 CHECK IF RUNGE–KUTTA ORDER CONDITIONS FOR FORWARD EULER, BACKWARD EULER, MIDPOINT, TRAPEZOIDAL METHODS ARE FULFILLED (✓) OR VIOLATED (×)

	Forward Euler	Backward Euler	Midpoint	Trapezoidal
Node-Conditions	✓	✓	✓	✓
Condition (Order 1)	✓	✓	✓	✓
Condition (Order 2)	×	×	✓	✓
Conditions (Order 3)	×	×	×	×

conditions are summarized in Table 2.2. Based on the provided results, the forward and backward Euler methods both satisfy the node conditions and the first-order condition. The midpoint and trapezoidal methods both additionally satisfy the second-order condition.

Observe that, in contrast to the explicit forward Euler and Heun's method, all implicit methods introduced so far (backward Euler method, midpoint rule, and trapezoidal method) exhibit at least one entry on the upper triangular part of the matrix **A**. In general, from the array representation of Runge–Kutta methods, it becomes immediately apparent if a method is explicit (**A** is strictly lower triangular) or implicit.

Engineering Note 2.2.1: Runge–Kutta Step Function Evaluation

In the case of explicit methods, the computation of the solution in the step function can be directly implemented for general methods using a sequential construction of the stages as shown in Listing 2.1.

Listing 2.1 Evaluation of the step function for general explicit Runge–Kutta methods.

```
%%%%%%%%%%%%%%%%%%%%%%%%%%%%%%%%%%%%%%%%%%%%%%%%
% Step Function (Runge-Kutta, Explicit)
%%%%%%%%%%%%%%%%%%%%%%%%%%%%%%%%%%%%%%%%%%%%%%%%
% <INPUTS>
% > x_old [m x 1]: State vector at old time point.
% > t_old [1 x 1]: Old time point.
% > f     [1 x 1]: F-handle (x_dot = f(x,t)).
% > h     [1 x 1]: Step-size h.
% > A     [s x s]: Butcher tableau coefficient matrix A.
% > b     [1 x s]: Butcher tableau coefficient vector b.
% > c     [s x 1]: Butcher tableau coefficient vector c.
%
```

```
% <OUTPUTS>
% > x_new [m x 1]: State vector at t_old + h.
%
% <NOTES>
% > the Butcher tableau (explicit) is of the following
    form:
%   c_1 |   0    ...    0
%   c_2 | a_2_1 ...    0
%    .  |   .           .
%    .  |   .           .
%    .  |   .           .
%   c_s | a_s_1 ... a_s_s
%   ------------------------
%       |  b_1  ...  b_s
%%%%%%%%%%%%%%%%%%%%%%%%%%%%%%%%%%%%%%%%%%%%%%%%%%%%%%%%%%%%
function x_new = psi(x_old, t_old, f, h, A, b, c)
    s = numel(c);
    X = repmat(x_old, 1, s);
    K = zeros(size(X));
    for j = 1:s
        for q = 1:j-1
            X(:,j) = X(:,j) + h * A(j,q) * K(:,q);
        end
        K(:,j) = f(X(:,j), t_old + c(j,1) * h);
    end
    x_new = x_old;
    for q = 1:s
        x_new = x_new + h * b(1,q) * K(:,q);
    end
end
```

Note that the nested loop for the computation of the stages only iterates over the lower triangular elements of the coefficient matrix $\mathbf{A}$ as all elements on the diagonal and above are zero for explicit methods.

For implicit methods, additional effort is required to solve the nonlinear system of equations for each integration step, which is usually accomplished using numerical root-finding methods. In the case of optimal control methods, these implicit equations are, for the most part, embedded in the optimization problem resulting from the transcription process (i.e., the equations are solved directly by the Newton-type iterations of the resulting nonlinear program). One reason for employing this strategy is that it avoids internal iterations introducing numerical noise in function evaluations; after all, the implicit equations can internally only be solved to a certain tolerance [20].

2.2.2 COLLOCATION METHODS

Collocation Methods

Collocation methods construct piecewise polynomials $\boldsymbol{\sigma}_i : \mathcal{I}_i \to \mathbb{R}^{n_x}$, $i = 0, \ldots, f-1$ approximating the solution of an initial value problem on subintervals $\mathcal{I}_i = [t_i, t_{i+1}]$ based on the collocation conditions

$$\dot{\boldsymbol{\sigma}}_i\left(t_{i,j}\right) = \mathbf{f}\left[\boldsymbol{\sigma}_i\left(t_{i,j}\right), t_{i,j}\right]$$
$$\boldsymbol{\sigma}_i\left(t_i\right) = \tilde{\mathbf{x}}\left(t_i\right)$$

with $t_{i,j} = t_i + hc_j$ and c_j, $j = 1, \ldots, s$ given. The collocation conditions in integral form are

$$\boldsymbol{\sigma}_i\left(t_{i,j}\right) = \tilde{\mathbf{x}}\left(t_i\right) + \int_{t_i}^{t_{i,j}} \mathbf{f}\left[\boldsymbol{\sigma}_i\left(t\right), t\right] dt$$

Collocation methods are the basis for a considerable number of numerical optimal control solvers (see, e.g., Betts [20], Becerra [31], Patterson and Rao [33], and Rieck [34]). It can be shown that collocation methods in the form used in this book belong to the class of implicit Runge–Kutta methods [44], which implies that the general properties and notions discussed in the last section are directly applicable. These collocation methods are in practice based either on a global approximation using schemes with sufficiently high order or in so-called *segmented form* using several local approximations of lower order.

It should be noted that these segments do not need to be necessarily of the same length or be discretized with the same method. Certain types of mesh refinement methods exploit these degrees of freedom to either refine the order of the collocation method for each segment or to split segments at particular points [45]. In this book, however, for notational simplicity, we assume the same segment length h and the same method/order for all segments. Moreover, we focus primarily on segmented forms with a large number of segments and methods of low order. These types of schemes ultimately yield highly decoupled nonlinear programming problems, which facilitates the numerical solution for specialized solvers able to exploit the resulting sparse structures.

Collocation methods come in two different flavors: differential and integral forms. For differential methods (see Fig. 2.2), the difference ("defect") between the derivative of the polynomial approximation and the right-hand side evaluated

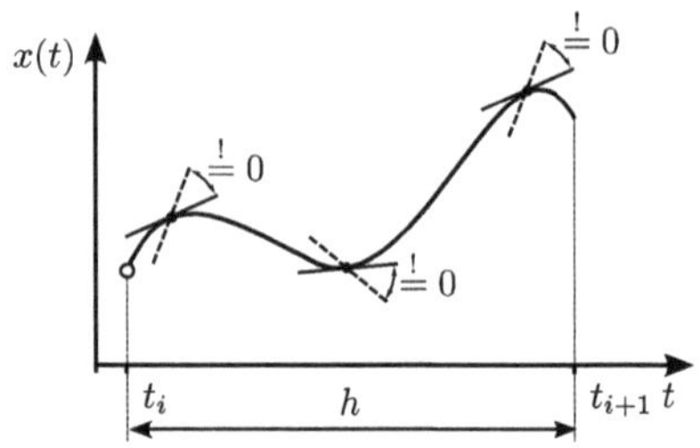

Fig. 2.2 Collocation defects (differential) for a polynomial segment.

with the polynomial approximation at the collocation points $t_{i,j}$, $j = 1, \ldots, s$ is required to be zero:

$$\dot{\boldsymbol{\sigma}}_i\left(t_{i,j}\right) - \mathbf{f}\left[\boldsymbol{\sigma}_i\left(t_{i,j}\right), t_{i,j}\right] = \mathbf{0}$$

In the second form, which is the form primarily used in this book, the defect is formulated in an integral sense:

$$\boldsymbol{\sigma}_i\left(t_{i,j}\right) - \tilde{\mathbf{x}}\left(t_i\right) - \int_{t_i}^{t_{i,j}} \mathbf{f}\left[\boldsymbol{\sigma}_i\left(t\right), t\right] dt = \mathbf{0}$$

The set of collocation methods considered in the following is based on the widely used Legendre–Gauss points. Readers may be familiar with Legendre–Gauss points in the context of quadrature (numerical integration) methods.

There exist several different types of point sets in this class that can be categorized based on the points required to be included for an interval.

- *Gauss* (none of the endpoints): Gauss points represent the optimal choice regarding the approximation order if no restriction is imposed on the placement of the points. With s Gauss points, polynomials up to a degree of $2s - 1$ can be integrated exactly by quadrature.
- *Radau* (one of the interval endpoints): Radau points represent the optimal choice regarding the approximation order if one of the endpoints needs to be included. With s Radau points, polynomials up to a degree of $2s - 2$ can be integrated exactly by quadrature (i.e., one degree less compared to Gauss points).
- *Lobatto* (both endpoints): Lobatto points represent the optimal choice regarding the approximation order if both endpoints need to be included. With s Lobatto points, polynomials up to a degree of $2s - 3$ can be integrated exactly by quadrature (i.e., two degrees less compared to Gauss points).

Note that there are two different possible sets of points for Radau methods, the one including the left endpoint and the one including the right. The set in which the right endpoint of the interval is included is usually termed the *flipped*

TABLE 2.3 THE FIRST GAUSS, RADAU (FLIPPED), AND LOBATTO POINTS ON THE INTERVAL [0, 1]

Gauss	Radau	Lobatto
$\frac{1}{2}$	1	–
$\frac{1}{2} - \frac{\sqrt{3}}{6}$	$\frac{1}{3}$	0
$\frac{1}{2} + \frac{\sqrt{3}}{6}$	1	1
$\frac{1}{2} - \frac{\sqrt{15}}{10}$	$\frac{2}{5} - \frac{\sqrt{6}}{10}$	0
$\frac{1}{2}$	$\frac{2}{5} + \frac{\sqrt{6}}{10}$	$\frac{1}{2}$
$\frac{1}{2} + \frac{\sqrt{15}}{10}$	1	1
$\vdots$	$\vdots$	$\vdots$

set. The first points for the Gauss, Radau (flipped), and Lobatto points are given in Table 2.3.

Observe that for Lobatto methods, the left and right interval endpoints must always be included. A direct implication of this requirement is that the global polynomial approximation is not only continuous but also continuously differentiable. Clearly, if for each segment the collocation conditions ensure that the derivative of the piecewise polynomial matches the right-hand side and the polynomials are continuous across segment boundaries, then also the derivatives of the polynomial segments must match at these boundaries. As mentioned at the beginning of this section, collocation methods are a subset of Runge–Kutta methods and can, as such, be compactly expressed in the form of Butcher tableaus introduced in the last section. In most optimal control applications, the implicit collocation equations are, however, not solved internally but directly embedded in the transcription method. For this purpose, additional state variables are introduced in the optimization method as decision variables, and the collocation conditions are imposed as equality constraints. Moreover, the polynomial approximation directly provides the possibility to return a dense output (i.e., an approximation to the solution of the initial value problem in between grid points).

Let us illustrate the construction of collocation methods using a Lagrange interpolant for the derivative values at distinct collocation points c_j, $j = 1, \ldots, s$. The Runge–Kutta coefficients $a_{j,q}$, b_q, $j, q = 1, \ldots, s$ for a particular method can be obtained based on the integration of this polynomial [44].

Let the Lagrange interpolant for a general function $f : \mathbb{R} \to \mathbb{R}$ be defined as

$$f(t) \approx \sum_{q=1}^{s} f(t_q)\, l_q(t)$$

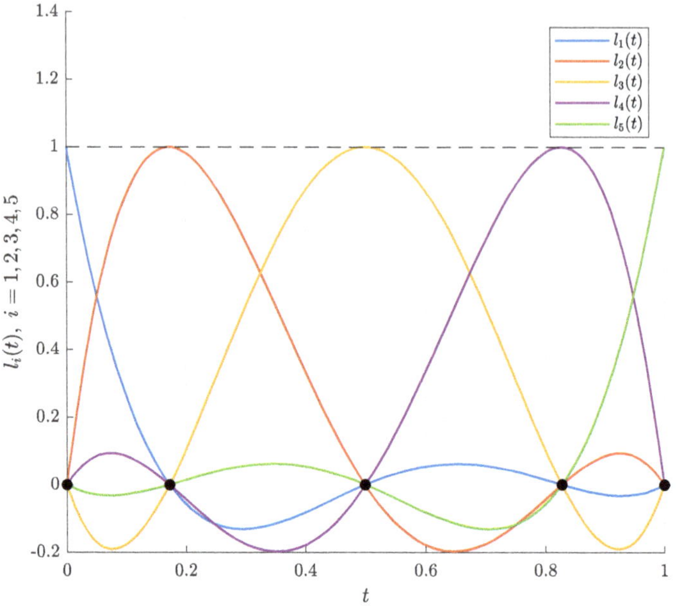

Fig. 2.3 Illustration of five Lagrange basis functions $l_1, \ldots, l_5$ on [0, 1].

with basis functions $l_q : \mathbb{R} \to \mathbb{R}$:

$$l_q(t) = \prod_{\substack{j=1 \\ j \neq q}}^{s} \frac{t - t_j}{t_q - t_j}$$

Each Lagrange basis function l_q has the property that it is one at knot point t_q and zero at all others (see Fig. 2.3 for an illustration):

$$l_q(t_j) = \begin{cases} 1, & q = j \\ 0, & \text{otherwise} \end{cases}$$

Using these basis functions, the procedure regarding integral collocation methods is to first formulate the Lagrange interpolant for the derivative values between collocation points $t_{i,q} = t_i + hc_q$, $q = 1, \ldots, s$ with c_q given:

$$\dot{\boldsymbol{\sigma}}_i(t) = \sum_{q=1}^{s} \mathbf{f}\left[\boldsymbol{\sigma}_i\left(t_{i,q}\right), t_{i,q}\right] l_q(t), \qquad t \in [t_i, t_{i+1}]$$

The integration of this interpolant with constant $\tilde{\mathbf{x}}(t_i)$ yields

$$\boldsymbol{\sigma}_i(t) = \tilde{\mathbf{x}}(t_i) + \sum_{q=1}^{s} \mathbf{f}\left[\boldsymbol{\sigma}_i\left(t_{i,q}\right), t_{i,q}\right] \int_{t_i}^{t} l_q(\theta)d\theta, \qquad t \in [t_i, t_{i+1}]$$

An evaluation of the interpolant at each collocation point $t_{i,j} = t_i + hc_j$, $j = 1, \ldots, s$ then allows for an identification of the Runge–Kutta coefficients $a_{j,q}$, $j, q = 1, \ldots, s$:

$$\boldsymbol{\sigma}_i\left(t_{i,j}\right) = \tilde{\mathbf{x}}(t_i) + \sum_{q=1}^{s} \underbrace{\mathbf{f}\left[\boldsymbol{\sigma}_i\left(t_{i,q}\right), t_{i,q}\right]}_{\mathbf{k}_q} \underbrace{\int_{t_i}^{t_{i,j}} l_q(t)dt}_{ha_{j,q}}$$

Similarly, the evaluation of the interpolant at the interval endpoint $t_{i+1} = t_i + h$ allows for an identification of the Runge–Kutta coefficients b_q, $q = 1, \ldots, s$:

$$\tilde{\mathbf{x}}(t_{i+1}) = \boldsymbol{\sigma}_i(t_{i+1}) = \tilde{\mathbf{x}}(t_i) + \sum_{q=1}^{s} \underbrace{\mathbf{f}\left[\boldsymbol{\sigma}_i\left(t_{i,q}\right), t_{i,q}\right]}_{\mathbf{k}_q} \underbrace{\int_{t_i}^{t_{i+1}} l_q(t)dt}_{hb_q}$$

Case 1: Let us return to the examples given in the introduction of this chapter, and use a rectangular-shaped approximation of the integrand. The interpolating polynomial for the state derivatives in this case is the constant function

$$l_1(t) = 1$$

and the semidefinite integral yields

$$\int_{t_i}^{t} l_1(\theta)\, d\theta = t - t_i$$

resulting in the dense output:

$$\boldsymbol{\sigma}_i(t) = \tilde{\mathbf{x}}(t_i) + \mathbf{f}\left[\boldsymbol{\sigma}_i\left(t_{i,1}\right), t_{i,1}\right](t - t_i)$$

At the point $t_{i,1} = t_i + hc_1$, we get

$$\boldsymbol{\sigma}_i\left(t_{i,1}\right) = \tilde{\mathbf{x}}(t_i) + \mathbf{f}\left[\boldsymbol{\sigma}_i\left(t_{i,1}\right), t_{i,1}\right] \underbrace{\left(t_{i,1} - t_i\right)}_{hc_1}$$

and, similarly, at $t_{i+1} = t_i + h$:

$$\tilde{\mathbf{x}}(t_{i+1}) = \boldsymbol{\sigma}_i(t_{i+1}) = \tilde{\mathbf{x}}(t_i) + \mathbf{f}\left[\boldsymbol{\sigma}_i(t_{i,1}), t_{i,q}\right] \underbrace{(t_{i+1} - t_i)}_{h}$$

After a division of the identified coefficients by h, the following Butcher tableau parameterized by c_1 is obtained:

$$\begin{array}{c|c} c_1 & c_1 \\ \hline & 1 \end{array}$$

For example, an insertion of the first (flipped) Radau point $c_1 = 1$ from the Legendre–Gauss points provided in Table 2.3 yields the Butcher tableau for the backward Euler method:

$$\begin{array}{c|c} 1 & 1 \\ \hline & 1 \end{array}$$

May be skipped (↴) at first read.

Case 2: The Lagrange basis functions for the interpolation of the state derivatives with two points are

$$l_1(t) = \frac{(t - t_i) - c_2 h}{h(c_1 - c_2)}$$

$$l_2(t) = \frac{(t - t_i) - c_1 h}{h(c_2 - c_1)}$$

and the semidefinite integrals yield

$$\int_{t_i}^{t} l_1(\theta)\, d\theta = \frac{(t - t_i)\left(t_i - t + 2c_2 h\right)}{2h(c_2 - c_1)}$$

$$\int_{t_i}^{t} l_2(\theta)\, d\theta = \frac{(t - t_i)\left(t_i - t + 2c_1 h\right)}{2h(c_1 - c_2)}$$

resulting in the dense output:

$$\boldsymbol{\sigma}_i(t) = \tilde{\mathbf{x}}(t_i) + \mathbf{f}\left[\boldsymbol{\sigma}_i\left(t_{i,1}\right), t_{i,1}\right] \frac{(t - t_i)\left(t_i - t + 2c_2 h\right)}{2h(c_2 - c_1)}$$

$$+\,\mathbf{f}\left[\boldsymbol{\sigma}_i\left(t_{i,2}\right),t_{i,2}\right]\frac{\left(t-t_i\right)\left(t_i-t+2c_1h\right)}{2h\left(c_1-c_2\right)}$$

From an evaluation of the polynomial at the collocation point $t_{i,1} = t_i + hc_1$, we obtain:

$$\boldsymbol{\sigma}_i\left(t_{i,1}\right) = \tilde{\mathbf{x}}\left(t_i\right) + \mathbf{f}\left[\boldsymbol{\sigma}_i\left(t_{i,1}\right),t_{i,1}\right]\overbrace{\frac{\left(t_{i,1}-t_i\right)\left(t_i-t_{i,1}+2c_2h\right)}{2h\left(c_2-c_1\right)}}^{h\frac{c_1\left(c_1-2c_2\right)}{2\left(c_1-c_2\right)}}$$
$$+\,\mathbf{f}\left[\boldsymbol{\sigma}_i\left(t_{i,2}\right),t_{i,2}\right]\underbrace{\frac{\left(t_{i,1}-t_i\right)\left(t_i-t_{i,1}+2c_1h\right)}{2h\left(c_1-c_2\right)}}_{h\frac{{c_1}^2}{2\left(c_1-c_2\right)}}$$

Similarly, at $t_{i,2} = t_i + hc_2$,

$$\boldsymbol{\sigma}_i\left(t_{i,2}\right) = \tilde{\mathbf{x}}\left(t_i\right) + \mathbf{f}\left[\boldsymbol{\sigma}_i\left(t_{i,1}\right),t_{i,1}\right]\overbrace{\frac{\left(t_{i,2}-t_i\right)\left(t_i-t_{i,2}+2c_2h\right)}{2h\left(c_2-c_1\right)}}^{h\frac{{c_2}^2}{2\left(c_2-c_1\right)}}$$
$$+\,\mathbf{f}\left[\boldsymbol{\sigma}_i\left(t_{i,2}\right),t_{i,2}\right]\underbrace{\frac{\left(t_{i,2}-t_i\right)\left(t_i-t_{i,2}+2c_1h\right)}{2h\left(c_1-c_2\right)}}_{h\frac{c_2\left(2c_1-c_2\right)}{2\left(c_1-c_2\right)}}$$

Finally, the evaluation at the next point $t_{i+1} = t_i + h$ yields,

$$\boldsymbol{\sigma}_i\left(t_{i+1}\right) = \tilde{\mathbf{x}}\left(t_i\right) + \mathbf{f}\left[\boldsymbol{\sigma}_i\left(t_{i,1}\right),t_{i,1}\right]\overbrace{\frac{\left(t_{i+1}-t_i\right)\left(t_i-t_{i+1}+2c_2h\right)}{2h\left(c_2-c_1\right)}}^{h\frac{2c_2-1}{2\left(c_2-c_1\right)}}$$
$$+\,\mathbf{f}\left[\boldsymbol{\sigma}_i\left(t_{i+1}\right),t_{i+1}\right]\underbrace{\frac{\left(t_{i+1}-t_i\right)\left(t_i-t_{i+1}+2c_1h\right)}{2h\left(c_1-c_2\right)}}_{h\frac{2c_1-1}{2\left(c_1-c_2\right)}}$$

As in the first example, we can divide the identified coefficients by h to obtain the following Butcher tableau parameterized by c_1 and c_2:

$$
\begin{array}{c|cc}
c_1 & \frac{c_1(c_1-2c_2)}{2(c_1-c_2)} & \frac{c_1{}^2}{2(c_1-c_2)} \\
c_2 & \frac{c_2{}^2}{2(c_2-c_1)} & \frac{c_2(2c_1-c_2)}{2(c_1-c_2)} \\
\hline
 & \frac{2c_2-1}{2(c_2-c_1)} & \frac{2c_1-1}{2(c_1-c_2)}
\end{array}
$$

If we insert, for example, the first two Lobatto points, $c_1 = 0$ and $c_2 = 1$, from the Legendre–Gauss points provided in Table 2.3, we obtain the Butcher tableau for the trapezoidal method:

$$
\begin{array}{c|cc}
0 & 0 & 0 \\
1 & 1/2 & 1/2 \\
\hline
 & 1/2 & 1/2
\end{array}
$$

The case with three collocation points that, for example, yields the Hermite–Simpson collocation method is provided in the supplemental material in Appendix B.

2.3 PARAMETRIC INITIAL VALUE PROBLEMS

Parametric Simulation Problem

Let the simulation problem depend on a parameter vector $\mathbf{z} \in \mathbb{R}^n$:

$$\dot{\mathbf{x}}(t, \mathbf{z}) = \mathbf{f}[\mathbf{x}(t, \mathbf{z}), t, \mathbf{z}]$$

$$\mathbf{x}(t_0, \mathbf{z}) = \boldsymbol{\gamma}(\mathbf{z})$$

with $\mathbf{f} : \mathbb{R}^{n_x} \times \mathbb{R} \times \mathbb{R}^n \to \mathbb{R}^{n_x}$ and the function $\boldsymbol{\gamma} : \mathbb{R}^n \to \mathbb{R}^{n_x}$ defining the initial condition. In clearance applications, the parameter vector $\mathbf{z}$ collects static parameters as well as control coefficients, and the clearance criterion $J : \mathbb{R}^{n_x} \times \mathbb{R}^n \to \mathbb{R}$ is of the form:

$$y = J[\mathbf{x}(t, \mathbf{z}), \mathbf{z}] \tag{2.8}$$

Find the derivatives of the criterion at some $t = t_f$ with respect to all elements in the parameter vector $z \in \mathbf{z}$ at a nominal $\mathbf{z} = \tilde{\mathbf{z}}$.

Recall that the key concept in formulating flight control law clearance problems as optimization problems is the use of minimization methods to search for a violation of the criterion. Under this approach, the criterion is introduced into the objective function to be minimized.

The evaluation of this criterion involves the simulation of the aircraft closed-loop system depending on parameters within the model (e.g., aerodynamic coefficients, weight and weight distribution), the control inputs entering the model (e.g., the stick or pedal inputs), or parameters related to the initial condition (e.g., the static and dynamic pressure at a reference point). As discussed extensively throughout Chapter 1, optimization methods require first- and second-order derivative information for the statement of the optimality conditions as well as the identification of a suitable search direction in each iteration. Methods for parametric simulation problems can be used to efficiently compute these derivatives.

Observe that the dependencies of the criterion (2.8) on the parameter vector $\mathbf{z}$ are twofold:

$$y = J[\overbrace{\mathbf{x}(t, \mathbf{z})}^{\text{Indirect}}, \underbrace{\mathbf{z}}_{\text{Direct}}] \tag{2.9}$$

First, the criterion depends directly on $\mathbf{z}$. Second, the state vector at the final time point is influenced indirectly by the parameter vector as the initial condition and the differential equation of the IVP are both allowed to be parameter dependent.

In the majority of cases, the indirect dependence represents the more challenging part regarding the derivative computation. This is contributed to the fact that the parameter sensitivities of the state vector obtained from the solution of the parametric initial value problem need to be taken into account.

A straightforward method to obtain the derivatives at some $t = t_f$ is to perform a finite difference approximation for the first derivatives as

$$\left.\frac{dJ}{dz_j}\right|_{t=t_f} \approx \frac{J|_{t=t_f,\mathbf{z}=\tilde{\mathbf{z}}+\mathbf{e}_j\delta_j} - J|_{t=t_f,\mathbf{z}=\tilde{\mathbf{z}}-\mathbf{e}_j\delta_j}}{2\delta_j} \tag{2.10}$$

and for the mixed derivatives

$$\left.\frac{d^2J}{dz_j dz_k}\right|_{t=t_f} \approx \frac{\begin{array}{l} J|_{t=t_f,\mathbf{z}=\tilde{\mathbf{z}}+\mathbf{e}_j\delta_j+\mathbf{e}_k\delta_k} - J|_{t=t_f,\mathbf{z}=\tilde{\mathbf{z}}-\mathbf{e}_j\delta_j+\mathbf{e}_k\delta_k} - J|_{t=t_f,\mathbf{z}=\tilde{\mathbf{z}}+\mathbf{e}_j\delta_j-\mathbf{e}_k\delta_k} \\ + J|_{t=t_f,\mathbf{z}=\tilde{\mathbf{z}}-\mathbf{e}_j\delta_j-\mathbf{e}_k\delta_k} \end{array}}{4\delta_j\delta_k} \tag{2.11}$$

with suitable step-sizes $\delta_j \in \mathbb{R}^+$ and $\delta_k \in \mathbb{R}^+$. As discussed in Sec. 1.4.3, there exist certain drawbacks of this strategy regarding the achievable accuracy due to truncation and round-off errors. Moreover, note that in the case of our clearance applications, the solution of additional initial value problems is required to evaluate the perturbed criteria in Eqs. (2.10) and (2.11). As shown in the sequel, there are strategies commonly employed in numerical optimal control methods that can be applied to the clearance task to mitigate these drawbacks. These strategies represent the forward and backward derivative procedure to obtain first- and second-order derivatives for a clearance criterion of the general form [Eq. (2.9)].

In the context of flight control law clearance, it is useful to partition the general parameter vector $\mathbf{z}$ as

$$\mathbf{z}^T = \left[\mathbf{p}^T, \mathbf{w}_1^T, \ldots, \mathbf{w}_{n_u}^T\right] \tag{2.12}$$

Here, the vectors $\mathbf{w}_1, \ldots, \mathbf{w}_{n_u}$ represent control coefficients related to parameterizations of the individual inputs collected in the control vector:

$$\mathbf{u}\left(t, \mathbf{w}_1, \ldots, \mathbf{w}_{n_u}\right) = \begin{bmatrix} u_1\left(t, \mathbf{w}_1\right) \\ \vdots \\ u_{n_u}\left(t, \mathbf{w}_{n_u}\right) \end{bmatrix}$$

The vector $\mathbf{p} \in \mathbb{R}^{n_p}$ collects all other static parameters affecting the initial condition as well as the closed-loop model. For a control parametrization with the coefficient vectors $\mathbf{w}_1, \ldots, \mathbf{w}_{n_u}$, the parametrization is of the form

$$\begin{aligned} u_1\left(t, \mathbf{w}_1\right) &= \mathbf{w}_1^T \mathbf{b}_1\left(t\right) \\ &\vdots \\ u_{n_u}\left(t, \mathbf{w}_{n_u}\right) &= \mathbf{w}_{n_u}^T \mathbf{b}_{n_u}\left(t\right) \end{aligned} \tag{2.13}$$

with the vectors $\mathbf{b}_1, \ldots, \mathbf{b}_{n_u}$ collecting basis functions. Note that each control is allowed to possess an individual parametrization.

A common choice of basis functions in optimal control applications are B-splines. These include the important special cases of a piecewise constant (degree-zero B-spline) and piecewise linear parameterization (degree-one B-spline). Moreover, B-splines allow for the use of nonuniform grids and arbitrary smoothness of the control interpolant. The B-spline basis functions up to degree $p = 3$ are illustrated in Fig. 2.4. As can be seen in Fig. 2.4, B-spline basis functions possess local support, meaning that they are nonzero only over a limited range. This property offers advantages in numerical optimal control methods. Essentially, the local support property ensures that the influence of a particular control coefficient is localized, thus reducing coupling within the problem, which is a desirable property facilitating the numerical solution using optimization algorithms [21].

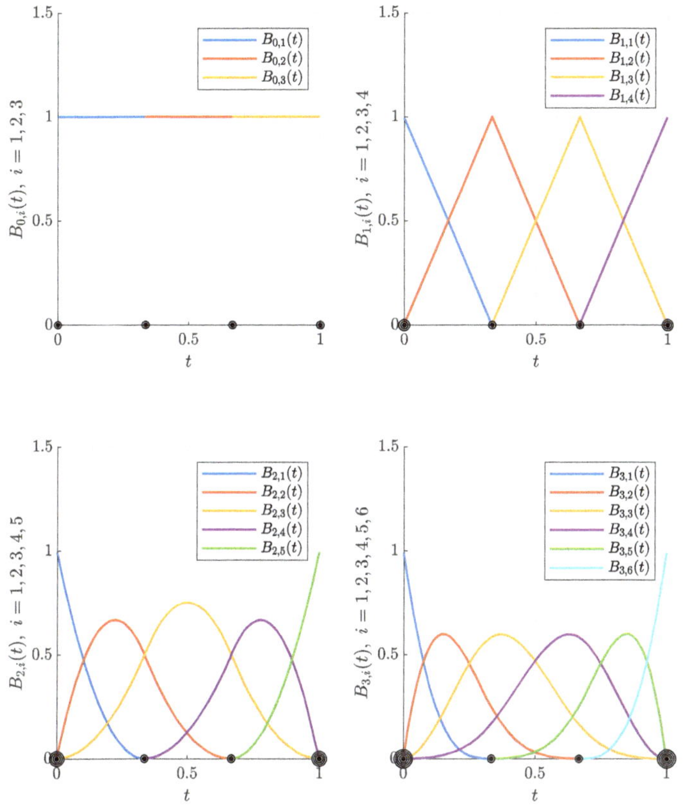

Fig. 2.4 B-Spline basis functions with knots 0, 1/3, 2/3, 1 for degrees $p = 0$ to $p = 3$ (top left to bottom right). Knot multiplicity is illustrated using black circles.

A B-spline parameterization of degree p can be defined based on a *clamped* knot-grid $\mathcal{K}$ of cardinality $M + 2p$ as follows [46]:

$$\mathcal{K} = \{\underbrace{k_1, \ldots, k_1}_{p+1}, \underbrace{k_2, \ldots, k_{M-1}}_{M-2}, \underbrace{k_M, \ldots, k_M}_{p+1}\}$$

with elements $k_1 \leqslant k_2 \leqslant \ldots \leqslant k_{M-1} \leqslant k_M$. Observe that for the clamped knot grid, the knots at the beginning and the end have multiplicity $p + 1$. A recursive definition of the basis functions $B_{p,i}(t)$, $i = 1, \ldots, M + p - 1$

May be skipped at first read.

is given as follows [21]:

$$B_{p,i}(t) = \frac{t - \mathcal{K}_i}{\mathcal{K}_{i+p} - \mathcal{K}_i} B_{p-1,i}(t) + \frac{\mathcal{K}_{i+p+1} - t}{\mathcal{K}_{i+p+1} - \mathcal{K}_{i+1}} B_{p-1,i+1}(t)$$

$$B_{0,i}(t) = \begin{cases} 1, & t \in (\mathcal{K}_i, \mathcal{K}_{i+1}) \\ 0, & \text{otherwise} \end{cases}$$

and the definition of the quotient $0/0 = 0$. Moreover, the *local support* property of the basis functions is expressed for $p > 0$ as

$$B_{p,i}(t) \begin{cases} > 0, & t \in \left(\mathcal{K}_i, \mathcal{K}_{i+p+1}\right) \\ = 0, & \text{otherwise} \end{cases}$$

Using the partitioning of the parameter vector $\mathbf{z}$ from Eq. (2.12) as well as a control parameterization [Eq. (2.13)], the right-hand side can be expressed in the common form:

$$\dot{\mathbf{x}}(t) = \bar{\mathbf{f}}\left[\mathbf{x}(t), \mathbf{u}\left(t, \mathbf{w}_1, \ldots, \mathbf{w}_{n_u}\right), \mathbf{p}\right] \tag{2.14}$$

Before we proceed with the derivative computation for the clearance criterion, let us, for completeness and as a basis for the following discussion, restate the general form of Runge–Kutta one-step methods in parameter-dependent form. The step function in this case is given as

$$\tilde{\mathbf{x}}(t_{i+1}, \mathbf{z}) = \boldsymbol{\psi}\left[\tilde{\mathbf{x}}(t_i, \mathbf{z}), t_i, \mathbf{z}\right]$$

with

$$\boldsymbol{\psi}\left[\tilde{\mathbf{x}}(t_i, \mathbf{z}), t_i, \mathbf{z}\right] = \tilde{\mathbf{x}}(t_i, \mathbf{z}) + h \sum_{q=1}^{s} b_q \mathbf{k}_q \left[\tilde{\mathbf{x}}(t_i, \mathbf{z}), t_i, \mathbf{z}\right]$$

and the stage derivatives are defined by

$$\mathbf{k}_j\left[\tilde{\mathbf{x}}(t_i, \mathbf{z}), t_i, \mathbf{z}\right] = \mathbf{f}\left(\tilde{\mathbf{x}}(t_i, \mathbf{z}) + h \sum_{q=1}^{s} a_{j,q} \mathbf{k}_q \left[\tilde{\mathbf{x}}(t_i, \mathbf{z}), t_i, \mathbf{z}\right], t_i + c_j h, \mathbf{z}\right),$$

$$j = 1, \ldots, s$$

Moreover, the approximation of the solution starts at the parameter-dependent initial condition $\tilde{\mathbf{x}}(t_0, \mathbf{z}) = \boldsymbol{\gamma}(\mathbf{z})$.

2.3.1 FORWARD DERIVATIVES

Clearance Criterion Derivatives (Forward)

The derivatives of the clearance criterion J with respect to a static parameter or control coefficient $z \in \mathbf{z}$ can be obtained from

$$\left.\frac{dJ}{dz}\right|_{t=t_f} = \left[\frac{\partial J}{\partial z} + \frac{\partial J}{\partial \tilde{\mathbf{x}}}\frac{d\tilde{\mathbf{x}}}{dz}\right]_{t=t_f} \tag{2.15}$$

with the derivatives

$$\left.\frac{d\tilde{\mathbf{x}}}{dz}\right|_{t=t_f}$$

computed as

$$\left.\frac{d\tilde{\mathbf{x}}}{dz}\right|_{t=t_{i+1}} = \left[\frac{\partial \boldsymbol{\psi}}{\partial z} + \frac{\partial \boldsymbol{\psi}}{\partial \tilde{\mathbf{x}}}\frac{d\tilde{\mathbf{x}}}{dz}\right]_{t=t_i}, \ i = 0, \ldots, f-1 \tag{2.16}$$

starting with

$$\left.\frac{d\tilde{\mathbf{x}}}{dz}\right|_{t=t_0} = \frac{\partial \boldsymbol{\gamma}}{\partial z}$$

In the following, the procedure for obtaining the derivatives of the clearance criterion is discussed for a single element $z \in \mathbf{z}$ as this procedure is equally valid for all static parameters or control coefficients collected in the parameter vector $\mathbf{z}$. Recall that the state approximations $\tilde{\mathbf{x}}(t_i, \mathbf{z})$, $i = 1, \ldots, f$ are obtained from an application of the Runge–Kutta one-step method starting at $\tilde{\mathbf{x}}(t_0; \mathbf{z}) = \boldsymbol{\gamma}(\mathbf{z})$. Thus, the steps for the evaluation of the clearance criterion depending on $\mathbf{z}$ are

$$\begin{aligned}
\tilde{\mathbf{x}}(t_0, \mathbf{z}) &= \boldsymbol{\gamma}(\mathbf{z}) \\
\tilde{\mathbf{x}}(t_1, \mathbf{z}) &= \boldsymbol{\psi}\left[\tilde{\mathbf{x}}(t_0, \mathbf{z}), t_0, \mathbf{z}\right] \\
\tilde{\mathbf{x}}(t_2, \mathbf{z}) &= \boldsymbol{\psi}\left[\tilde{\mathbf{x}}(t_1, \mathbf{z}), t_1, \mathbf{z}\right] \\
&\ \vdots \\
\tilde{\mathbf{x}}(t_{f-1}, \mathbf{z}) &= \boldsymbol{\psi}\left[\tilde{\mathbf{x}}(t_{f-2}, \mathbf{z}), t_{f-2}, \mathbf{z}\right] \\
\tilde{\mathbf{x}}(t_f, \mathbf{z}) &= \boldsymbol{\psi}\left[\tilde{\mathbf{x}}(t_{f-1}, \mathbf{z}), t_{f-1}, \mathbf{z}\right] \\
y &= J\left[\tilde{\mathbf{x}}(t_f, \mathbf{z}), \mathbf{z}\right]
\end{aligned} \tag{2.17}$$

The procedure for the computation of the derivatives can be directly derived through a successive application of the chain rule to each of these steps. Note that this approach is equivalent to the employment of the forward derivative method from Sec. 1.4.1.2 to the evaluation procedure of the criterion.

May be skipped (↴) at first read.

Let us visualize the evaluation of the criterion using a subsystem-based representation with $\boldsymbol{\psi}_i = \boldsymbol{\psi}\left[\tilde{\mathbf{x}}(t_i, \mathbf{z}), t_i, \mathbf{z}\right]$ and $\tilde{\mathbf{x}}_i = \tilde{\mathbf{x}}(t_i, \mathbf{z})$:

$$\overset{\downarrow \mathbf{z}}{\boxed{\gamma}} \overset{\tilde{\mathbf{x}}_0}{\rightarrow} \overset{\downarrow \mathbf{z}}{\boxed{\boldsymbol{\psi}_0}} \overset{\tilde{\mathbf{x}}_1}{\rightarrow} \overset{\downarrow \mathbf{z}}{\boxed{\boldsymbol{\psi}_1}} \overset{\tilde{\mathbf{x}}_2}{\rightarrow} \ldots \overset{\tilde{\mathbf{x}}_{f-1}}{\rightarrow} \overset{\downarrow \mathbf{z}}{\boxed{\boldsymbol{\psi}_{f-1}}} \overset{\tilde{\mathbf{x}}_f}{\rightarrow} \overset{\downarrow \mathbf{z}}{\boxed{J}} \overset{y}{\rightarrow} \tag{2.18}$$

Using the notions from Sec. 1.4.3 with $f+2$ subsystems of the form $\mathbf{v}_1 = \mathbf{f}_1(\cdot), \ldots, \mathbf{v}_{f+2} = \mathbf{f}_{f+2}(\cdot)$ as well as the associated index sets $\mathcal{F}_1, \ldots, \mathcal{F}_{f+2}$ defined as

$$\begin{aligned}
&\mathbf{z} = \mathbf{v}_0 \\
&\tilde{\mathbf{x}}_0 = \mathbf{v}_1 = \underbrace{\mathbf{f}_1}_{\gamma}(\mathbf{v}_0), && \mathcal{F}_1 = \{0\} \\
&\tilde{\mathbf{x}}_1 = \mathbf{v}_2 = \underbrace{\mathbf{f}_2}_{\boldsymbol{\psi}_0}(\mathbf{v}_0, \mathbf{v}_1), && \mathcal{F}_2 = \{0, 1\} \\
&\tilde{\mathbf{x}}_2 = \mathbf{v}_3 = \underbrace{\mathbf{f}_3}_{\boldsymbol{\psi}_1}(\mathbf{v}_0, \mathbf{v}_2), && \mathcal{F}_3 = \{0, 2\} \\
&\vdots \\
&\tilde{\mathbf{x}}_f = \mathbf{v}_{f+1} = \underbrace{\mathbf{f}_{f+1}}_{\boldsymbol{\psi}_{f-1}}(\mathbf{v}_0, \mathbf{v}_f), && \mathcal{F}_{f+1} = \{0, f\} \\
&y = \mathbf{v}_{f+2} = \underbrace{\mathbf{f}_{f+2}}_{J}(\mathbf{v}_0, \mathbf{v}_{f+1}), && \mathcal{F}_{f+2} = \{0, f+1\}
\end{aligned}$$

this representation becomes

$$\overset{\downarrow \mathbf{v}_0}{\boxed{\mathbf{f}_1}} \overset{\mathbf{v}_1}{\rightarrow} \overset{\downarrow \mathbf{v}_0}{\boxed{\mathbf{f}_2}} \overset{\mathbf{v}_2}{\rightarrow} \overset{\downarrow \mathbf{v}_0}{\boxed{\mathbf{f}_3}} \overset{\mathbf{v}_3}{\rightarrow} \ldots \overset{\mathbf{v}_f}{\rightarrow} \overset{\downarrow \mathbf{v}_0}{\boxed{\mathbf{f}_{f+1}}} \overset{\mathbf{v}_{f+1}}{\rightarrow} \overset{\downarrow \mathbf{v}_0}{\boxed{\mathbf{f}_{f+2}}} \overset{\mathbf{v}_{f+2}}{\rightarrow}$$

Then, the application of the forward derivative method [Eq. (1.15)] directly yields the derivatives of the criterion:

$$\begin{aligned}
\mathbf{s}_0 &= \frac{\partial \mathbf{z}}{\partial z} \\
\mathbf{s}_1 &= \frac{\partial \mathbf{f}_1}{\partial \mathbf{v}_0}\mathbf{s}_0 \\
\mathbf{s}_2 &= \frac{\partial \mathbf{f}_2}{\partial \mathbf{v}_0}\mathbf{s}_0 + \frac{\partial \mathbf{f}_2}{\partial \mathbf{v}_1}\mathbf{s}_1 \\
\mathbf{s}_3 &= \frac{\partial \mathbf{f}_3}{\partial \mathbf{v}_0}\mathbf{s}_0 + \frac{\partial \mathbf{f}_3}{\partial \mathbf{v}_2}\mathbf{s}_2 \\
&\vdots \\
\mathbf{s}_{f+1} &= \frac{\partial \mathbf{f}_{f+1}}{\partial \mathbf{v}_0}\mathbf{s}_0 + \frac{\partial \mathbf{f}_{f+1}}{\partial \mathbf{v}_f}\mathbf{s}_f \\
\mathbf{s}_{f+2} &= \frac{\partial \mathbf{f}_{f+2}}{\partial \mathbf{v}_0}\mathbf{s}_0 + \frac{\partial \mathbf{f}_{f+2}}{\partial \mathbf{v}_{f+1}}\mathbf{s}_{f+1}
\end{aligned}$$

Observe, in particular, that $\mathbf{s}_0$ is a vector with only a single (unit) entry at the index of the element z in $\mathbf{z}$ and that the preceding equations are equivalent to Eqs. (2.15) and (2.16) with $\mathbf{s}_i = \partial\tilde{\mathbf{x}}/\partial z|_{t=t_i}$ for $i = 1, \ldots, f+1$ as well as

$$\begin{aligned}
\frac{\partial \mathbf{f}_1}{\partial \mathbf{v}_0}\mathbf{s}_0 &= \frac{\partial \boldsymbol{\gamma}}{\partial z} \\
\frac{\partial \mathbf{f}_i}{\partial \mathbf{v}_0}\mathbf{s}_0 &= \left.\frac{\partial \boldsymbol{\psi}}{\partial z}\right|_{t=t_{i-2}}, \quad i = 2, \ldots, f+1 \\
\frac{\partial \mathbf{f}_{f+2}}{\partial \mathbf{v}_0}\mathbf{s}_0 &= \frac{\partial J}{\partial z}
\end{aligned}$$

Apparently, all steps in the procedure [Eq. (2.16)] depend on the partial derivatives with respect to z, implying that they need to be repeated for each static parameter and control coefficient $z \in \mathbf{z}$ if we seek to obtain all derivatives for the clearance criterion. As discussed in the following section, this effort for the computation of first-order derivatives can be significantly reduced using the backward derivative method [Eq. (1.16)]. In general, this backward method is efficient compared to the forward method if there exist few outputs and a large number of independent inputs. In the optimal control formulation for flight control law testing, we have a single scalar criterion (output) and usually a considerably larger number of static parameters and control coefficients as independent inputs. As such, we can expect that the backward method performs more efficiently compared to the forward method regarding the computation of the first-order derivatives. It should be noted, however, that the forward method is not without use in clearance applications. As shown in Sec. 2.3.3, the forward procedure can be used for the efficient computation of mixed (second-order) derivatives.

Before proceeding to the backward derivative method, let us provide another point of view regarding the forward procedure. If we rewrite the steps of the forward derivative propagation using the increment function $\boldsymbol{\phi}$ of the Runge–Kutta method instead of the step function $\boldsymbol{\psi}$ [see Eq. (2.17)],

$$\mathbf{s}_{i+1} = \mathbf{s}_i + \left.\frac{\partial \boldsymbol{\phi}}{\partial \tilde{\mathbf{x}}}\right|_{t=t_i} \mathbf{s}_i + \left.\frac{\partial \boldsymbol{\phi}}{\partial z}\right|_{t=t_i}$$

with initial condition $\mathbf{s}_0 = \partial \boldsymbol{\gamma}/\partial z$, it can be identified as the forward sensitivity differential equation

$$\dot{\mathbf{s}}(t) = \frac{\partial \mathbf{f}}{\partial \mathbf{x}} \mathbf{s}(t) + \frac{\partial \mathbf{f}}{\partial z} \tag{2.19}$$

with initial condition $\mathbf{s}(t_0) = \partial \boldsymbol{\gamma}/\partial z$ solved with the same Runge–Kutta one-step method as the initial value problem for the states [35]. Assuming differentiability, Eq. (2.19) can be derived in the continuous case by reversing the order of time and parameter differentiation [21]:

$$\frac{d}{dt}\left(\frac{\partial \mathbf{x}}{\partial z}\right) = \underbrace{\frac{\partial}{\partial z}\left(\frac{d\mathbf{x}}{dt}\right)}_{\dot{\mathbf{s}}(t)} = \frac{\partial \mathbf{f}}{\partial \mathbf{x}} \underbrace{\frac{\partial \mathbf{x}}{\partial z}}_{\mathbf{s}(t)} + \frac{\partial \mathbf{f}}{\partial z}$$

Thus, the derivative computation procedure of the clearance criterion based on the forward method essentially requires the solution of ***one*** *initial value problem* ***for each parameter*** (i.e., each static parameter or control coefficient in our application).

2.3.2 BACKWARD DERIVATIVES

Clearance Criterion Derivatives (Backward)

The derivatives of the clearance criterion J with respect to a static parameter or control coefficient $z \in \mathbf{z}$ can be obtained from

$$\left.\frac{dJ}{dz}\right|_{t=t_f} = \left.\frac{\partial J}{\partial z}\right|_{t=t_f} + \boldsymbol{\nu}_1^T \frac{\partial \boldsymbol{\gamma}}{\partial z} + \sum_{i=1}^{f} \boldsymbol{\nu}_{i+1}^T \left.\frac{\partial \boldsymbol{\psi}}{\partial z}\right|_{t=t_{i-1}} \tag{2.20}$$

with

$$\boldsymbol{\nu}_{f+1}^T = \frac{\partial J}{\partial \tilde{\mathbf{x}}}|_{t=t_f}$$

and

$$\boldsymbol{\nu}_{i-1}^T = \boldsymbol{\nu}_i^T \left.\frac{\partial \boldsymbol{\psi}}{\partial \tilde{\mathbf{x}}}\right|_{t=t_{i-2}}, \; i = f+1, \ldots, 2 \qquad (2.21)$$

Let us first develop an intuition regarding the interpretation of the derivative propagation [Eq. (2.21)] from the backward procedure. The computation starts with the assignment of the partial derivatives of the criterion with respect to the states at the final time point to $\boldsymbol{\nu}_{f+1}$. As such, $\boldsymbol{\nu}_{f+1}$ represents the sensitivity of the criterion with respect to the terminal states. Now, observe that Eq. (2.21) is essentially an application of the chain rule in a backward sense. The first step

$$\boldsymbol{\nu}_f^T = \underbrace{\boldsymbol{\nu}_{f+1}^T}_{\left.\frac{\partial J}{\partial \tilde{\mathbf{x}}}\right|_{t=t_f}} \left.\frac{\partial \boldsymbol{\psi}}{\partial \tilde{\mathbf{x}}}\right|_{t=t_{f-1}}$$

expresses that $\boldsymbol{\nu}_f$ represents the sensitivity with respect to the states at the second last time point as it is defined as *the partial derivative of the criterion with respect to the states at the last time point times the derivative of the step function (providing the states at the last time point) with respect to the states at the second-last time point.* This backward propagation can be repeated in exactly the same form for $\boldsymbol{\nu}_{f-1}, \boldsymbol{\nu}_{f-2}, \ldots$ until we arrive at $\boldsymbol{\nu}_1$ representing the sensitivity of the criterion with respect to the states at the initial time point. As such, in Eq. (2.20), the term

$$\boldsymbol{\nu}_1^T \frac{\partial \boldsymbol{\gamma}}{\partial z}$$

captures the sensitivity of the criterion with respect to the parameter through the parameter dependency in the initial state function as it is defined as *the sensitivity of the criterion with respect to the states at the initial time point times the derivative of the function returning the initial states with respect to the parameter.* Similarly, in Eq. (2.20), each term

$$\boldsymbol{\nu}_{i+1}^T \left.\frac{\partial \boldsymbol{\psi}}{\partial z}\right|_{t=t_{i-1}}$$

captures the sensitivity of the criterion with respect to the parameter through the parameter dependency in some of the other states at t_i, $i > 0$ as it is defined as *the sensitivity of criterion with respect to the states at this time point times the derivative of the step function (providing the states at this time point) with respect to the parameter.* Thus, the discussed terms appearing in Eq. (2.20) collect all *indirect* influences of the parameter on the criterion through the state dependency.

May be skipped (↴) at first read.

Let us develop the backward procedure using the subsystem representation (2.18) based on the adjoints $\boldsymbol{\nu}_i$, $i = 1, \ldots, f+1$ associated with the corresponding vectors $\tilde{\mathbf{x}}_i$:

$$\overset{\downarrow \mathbf{z}}{\boxed{\gamma}} \underset{\boldsymbol{\nu}_1}{\overset{\tilde{\mathbf{x}}_0}{\rightleftharpoons}} \overset{\downarrow \mathbf{z}}{\boxed{\boldsymbol{\psi}_0}} \underset{\boldsymbol{\nu}_2}{\overset{\tilde{\mathbf{x}}_1}{\rightleftharpoons}} \overset{\downarrow \mathbf{z}}{\boxed{\boldsymbol{\psi}_1}} \underset{\boldsymbol{\nu}_3}{\overset{\tilde{\mathbf{x}}_2}{\rightleftharpoons}} \cdots \underset{\boldsymbol{\nu}_f}{\overset{\tilde{\mathbf{x}}_{f-1}}{\rightleftharpoons}} \overset{\downarrow \mathbf{z}}{\boxed{\boldsymbol{\psi}_{f-1}}} \underset{\boldsymbol{\nu}_{f+1}}{\overset{\tilde{\mathbf{x}}_f}{\rightleftharpoons}} \overset{\downarrow \mathbf{z}}{\boxed{J}} \overset{y}{\rightarrow}$$

Recall that the backward derivative method introduced in Sec. 1.4.1 propagates derivative information in the opposite sense compared to the direction of the evaluation procedure. Again, using the notation from Sec. 1.4.3 and together with the adjoints $\boldsymbol{\nu}_0$ and $\boldsymbol{\nu}_{f+2}$ for the initial and final assignment, we arrive at the following representation:

$$\overbrace{\overset{\downarrow \mathbf{v}_0}{\boxed{\mathbf{f}_1}} \underset{\boldsymbol{\nu}_1}{\overset{\mathbf{v}_1}{\rightleftharpoons}} \overset{\downarrow \mathbf{v}_0}{\boxed{\mathbf{f}_2}} \underset{\boldsymbol{\nu}_2}{\overset{\mathbf{v}_2}{\rightleftharpoons}} \overset{\downarrow \mathbf{v}_0}{\boxed{\mathbf{f}_3}} \underset{\boldsymbol{\nu}_3}{\overset{\mathbf{v}_3}{\rightleftharpoons}} \cdots \underset{\boldsymbol{\nu}_f}{\overset{\mathbf{v}_f}{\rightleftharpoons}} \overset{\downarrow \mathbf{v}_0}{\boxed{\mathbf{f}_{f+1}}} \underset{\boldsymbol{\nu}_{f+1}}{\overset{\mathbf{v}_{f+1}}{\rightleftharpoons}} \overset{\downarrow \mathbf{v}_0}{\boxed{\mathbf{f}_{f+2}}} \underset{\boldsymbol{\nu}_{f+2}}{\overset{\mathbf{v}_{f+2}}{\rightleftharpoons}}}^{\boldsymbol{\nu}_0}$$

We define the following index sets for the backward procedure:

$$\begin{aligned} \mathcal{B}_{f+1} &= \{f+2\} \\ \mathcal{B}_f &= \{f+1\} \\ &\vdots \\ \mathcal{B}_1 &= \{2\} \\ \mathcal{B}_0 &= \{1, 2, \ldots, f+1, f+2\} \end{aligned}$$

Recall from Sec. 1.4.3 that an index j is included in an index set $\mathcal{B}_i$ if $\mathbf{f}_j(\cdot)$ depends on $\mathbf{v}_i$. With these definitions, we can formally write the backward procedure starting with $\boldsymbol{\nu}_{f+2} = 1$:

$$\boldsymbol{\nu}_{f+1}^T = \underbrace{\boldsymbol{\nu}_{f+2}^T}_{1} \frac{\partial \mathbf{f}_{f+2}}{\partial \mathbf{v}_{f+1}} = \underbrace{\frac{\partial \mathbf{f}_{f+2}}{\partial \mathbf{v}_{f+1}}}_{\left.\frac{\partial J}{\partial \tilde{\mathbf{x}}}\right|_{t=t_f}}$$

$$\boldsymbol{\nu}_f^T = \boldsymbol{\nu}_{f+1}^T \underbrace{\frac{\partial \mathbf{f}_{f+1}}{\partial \mathbf{v}_f}}_{\left.\frac{\partial \boldsymbol{\psi}}{\partial \tilde{\mathbf{x}}}\right|_{t=t_{f-1}}}$$

$$\vdots$$

$$\boldsymbol{\nu}_1^T = \boldsymbol{\nu}_2^T \underbrace{\frac{\partial \mathbf{f}_2}{\partial \mathbf{v}_1}}_{\left.\frac{\partial \boldsymbol{\psi}}{\partial \tilde{\mathbf{x}}}\right|_{t=t_0}}$$

$$\underbrace{\boldsymbol{\nu}_0^T}_{\left.\frac{dJ}{dz}\right|_{t=t_f}} = \boldsymbol{\nu}_1^T \underbrace{\frac{\partial \mathbf{f}_1}{\partial \mathbf{v}_0}}_{\frac{\partial \boldsymbol{\gamma}}{\partial z}} + \underbrace{\boldsymbol{\nu}_2^T \frac{\partial \mathbf{f}_2}{\partial \mathbf{v}_0} + \ldots + \boldsymbol{\nu}_{f+1}^T \frac{\partial \mathbf{f}_{f+1}}{\partial \mathbf{v}_0}}_{\sum_{i=1}^{f} \boldsymbol{\nu}_{i+1} \left.\frac{\partial \boldsymbol{\psi}}{\partial z}\right|_{t=t_{i-1}}} + \underbrace{\boldsymbol{\nu}_{f+2}^T}_{1} \underbrace{\frac{\partial \mathbf{f}_{f+2}}{\partial \mathbf{v}_0}}_{\left.\frac{\partial J}{\partial z}\right|_{t=t_f}}$$

Note that by inserting the function definitions (see braces in the preceding equations), we directly obtain the backward derivative procedure for the clearance criterion.

As for the forward procedure, Eqs. (2.21) propagating the derivative information with respect to the states can be interpreted as the solution to an initial value problem. Rewriting Eqs. (2.21) in terms of the increment function [Eq. (2.3)] and the definition $\boldsymbol{\lambda}_i = -\boldsymbol{\nu}_{i+1}$, $i = 1, \ldots, f$ yields the discrete adjoint equation:

$$\boldsymbol{\lambda}_{i-1}^T = \boldsymbol{\lambda}_i^T + h\boldsymbol{\lambda}_i^T \left.\frac{\partial \boldsymbol{\phi}}{\partial \tilde{\mathbf{x}}}\right|_{t=t_i}$$

with the transversality condition $\boldsymbol{\lambda}_f^T = -\partial J / \partial \tilde{\mathbf{x}}|_{t=t_f}$ [21], which constitutes a discrete form of the continuous adjoint equation:

$$\dot{\boldsymbol{\lambda}}^T(t) = -\boldsymbol{\lambda}^T \frac{\partial \mathbf{f}}{\partial \mathbf{x}}$$

with final condition

$$\boldsymbol{\lambda}^T(t_f) = -\frac{\partial J}{\partial \mathbf{x}}|_{t=t_f}$$

Observe that in the case of the backward sensitivity method the backpropagation of the sensitivities with respect to the states [Eq. (2.21)] does not depend on the partial derivatives with respect to z. Thus, the computational effort required for the backward procedure can be interpreted in the sense that, as opposed to the forward procedure, it requires the solution of ***one** initial value problem* ***for all parameters*** (static parameters and control coefficients).

Engineering Note 2.3.1: Runge–Kutta Step Function Derivatives

Note that throughout the sections for the forward and backward procedure it is assumed that the partial derivatives of the step function are available. However, during the flight control law development process, typically only linearization routines for the closed-loop model $\bar{\mathbf{f}}$ of the form of Eq. (2.14) are at our disposal. These linearization routines can be used to obtain the partial derivatives of the state derivatives with respect to the states $\partial\bar{\mathbf{f}}/\partial\tilde{\mathbf{x}}$, static parameters $\partial\bar{\mathbf{f}}/\partial\mathbf{p}$, and controls $\partial\bar{\mathbf{f}}/\partial\mathbf{u}$. In this case, the partial derivatives with respect to the parameter vector $\mathbf{z}$ defined in Eq. (2.12) are derived as

$$\frac{\partial\mathbf{f}}{\partial\mathbf{z}} = \frac{\partial\bar{\mathbf{f}}}{\partial\mathbf{p}}\frac{\partial\mathbf{p}}{\partial\mathbf{z}} + \sum_{r=1}^{n_u}\frac{\partial\bar{\mathbf{f}}}{\partial u_r}\frac{\partial u_r}{\partial\mathbf{w}_r}\frac{\partial\mathbf{w}_r}{\partial\mathbf{z}}$$

Note that the matrix $\partial\mathbf{p}/\partial\mathbf{z} \in \mathbb{R}^{n_p\times n}$ is sparse, and it can be defined by the COO vector triplet (vector of row indices, vector of column indices, value vector)

$$\left(\left[1,\ldots,n_p\right];\left[1,\ldots,n_p\right];\mathbf{1}\right)$$

where $\mathbf{1}$ is a vector of appropriate size with all entries one. Similarly, the matrices $\partial\mathbf{w}_r/\partial\mathbf{z} \in \mathbb{R}^{n_{w,r}\times n}$, $r = 1,\ldots,n_u$ are sparse with coordinates:

$$\left(\left[1,\ldots,n_{w,r}\right];\left[n_p + 1 + \sum_{v=1}^{r-1} n_{w,v},\ldots,n_p + \sum_{v=1}^{r} n_{w,v}\right];\mathbf{1}\right)$$

Note further that from Eq. (2.13) it is readily apparent that the partial derivatives $\partial u_r/\partial\mathbf{w}_r$ of the control function with respect to the control coefficients are the individual basis functions evaluated at a particular time point.

Thus, for an explicit Runge–Kutta integration the derivatives of the step function with respect to the states and parameters (see Listing 2.2) can directly be obtained by the application of the forward derivative method to the function given in Listing 2.1.

Listing 2.2 Evaluation of the step function for general explicit Runge–Kutta methods including derivatives.

```
%%%%%%%%%%%%%%%%%%%%%%%%%%%%%%%%%%%%%%%%%%%%%%%%%%%%%
% Step Function Derivatives (Runge-Kutta, Explicit)
%%%%%%%%%%%%%%%%%%%%%%%%%%%%%%%%%%%%%%%%%%%%%%%%%%%%%
% <INPUTS>
```

```
% > x_old [m x 1]: State vector at old time point.
% > t_old [1 x 1]: Old time point.
% > f     [1 x 1]: F-handle [x_dot,x_dot_x,x_dot_z]=f(x,t
    ,z)
% > h     [1 x 1]: Step-size h.
% > A     [s x s]: Butcher tableau coefficient matrix A.
% > b     [1 x s]: Butcher tableau coefficient vector b.
% > c     [s x 1]: Butcher tableau coefficient vector c.
% > z     [n x 1]: Parameter vector.
%
% <OUTPUTS>
% > x_new   [m x 1]: State vector at t_old + h.
% > x_new_x [m x m]: Derivative of x_new w.r.t. states.
% > x_new_z [m x n]: Derivative of x_new w.r.t.
    parameters.
%%%%%%%%%%%%%%%%%%%%%%%%%%%%%%%%%%%%%%%%%%%%%%%%%%%%%%%%%%%
function [x_new,x_new_x,x_new_z]=psi(x_old,t_old,f,h,A,b,
    c,z)
    s = numel(c);
    n = numel(z);
    m = numel(x_old);
    X = repmat(x_old, 1, s);
    K = zeros(size(X));
    K_x = cell(s,1);
    K_z = cell(s,1);
    for j=1:s
        K_x{j} = eye(m);
        K_z{j} = zeros(m, n);
        for q=1:j-1
            X(:,j) = X(:,j) + h * A(j,q) * K(:,q);
            K_x{j} = K_x{j} + h * A(j,q) * K_x{q};
            K_z{j} = K_z{j} + h * A(j,q) * K_z{q};
        end
        [x_dot, x_dot_x, x_dot_z] = f(X(:,j), t_old + h *
             c(j),z);
        K(:,j) = x_dot;
        K_x{j} = x_dot_x * K_x{j};
        K_z{j} = x_dot_z + x_dot_x * K_z{j};
    end
    x_new   = x_old;
    x_new_x = eye(m);
    x_new_z = zeros(m, n);
```

```
    for q=1:s
        x_new   = x_new   + h * b(1,q) * K(:,q);
        x_new_x = x_new_x + h * b(1,q) * K_x{q};
        x_new_z = x_new_z + h * b(1,q) * K_z{q};
    end
end
```

2.3.3 MIXED DERIVATIVES

Clearance Criterion Derivatives (Mixed)

The mixed derivatives of the clearance criterion with respect to parameters $z_j \in \mathbf{z}$ and $z_k \in \mathbf{z}$ can be obtained from

$$\left.\frac{d^2 J}{dz_j dz_k}\right|_{t=t_f} = \left[\frac{\partial^2 J}{\partial z_k \partial \tilde{\mathbf{x}}}\frac{d\tilde{\mathbf{x}}}{dz_j} + \frac{\partial^2 J}{\partial z_k \partial z_j}\right]_{t=t_f} + \boldsymbol{\nu}_1^T \frac{\partial^2 \boldsymbol{\gamma}}{\partial z_k \partial z_j} + \boldsymbol{\mu}_1^T \frac{\partial \boldsymbol{\gamma}}{\partial z_k} + \tag{2.22}$$

$$\sum_{i=1}^{f}\left(\boldsymbol{\nu}_{i+1}^T\left[\frac{\partial^2 \boldsymbol{\psi}}{\partial z_k \partial z_j} + \frac{\partial^2 \boldsymbol{\psi}}{\partial z_k \partial \tilde{\mathbf{x}}}\frac{d\tilde{\mathbf{x}}}{dz_j}\right]_{t=t_{i-1}} + \boldsymbol{\mu}_{i+1}^T \left.\frac{\partial \boldsymbol{\psi}}{\partial z_k}\right|_{t=t_{i-1}}\right)$$

with

$$\boldsymbol{\mu}_i = \frac{d\boldsymbol{\nu}_i}{dz_j}$$

from

$$\boldsymbol{\mu}_{f+1}^T = \left[\frac{\partial^2 J}{\partial \tilde{\mathbf{x}} \partial z_j} + \frac{\partial^2 J}{\partial \tilde{\mathbf{x}}^2}\frac{d\tilde{\mathbf{x}}}{dz_j}\right]_{t=t_f}$$

and

$$\boldsymbol{\mu}_{i-1}^T = \boldsymbol{\mu}_i^T \left.\frac{\partial \boldsymbol{\psi}}{\partial \tilde{\mathbf{x}}}\right|_{t=t_{i-2}} + \boldsymbol{\nu}_i^T\left[\frac{\partial^2 \boldsymbol{\psi}}{\partial \tilde{\mathbf{x}} \partial z_j} + \sum_{r=1}^{n_x}\frac{\partial^2 \boldsymbol{\psi}}{\partial \tilde{\mathbf{x}} \partial x_r}\frac{d\tilde{x}_r}{dz_j}\right]_{t=t_{i-2}},$$

$$i = f+1, \ldots, 2 \tag{2.23}$$

All $\boldsymbol{\nu}_i$, $i = 1, \ldots, f+1$ and $\partial\tilde{\mathbf{x}}/\partial z_j|_{t=t_i}$, $i = 0, \ldots, f$ are available from the backward, respectively, forward derivative procedure.

The equations regarding the mixed derivatives of the clearance criterion with respect to parameters z_j and z_k are directly derived by a differentiation of the backward procedure presented in the preceeding section for a parameter z_k:

$$
\begin{aligned}
\boldsymbol{\nu}_{f+1}^T &= \left.\frac{\partial J}{\partial \tilde{\mathbf{x}}}\right|_{t=t_f} \\
\boldsymbol{\nu}_f^T &= \boldsymbol{\nu}_{f+1}^T \left.\frac{\partial \boldsymbol{\psi}}{\partial \tilde{\mathbf{x}}}\right|_{t=t_{f-1}} \\
&\vdots \\
\boldsymbol{\nu}_1^T &= \boldsymbol{\nu}_2^T \left.\frac{\partial \boldsymbol{\psi}}{\partial \tilde{\mathbf{x}}}\right|_{t=t_0} \\
\left.\frac{dJ}{dz_k}\right|_{t=t_f} &= \left.\frac{\partial J}{\partial z_k}\right|_{t=t_f} + \boldsymbol{\nu}_1^T \frac{\partial \boldsymbol{\gamma}}{\partial z_k} + \sum_{i=1}^{f} \boldsymbol{\nu}_{i+1}^T \left.\frac{\partial \boldsymbol{\psi}}{\partial z_k}\right|_{t=t_{i-1}}
\end{aligned}
$$

with respect to a parameter z_j. Note that the state derivatives $\partial \tilde{\mathbf{x}}/\partial z_j|_{t=t_i}$ are all available from the forward method for the first derivatives with respect to z_j.

It is interesting to observe that this direct differentiation essentially constitutes an application of the general forward derivative procedure to the backward procedure derived for the clearance criterion. The strategy followed here can be interpreted as an application of the general approach outlined in Sec. 1.4.1 for obtaining higher-order derivatives (i.e., the repeated application of the derivative methods to the resulting procedures from the forward or backward method).

Regarding the computational effort, note that, considering symmetry, there are $n(n+1)/2$ potentially distinct mixed derivatives of the criterion with respect to n parameters. Thus, the finite difference method [Eq. (2.11)] requires a total number of $2n(n+1)$ vector-valued state propagations (simulations) to obtain the respective perturbed values and, ultimately, approximations to the derivatives.

The procedure for evaluating all mixed derivatives outlined in this section depends on the evaluation of the backward and forward procedures. Hence, to evaluate all terms in Eq. (2.22), it is required to perform

- one vector-valued derivative propagation from the backward procedure (2.21) to obtain all $\boldsymbol{\nu}_i$; and
- n vector-valued derivative propagations from the forward procedure (2.16) to obtain all $\partial \tilde{\mathbf{x}}/\partial z_j|_{t=t_i}$, $j = 1, \ldots, n$; and
- n vector-valued derivative propagations from the forward procedure applied to the backward procedure [Eq. (2.23)] to obtain all $\boldsymbol{\mu}_i = \partial \boldsymbol{\nu}_i/\partial z_j$, $j = 1, \ldots, n$.

This amounts to $1 + 2n$ vector-valued derivative propagations to obtain the exact Hessian of the clearance criterion. Note that also the Jacobian can be efficiently obtained by evaluating either Eq. (2.20) or Eq. (2.15) as a by-product.

It should be mentioned that the outlined procedure is efficient, if the method of choice is the integration in a "single sweep" (single shooting) resulting in a clearance optimal control problem in which we seek to minimize a criterion *without additional constraint functions.* For general optimal control problems, the procedure needs to be repeated for all such functions rendering it less attractive from an efficiency perspective for an increasing number of constraints. However, in this general setting, we prefer anyhow (e.g., due to sparsity considerations) to employ the segmented collocation approaches introduced in Sec. 2.2.2 as a basis for the majority of optimal control-based clearance methods presented throughout Part II.

CHAPTER 3

Optimal Control

3.1 OPTIMAL CONTROL PROBLEMS

Clearance Optimal Control Problem

Find the piecewise continuous control history $\underline{\mathbf{u}} \leqslant \mathbf{u}(t) \leqslant \overline{\mathbf{u}}$ for $t \in [t_0, t_f]$ with $\mathbf{u}(t)$ bounded by $\underline{\mathbf{u}}, \overline{\mathbf{u}} \in \mathbb{R}^{n_u}$, the corresponding state history $\underline{\mathbf{x}} \leqslant \mathbf{x}(t) \leqslant \overline{\mathbf{x}}$ bounded by $\underline{\mathbf{x}}, \overline{\mathbf{x}} \in \mathbb{R}^{n_x}$, and the static parameters $\underline{\mathbf{p}} \leqslant \mathbf{p} \leqslant \overline{\mathbf{p}}$ bounded by $\underline{\mathbf{p}}, \overline{\mathbf{p}} \in \mathbb{R}^{n_p}$ that minimize the state- and parameter-dependent endpoint (Mayer-type) cost function $J : \mathbb{R}^{n_x} \times \mathbb{R}^{n_p} \to \mathbb{R}$:

$$
\begin{aligned}
\underset{\mathbf{u}(t)}{\text{minimize}} \quad & J\left[\mathbf{x}\left(t_f\right), \mathbf{p}\right] \\
\text{subject to} \quad & \dot{\mathbf{x}}(t) = \mathbf{f}\left[\mathbf{x}(t), \mathbf{u}(t), \mathbf{p}\right] \\
& \boldsymbol{\gamma}\left[\mathbf{x}(t_0), \mathbf{p}\right] = \mathbf{0} \\
& \underline{\mathbf{p}} \leqslant \mathbf{p} \leqslant \overline{\mathbf{p}} \\
& \underline{\mathbf{x}} \leqslant \mathbf{x}(t) \leqslant \overline{\mathbf{x}} \\
& \underline{\mathbf{u}} \leqslant \mathbf{u}(t) \leqslant \overline{\mathbf{u}} \\
& t \in \left[t_0, t_f\right], \; t_0, t_f \text{ fixed}
\end{aligned}
\tag{3.1}
$$

The optimal control problem is subject to a first-order ordinary differential equation and an initial boundary condition defined by $\mathbf{f} : \mathbb{R}^{n_x} \times \mathbb{R}^{n_u} \times \mathbb{R}^{n_p} \to \mathbb{R}^{n_x}$ and $\boldsymbol{\gamma} : \mathbb{R}^{n_x} \times \mathbb{R}^{n_p} \to \mathbb{R}^{n_\gamma}$.

In the third fundamental building block, we view the clearance problem from an optimal control perspective in continuous time. Similar to optimization problems, there exist several (often equivalent) types of optimal control problem formulations as well as transformations to map between those types. Hence, we first focus on the discussion of common notions used in optimal control theory and the classification of the clearance problem in this context. Following this classification, Sec. 3.2 then presents useful standard transformations that can be employed to widen the range of problems modeled by Eq. (3.1) or to represent the clearance problem in a more compact form.

In optimal control theory, there exist two main types of cost terms in standard problem formulations. The first type is referred to as a *Lagrange cost* and represents a running cost integrated over the time interval. The second type is referred to as a *Mayer cost* and represents a cost term associated with the endpoints of the time interval. If both cost types are introduced, the performance index is termed a *Bolza cost*. Observe that the clearance optimal control problem [Eq. (3.1)] features a cost depending on the states at the final time point. Thus, Eq. (3.1) is of *Mayer* type.

Constraints are categorized in a similar way as the cost (i.e., depending on whether they act on the whole interval or solely at the endpoints). If constraints are imposed along the trajectory, such constraints are referred to as *path constraints*. These path constraints can be further classified as either *purely state-dependent* path constraints or *mixed state-control* path constraints. Conversely, if constraints are associated with the states at the endpoints of the interval, such conditions are termed initial and/or final boundary conditions. Observe that in addition to the control constraints, Eq. (3.1) features purely state-dependent path constraints as well as an initial boundary condition. However, no boundary condition for the endpoint is required for our purposes.

The time horizon over which the optimal control problem is solved is referred to as the *phase time interval*. In general, this interval can be either fixed or free, meaning that the endpoints of the interval can be either set to a specific value or subject to optimization. In optimal control applications, it is often useful to introduce more than one phase and join these phases at the phase boundaries using appropriate linking constraints. This type of modeling leads to so-called *multiphase* problems. Though, throughout this book, and in particular in the case of the clearance optimal control problem [Eq. (3.1)], it suffices for us to only consider a single phase with a fixed time interval.

There exist two general approaches for solving optimal control problems. Under the first approach, the so-called *direct* approach, the optimal control problem is transcribed to a parameter optimization problem (see Chapter 1) using a suitable simulation method (see Chapter 2). Then, the optimality conditions of this parameter optimization problem are solved numerically. Under the second approach, the so-called *indirect* method, the optimality conditions are derived first for the problem in continuous time. Then, in a second step, these optimality conditions can be solved either numerically or by analytical means.

The derivation of the optimality conditions as well as the numerical solution under the indirect approach is, however, no easy task for realistic models in clearance applications. Besides the complexity of the aircraft closed-loop system, the derivation of these conditions for a particular problem requires knowledge regarding the sequence of arcs on which constraints are active/inactive. Moreover, the resulting problem to be solved represents a boundary value problem involving not only the state dynamics but also a second system, called the *costate* system. For a numerical solution, (good) initial guesses for both the states as well as the (rather nonintuitive) costates are required. A common strategy in practice is thus to first employ the direct approach to solve the problem numerically and use this solution as a basis for the solution using the indirect approach. We note, however, that in the majority of practical use cases, the discretized solution obtained from the direct approach suffices for our purposes. Nevertheless, we would like to emphasize that insights from the indirect approach can prove valuable also in our application. In particular, they allow us to analyze the structure of the optimal solution and exhibit the type of solutions expected for certain classes of clearance problems. Thus, even though this book focuses on the direct approach for solving the clearance optimal control problems numerically, the results from the indirect approach can be used to perform sanity checks of the numerical solution and reveal the structure of the worst-case control.

3.2 TRANSFORMED PROBLEMS

Transformations for Clearance Optimal Problems

Among others, the following modeling techniques can be used to generalize the formulations that can be modeled by the clearance optimal control problem [Eq. (3.1)] or to obtain a more compact representation:

Lagrange Cost: For $\min \int_{t_0}^{t_f} l\left(\mathbf{x}, \mathbf{u}, \mathbf{p}\right) dt$, introduce a state x_l with

$$\dot{x}_l = l\left(\mathbf{x}, \mathbf{u}, \mathbf{p}\right),\ x_l(t_0) = 0 \rightarrow \min x_l(t_f).$$

Parameters: For $\underline{\mathbf{p}} \leqslant \mathbf{p} \leqslant \overline{\mathbf{p}}$, add bounded states $\underline{\mathbf{p}} \leqslant \mathbf{x}_p \leqslant \overline{\mathbf{p}}$ with

$$\dot{\mathbf{x}}_p = \mathbf{0}.$$

Free Interval: For $t \in [t_0, t_f]$, t_0, t_f free, introduce parameters t_0, t_f and propagate in normalized time $\tau = (t - t_0)/(t_f - t_0) \in [0, 1]$:

$$\mathbf{x}' = \left(t_f - t_0\right) \mathbf{f}\left(\mathbf{x}, \mathbf{u}, \mathbf{p}\right)$$

Autonomization: Introduce a state x_t representing time with

$$\dot{x}_t = 1,\ x_t(t_0) = t_0$$

Path Constraints: For $\underline{\mathbf{g}} \leqslant \mathbf{g}(\mathbf{x}) \leqslant \overline{\mathbf{g}}$, add bounded states $\underline{\mathbf{g}} \leqslant \mathbf{x}_g \leqslant \overline{\mathbf{g}}$ with

$$\dot{\mathbf{x}}_g = \frac{\partial \mathbf{g}}{\partial \mathbf{x}} \mathbf{f}(\mathbf{x}, \mathbf{u}, \mathbf{p}), \ \mathbf{x}_g(t_0) = \mathbf{g}[\mathbf{x}(t_0)]$$

The clearance optimal control problem (3.1) accommodates a wider range of problems than it may appear at a first glance. In the following, several techniques are outlined allowing a transformation from more general formulations to a problem of the same type as Eq. (3.1) or to obtain an equivalent optimal control problem without the explicit appearance of the parameters.

Lagrange and Mayer Cost: If required, a cost term of Lagrange-type

$$\int_{t_0}^{t_f} l\left[\mathbf{x}(t), \mathbf{u}(t), \mathbf{p}\right] dt$$

can be modeled in Eq. (3.1).

A clearance problem featuring this cost type is transformed into a Mayer form by the introduction of an additional state $x_l \in \mathbb{R}$ propagated as $\dot{x}_l(t) = l\left[\mathbf{x}(t), \mathbf{u}(t), \mathbf{p}\right]$ with initial condition $x_l(t_0) = 0$:

$$\begin{aligned} &\underset{\mathbf{u}(t)}{\text{minimize}} && \int_{t_0}^{t_f} l\left[\mathbf{x}(t), \mathbf{u}(t), \mathbf{p}\right] dt \\ &\text{subject to} && \dot{\mathbf{x}}(t) = \mathbf{f}\left[\mathbf{x}(t), \mathbf{u}(t), \mathbf{p}\right] \\ &&& \boldsymbol{\gamma}\left[\mathbf{x}(t_0), \mathbf{p}\right] = \mathbf{0} \\ &&& \vdots \\ &&& t \in \left[t_0, t_f\right], \ t_0, t_f \text{ fixed} \end{aligned} \quad \rightarrow \quad \begin{aligned} &\underset{\mathbf{u}(t)}{\text{minimize}} && x_l(t_f) \\ &\text{subject to} && \dot{\mathbf{x}}(t) = \mathbf{f}\left[\mathbf{x}(t), \mathbf{u}(t), \mathbf{p}\right] \\ &&& \dot{x}_l(t) = l\left[\mathbf{x}(t), \mathbf{u}(t), \mathbf{p}\right] \\ &&& x_l(t_0) = 0 \\ &&& \boldsymbol{\gamma}\left[\mathbf{x}(t_0), \mathbf{p}\right] = \mathbf{0} \\ &&& \vdots \\ &&& t \in \left[t_0, t_f\right], \ t_0, t_f \text{ fixed} \end{aligned}$$

Note that the resulting problem represents a Mayer-type problem of the same form as Eq. (3.1) using an augmented state vector propagated as

$$\frac{d}{dt}\begin{bmatrix} \mathbf{x}(t) \\ x_l(t) \end{bmatrix} = \begin{bmatrix} \mathbf{f}\left[\mathbf{x}(t), \mathbf{u}(t), \mathbf{p}\right] \\ l\left[\mathbf{x}(t), \mathbf{u}(t), \mathbf{p}\right] \end{bmatrix}$$

and initial condition

$$\begin{bmatrix} \gamma\left[\mathbf{x}(t_0), \mathbf{p}\right] \\ x_l(t_0) \end{bmatrix} = \mathbf{0}$$

Parameters and States: All static parameters $\mathbf{p}$ can be equivalently modeled through the introduction of additional states $\mathbf{x}_p \in \mathbb{R}^{n_p}$. These states are propagated with a zero rate $\dot{\mathbf{x}}_p(t) = \mathbf{0}$ and are bounded by* $\underline{\mathbf{p}} \leqslant \mathbf{x}_p(t) \leqslant \overline{\mathbf{p}}$. Therefore, the optimal control problem is transformed as

$$\begin{aligned} &\underset{\mathbf{u}(t)}{\text{minimize}} && J\left[\mathbf{x}(t_f), \mathbf{p}\right] \\ &\text{subject to} && \dot{\mathbf{x}}(t) = \mathbf{f}\left[\mathbf{x}(t), \mathbf{u}(t), \mathbf{p}\right] \\ &&& \gamma\left[\mathbf{x}(t_0), \mathbf{p}\right] = \mathbf{0} \\ &&& \underline{\mathbf{p}} \leqslant \mathbf{p} \leqslant \overline{\mathbf{p}} \\ &&& \underline{\mathbf{x}} \leqslant \mathbf{x}(t) \leqslant \overline{\mathbf{x}} \\ &&& \vdots \\ &&& t \in \left[t_0, t_f\right],\ t_0, t_f \text{ fixed} \end{aligned} \quad \rightarrow \quad \begin{aligned} &\underset{\mathbf{u}(t)}{\text{minimize}} && J\left[\mathbf{x}(t_f), \mathbf{x}_p(t_f)\right] \\ &\text{subject to} && \dot{\mathbf{x}}(t) = \mathbf{f}\left[\mathbf{x}(t), \mathbf{u}(t), \mathbf{x}_p(t)\right] \\ &&& \dot{\mathbf{x}}_p(t) = \mathbf{0} \\ &&& \gamma\left[\mathbf{x}(t_0), \mathbf{x}_p(t_0)\right] = \mathbf{0} \\ &&& \underline{\mathbf{p}} \leqslant \mathbf{x}_p(t) \leqslant \overline{\mathbf{p}} \\ &&& \underline{\mathbf{x}} \leqslant \mathbf{x}(t) \leqslant \overline{\mathbf{x}} \\ &&& \vdots \\ &&& t \in \left[t_0, t_f\right],\ t_0, t_f \text{ fixed} \end{aligned}$$

Augmentation of the state vector by the states representing the parameters with the dynamics

$$\frac{d}{dt}\begin{bmatrix} \mathbf{x}(t) \\ \mathbf{x}_p(t) \end{bmatrix} = \begin{bmatrix} \mathbf{f}\left[\mathbf{x}(t), \mathbf{u}(t), \mathbf{x}_p(t)\right] \\ \mathbf{0} \end{bmatrix}$$

and the extension of the state constraints

$$\begin{bmatrix} \underline{\mathbf{x}} \\ \underline{\mathbf{p}} \end{bmatrix} \leqslant \begin{bmatrix} \mathbf{x}(t) \\ \mathbf{x}_p(t) \end{bmatrix} \leqslant \begin{bmatrix} \overline{\mathbf{x}} \\ \overline{\mathbf{p}} \end{bmatrix}$$

transforms Eq. (3.1) to a problem without the explicit appearance of the parameters. This more compact representation is the preferred form for the discussion of the optimality conditions of optimal control problems in the following section.

---

*If the initial condition allows inequality constraints, it is sufficient to impose $\underline{\mathbf{p}} \leqslant \mathbf{x}_p(t_0) \leqslant \overline{\mathbf{p}}$ or $\underline{\mathbf{p}} \leqslant \mathbf{x}_p(t_f) \leqslant \overline{\mathbf{p}}$.

May be skipped (↴) at first read.

Free to Fixed Time Interval: Problems with free initial time t_0 and/or final time t_f can be transformed into an equivalent problem defined for a fixed time interval. This transformation is achieved using a time transformation based on the introduction of t_0 and/or t_f as parameters. For this transformation, a new independent variable, the normalized time $\tau \in [0, 1]$, is defined:

$$\tau\,(t) = \frac{t - t_0}{t_f - t_0}$$

If this independent variable is used instead of the time t, the phase interval can be fixed to $[0, 1]$, and the dynamic system is propagated as

$$\mathbf{x}'\,(\tau) = \frac{d\mathbf{x}}{d\tau} = \frac{d\mathbf{x}}{dt}\frac{dt}{d\tau} = \frac{d\mathbf{x}}{dt}\left(t_f - t_0\right)$$

Thus, the problem is transformed as

$$\begin{aligned}
&\underset{\mathbf{u}(t)}{\text{minimize}} && J\left[\mathbf{x}\left(t_f\right),\mathbf{p}\right] & & \underset{\mathbf{u}(t)}{\text{minimize}} && J\left[\mathbf{x}\,(1)\,,\mathbf{p}\right]\\
&\text{subject to} && \dot{\mathbf{x}}\,(t) = \mathbf{f}\left[\mathbf{x}(t),\mathbf{u}(t),\mathbf{p}\right] & & \text{subject to} && \mathbf{x}'(\tau) = (t_f - t_0)\mathbf{f}[\mathbf{x}(\tau),\mathbf{u}(\tau),\mathbf{p}]\\
& && \boldsymbol{\gamma}\left[\mathbf{x}\,(t_0)\,,\mathbf{p}\right] = \mathbf{0} & \rightarrow & && \boldsymbol{\gamma}\left[\mathbf{x}\,(0)\,,\mathbf{p}\right] = \mathbf{0}\\
& && \underline{\mathbf{p}} \leqslant \mathbf{p} \leqslant \overline{\mathbf{p}} & & && \underline{\mathbf{p}} \leqslant \mathbf{p} \leqslant \overline{\mathbf{p}}\\
& && \underline{\mathbf{x}} \leqslant \mathbf{x}(t) \leqslant \overline{\mathbf{x}} & & && \underline{\mathbf{x}} \leqslant \mathbf{x}(\tau) \leqslant \overline{\mathbf{x}}\\
& && \underline{\mathbf{u}} \leqslant \mathbf{u}(t) \leqslant \overline{\mathbf{u}} & & && \underline{\mathbf{u}} \leqslant \mathbf{u}(\tau) \leqslant \overline{\mathbf{u}}\\
& && \underline{t}_0 \leqslant t_0 \leqslant \overline{t}_0 & & && \underline{t}_0 \leqslant t_0 \leqslant \overline{t}_0\\
& && \underline{t}_f \leqslant t_f \leqslant \overline{t}_f & & && \underline{t}_f \leqslant t_f \leqslant \overline{t}_f\\
& && t \in \left[t_0, t_f\right] & & && \tau \in [0, 1]
\end{aligned}$$

Note that with the definition of an augmented parameter vector with appropriate bounds

$$\begin{bmatrix}\underline{\mathbf{p}}\\ \underline{t}_0\\ \underline{t}_f\end{bmatrix} \leqslant \begin{bmatrix}\mathbf{p}\\ t_0\\ t_f\end{bmatrix} \leqslant \begin{bmatrix}\overline{\mathbf{p}}\\ \overline{t}_0\\ \overline{t}_f\end{bmatrix}$$

this problem represents a problem of the same type as Eq. (3.1) with the initial time fixed to zero and final time fixed to one.

Autonomization: Nonautonomous systems with an explicit dependence on the time t can, without loss of generality, be transformed into autonomous systems. Such a transformation is based on the introduction of a state $x_t \in \mathbb{R}$ propagated as $\dot{x}_t(t) = 1$ with initial condition $x_t(t_0) = t_0$:

$$\begin{aligned} &\underset{\mathbf{u}(t)}{\text{minimize}} && J\left[\mathbf{x}(t_f), \mathbf{p}\right] \\ &\text{subject to} && \dot{\mathbf{x}}(t) = \mathbf{f}\left[t, \mathbf{x}(t), \mathbf{u}(t), \mathbf{p}\right] \\ &&& \gamma\left[\mathbf{x}(t_0), \mathbf{p}\right] = \mathbf{0} \\ &&& \vdots \\ &&& t \in \left[t_0, t_f\right],\ t_0, t_f \text{ fixed} \end{aligned} \quad \rightarrow \quad \begin{aligned} &\underset{\mathbf{u}(t)}{\text{minimize}} && J\left[\mathbf{x}(t_f), \mathbf{p}\right] \\ &\text{subject to} && \dot{\mathbf{x}}(t) = \mathbf{f}\left[x_t(t), \mathbf{x}(t), \mathbf{u}(t), \mathbf{p}\right] \\ &&& \dot{x}_t(t) = 1 \\ &&& x_t(t_0) = t_0 \\ &&& \gamma\left[\mathbf{x}(t_0), \mathbf{p}\right] = \mathbf{0} \\ &&& \vdots \\ &&& t \in \left[t_0, t_f\right],\ t_0, t_f \text{ fixed} \end{aligned}$$

Augmenting the state vector by the state representing time with the dynamics

$$\frac{d}{dt}\begin{bmatrix} \mathbf{x}(t) \\ x_t(t) \end{bmatrix} = \begin{bmatrix} \mathbf{f}\left[x_t(t), \mathbf{x}(t), \mathbf{u}(t), \mathbf{p}\right] \\ 1 \end{bmatrix}$$

and extending the initial conditions

$$\begin{bmatrix} \gamma\left[\mathbf{x}(t_0), \mathbf{p}\right] \\ x_t(t_0) - t_0 \end{bmatrix} = \mathbf{0}$$

this problem represents an autonomous problem of the same type as Eq. (3.1).

General State-Dependent Path Constraints: Problems with purely state-dependent path constraints of the general form $\underline{\mathbf{g}} \leqslant \mathbf{g}[\mathbf{x}(t)] \leqslant \overline{\mathbf{g}}$, $\mathbf{g} : \mathbb{R}^{n_x} \rightarrow \mathbb{R}^{n_g}$ bounded by $\underline{\mathbf{g}}, \overline{\mathbf{g}} \in \mathbb{R}^{n_g}$ can be transformed into a problem featuring only box-bounded states.

Assuming sufficient differentiability, this transformation is achieved through the introduction of a state vector $\mathbf{x}_g(t) \in \mathbb{R}^{n_g}$ propagated as $\dot{\mathbf{x}}_g(t) = \partial\mathbf{g}/\partial\mathbf{x}\ \mathbf{f}\left[\mathbf{x}(t), \mathbf{u}(t), \mathbf{p}\right]$ with initial condition $\mathbf{x}_g(t_0) = \mathbf{g}\left[\mathbf{x}(t_0)\right]$:

$$
\begin{aligned}
&\underset{\mathbf{u}(t)}{\text{minimize}} && J\left[\mathbf{x}\left(t_f\right),\mathbf{p}\right] \\
&\text{subject to} && \dot{\mathbf{x}}(t) = \mathbf{f}\left[t,\mathbf{x}(t),\mathbf{u}(t),\mathbf{p}\right] \\
& && \boldsymbol{\gamma}\left[\mathbf{x}(t_0),\mathbf{p}\right] = \mathbf{0} \\
& && \vdots \\
& && t \in \left[t_0, t_f\right],\ t_0, t_f \text{ fixed}
\end{aligned}
\quad \rightarrow \quad
\begin{aligned}
&\underset{\mathbf{u}(t)}{\text{minimize}} && J\left[\mathbf{x}\left(t_f\right),\mathbf{p}\right] \\
&\text{subject to} && \dot{\mathbf{x}}(t) = \mathbf{f}\left[x_t(t),\mathbf{x}(t),\mathbf{u}(t),\mathbf{p}\right] \\
& && \dot{\mathbf{x}}_g(t) = \frac{\partial \mathbf{g}}{\partial \mathbf{x}}\mathbf{f}\left[\mathbf{x}(t),\mathbf{u}(t),\mathbf{p}\right] \\
& && \mathbf{x}_g(t_0) = \mathbf{g}\left[\mathbf{x}(t_0)\right] \\
& && \underline{\mathbf{g}} \leqslant \mathbf{x}_g(t) \leqslant \overline{\mathbf{g}} \\
& && \boldsymbol{\gamma}\left[\mathbf{x}(t_0),\mathbf{p}\right] = \mathbf{0} \\
& && \vdots \\
& && t \in \left[t_0, t_f\right],\ t_0, t_f \text{ fixed}
\end{aligned}
$$

Augmenting the state vector with appropriate bounds

$$
\begin{bmatrix} \underline{\mathbf{x}} \\ \underline{\mathbf{g}} \end{bmatrix} \leqslant \begin{bmatrix} \mathbf{x}(t) \\ \mathbf{x}_g(t) \end{bmatrix} \leqslant \begin{bmatrix} \overline{\mathbf{x}} \\ \overline{\mathbf{g}} \end{bmatrix}
$$

propagated by

$$
\frac{d}{dt}\begin{bmatrix} \mathbf{x}(t) \\ \mathbf{x}_g(t) \end{bmatrix} = \begin{bmatrix} \mathbf{f}\left[\mathbf{x}(t),\mathbf{u}(t),\mathbf{p}\right] \\ \frac{\partial \mathbf{g}}{\partial \mathbf{x}}\mathbf{f}\left[\mathbf{x}(t),\mathbf{u}(t),\mathbf{p}\right] \end{bmatrix}
$$

and the extension of the initial conditions

$$
\begin{bmatrix} \boldsymbol{\gamma}\left[\mathbf{x}(t_0),\mathbf{p}\right] \\ \mathbf{x}_g(t_0) - \mathbf{g}\left[\mathbf{x}(t_0)\right] \end{bmatrix} = \mathbf{0}
$$

this problem represents a problem of the same form as Eq. (3.1). Note that using the autonomization approach, this type of modeling can also be used to allow for time-dependent path constraints in Eq. (3.1) (including the case in which the bounds are time-varying).

For additional transformations regarding particular types of cost functions, please refer to the supplemental material provided in Appendix D.

Engineering Note 3.2.1: Transformations in Optimal Control Applications

The transformations presented in this section are useful to conceptually extend the generality of the clearance optimal control problem [Eq. (3.1)] and facilitate theoretical developments. In fact, such transformations are frequently used in practice for the implementation of certain problem types in optimal control solvers, in particular, if solvers do not natively support required modeling concepts or if the reformulated problem facilitates the convergence of the numerical solution.

The reader may have noticed that several of the presented transformations rely on the introduction of additional states. In this context, it should be mentioned, however, that this type of modeling introduces an overhead regarding the computational cost (hidden to some extent). Let us illustrate this point based on the introduction of the parameters as states. Under a direct solution approach, such a type of modeling would introduce a new decision variable for each state discretization time point and each parameter. Moreover, additional constraints (modeling the zero rates of these states) would explicitly link all decision variables associated with the same parameter. This strategy is rarely the way to go in practical applications, but solvers support more often than not the concept of a parameter and only introduce a single decision variable for each parameter. Similar considerations apply to the transformation of a Lagrange-type to a Mayer-type cost. Here, solvers supporting the concept of a Lagrange cost function usually solve the integral term internally using the quadrature method corresponding to the respective state integration method.

Nevertheless, besides the efficiency aspect, it is worth stressing that posing the problem in a different form through reformulations and problem transformations can have a *significant* effect on the performance of numerical solvers. This observation extends beyond the transformations presented in this section but in particular covers reformulations to structurally different problems while preserving the minimizers of the original problem. These kinds of transformations often make the difference if a formulation of the clearance problem can be analyzed properly as well as ultimately implemented and solved numerically in practice.

—

"Indeed, perhaps the most important issue needed to successfully solve a problem is the proper formulation" (Betts [20], Epilogue).

3.3 OPTIMALITY CONDITIONS

Optimal Control

The optimal control $\hat{\mathbf{u}}(t)$ is an admissible control globally minimizing the Hamiltonian $\mathcal{H}$:

$$\hat{\mathbf{u}}(t) = \arg \min_{\underline{\mathbf{u}} \leqslant \mathbf{u} \leqslant \overline{\mathbf{u}}} \mathcal{H}, \text{ a.e. in } [t_0, t_f]$$

Here, the Hamiltonian is evaluated with optimal states $\hat{\mathbf{x}}(t)$ and costates $\boldsymbol{\lambda}(t)$ evolving as

$$\frac{d\hat{\mathbf{x}}}{dt} = \left(\frac{\partial \mathcal{H}}{\partial \boldsymbol{\lambda}}\right)^T, \quad \frac{d\boldsymbol{\lambda}}{dt} = -\left(\frac{\partial \mathcal{H}}{\partial \mathbf{x}}\right)^T$$

For a control-affine system with control $\underline{u} \leqslant u \leqslant \overline{u}$, the switching function is defined as

$$\mathcal{S}(t) = \frac{\partial \mathcal{H}}{\partial u}$$

and its sign determines the optimal control $\hat{u}$ ($\mathcal{S} > 0 \rightarrow \hat{u} = \underline{u}$, $\mathcal{S} < 0 \rightarrow \hat{u} = \overline{u}$). If $\mathcal{S} = 0$ on a finite time interval, the *singular* control is determined by equating its time derivative with the first explicit appearance of the control to zero. Similarly, for problems featuring a purely state-dependent constraint

$$c\left[\mathbf{x}(t)\right] \leqslant 0$$

the *boundary* control holding the constraint on a boundary arc $c\left[\hat{\mathbf{x}}(t)\right] = 0$ is determined by equating its time derivative with the first explicit appearance of the control to zero.

The *minimum principle* [47][†] represents one of the cornerstones in optimal control theory. Unfortunately, the rigorous derivation of this principle can be considered rather technical and is beyond the scope of this book. Nevertheless, to provide the reader with an intuition regarding quantities such as the Hamiltonian as well as the costate system arising in the optimality conditions, we initiate our discussion based on a derivation of first-order necessary conditions using a variational approach. Hence, to begin with, let us consider the reduced problem

[†] The original work by Pontryagin and his coworkers considers a maximization problem. As the clearance optimal control problem constitutes (by our convention) a minimization problem, we use the term *minimum principle* throughout this book.

$$
\begin{aligned}
\underset{\mathbf{u}(t)}{\text{minimize}} \quad & J\left[\mathbf{x}\left(t_f\right)\right] \\
\text{subject to} \quad & \dot{\mathbf{x}}(t) = \mathbf{f}\left[\mathbf{x}(t), \mathbf{u}(t)\right] \\
& \mathbf{x}(t_0) = \mathbf{x}_0 \\
& t \in \left[t_0, t_f\right],\; t_0, t_f \text{ fixed}
\end{aligned} \tag{3.2}
$$

without control and state path constraints. The initial state is fixed to $\mathbf{x}_0$, and the final state is free. The *Hamiltonian* for this problem is defined as the product of the *costates* $\boldsymbol{\lambda}(t) \in \mathbb{R}^{n_x}$ and $\mathbf{f}$:

$$
\mathcal{H}[\boldsymbol{\lambda}(t), \mathbf{x}(t), \mathbf{u}(t)] = \boldsymbol{\lambda}^T(t)\, \mathbf{f}[\mathbf{x}(t), \mathbf{u}(t)] \tag{3.3}
$$

To simplify notation in the following, we omit, wherever possible, function arguments and use the short notations $\square_0 = \square(t_0)$ and $\square_f = \square\left(t_f\right)$.

For the derivation of the first-order necessary conditions of optimality, the augmented cost $\tilde{J}$ is introduced as

$$
\tilde{J} = J\left(\mathbf{x}_f\right) + \int_{t_0}^{t_f} \boldsymbol{\lambda}^T\left[\mathbf{f}(\mathbf{x}, \mathbf{u}) - \dot{\mathbf{x}}\right] dt
$$

May be skipped at first read.

by adjoining the dynamic constraints in differential form using the costates $\boldsymbol{\lambda}$ to the cost. The first variation of the augmented cost is

$$
\delta\tilde{J} = \frac{\partial J}{\partial \mathbf{x}_f}\delta\mathbf{x}_f + \int_{t_0}^{t_f}\left(\boldsymbol{\lambda}^T\frac{\partial \mathbf{f}}{\partial \mathbf{u}}\delta\mathbf{u} + \mathbf{f}^T\delta\boldsymbol{\lambda} - \dot{\mathbf{x}}^T\delta\boldsymbol{\lambda} - \boldsymbol{\lambda}^T\delta\dot{\mathbf{x}} + \boldsymbol{\lambda}^T\frac{\partial \mathbf{f}}{\partial \mathbf{x}}\delta\mathbf{x}\right) dt \tag{3.4}
$$

Integration by parts for the term $\boldsymbol{\lambda}^T\delta\dot{\mathbf{x}}$

$$
\int_{t_0}^{t_f} \boldsymbol{\lambda}^T\delta\dot{\mathbf{x}}dt = \boldsymbol{\lambda}_f^T\delta\mathbf{x}_f - \boldsymbol{\lambda}_0^T\delta\mathbf{x}_0 - \int_{t_0}^{t_f}\dot{\boldsymbol{\lambda}}^T\delta\mathbf{x}dt
$$

and insertion of these terms (considering only admissible variations, i.e., $\delta\mathbf{x}_0 = \mathbf{0}$) in Eq. (3.4) yields

$$
\begin{aligned}
\delta\tilde{J} = \left[\frac{\partial J}{\partial \mathbf{x}_f} - \boldsymbol{\lambda}_f^T\right]\delta\mathbf{x}_f + \int_{t_0}^{t_f}\Bigg(&\boldsymbol{\lambda}^T\frac{\partial \mathbf{f}}{\partial \mathbf{u}}\delta\mathbf{u} + \left[\mathbf{f}^T - \dot{\mathbf{x}}^T\right]\delta\boldsymbol{\lambda} \\
&+ \left[\boldsymbol{\lambda}^T\frac{\partial \mathbf{f}}{\partial \mathbf{x}} + \dot{\boldsymbol{\lambda}}^T\right]\delta\mathbf{x}\Bigg) dt
\end{aligned}
$$

Noting that from the definition of the Hamiltonian [Eq. (3.3)] we have

$$\frac{\partial \mathcal{H}}{\partial \mathbf{u}} = \boldsymbol{\lambda}^T \frac{\partial \mathbf{f}}{\partial \mathbf{u}}, \qquad \frac{\partial \mathcal{H}}{\partial \boldsymbol{\lambda}} = \mathbf{f}^T, \qquad \frac{\partial \mathcal{H}}{\partial \mathbf{x}} = \boldsymbol{\lambda}^T \frac{\partial \mathbf{f}}{\partial \mathbf{x}}$$

the first variation can be expressed as

$$\delta\tilde{J} = \left[\frac{\partial J}{\partial \mathbf{x}_f} - \boldsymbol{\lambda}_f^T\right]\delta\mathbf{x}_f + \int_{t_0}^{t_f}\left(\frac{\partial \mathcal{H}}{\partial \mathbf{u}}\delta\mathbf{u} + \left[\frac{\partial \mathcal{H}}{\partial \boldsymbol{\lambda}} - \dot{\mathbf{x}}^T\right]\delta\boldsymbol{\lambda} + \left[\frac{\partial \mathcal{H}}{\partial \mathbf{x}} + \dot{\boldsymbol{\lambda}}^T\right]\delta\mathbf{x}\right)dt$$

Equating this first variation of the augmented cost to zero (i.e., $\delta\tilde{J} = 0$) yields first-order necessary conditions of optimality.

The following equations need to be fulfilled at an optimal solution of Eq. (3.2):

$$\frac{d\mathbf{x}}{dt} = \left(\frac{\partial \mathcal{H}}{\partial \boldsymbol{\lambda}}\right)^T, \quad \mathbf{x}(t_0) = \mathbf{x}_0 \qquad \textit{(State Equation)}$$

$$\frac{d\boldsymbol{\lambda}}{dt} = -\left(\frac{\partial \mathcal{H}}{\partial \mathbf{x}}\right)^T, \quad \boldsymbol{\lambda}(t_f) = \left(\frac{\partial J}{\partial \mathbf{x}_f}\right)^T \qquad \textit{(Costate Equation)}$$

$$\frac{\partial \mathcal{H}}{\partial \mathbf{u}} = \mathbf{0} \qquad \textit{(Control Equation)}$$

Note that, in our discussion, the control vector has so far been considered unbounded. Therefore, all control variations are admissible, and the control equation can be interpreted as a necessary condition for a local extremal point of the Hamiltonian with regard to the control. Recall, however, that in our clearance problem, the controls are limited by

$$\underline{\mathbf{u}} \leqslant \mathbf{u} \leqslant \bar{\mathbf{u}}$$

For $\underline{\mathbf{u}} \leqslant \bar{\mathbf{u}}$, these box-bounds define a nonempty, closed, and convex control set $\mathcal{U}$. Under this constraint, clearly not all control variations are admissible. For this case, the minimum principle states that given the optimal states $\hat{\mathbf{x}}(t)$ the optimal control is the *admissible* control minimizing the Hamiltonian, that is,

$$\hat{\mathbf{u}}(t) = \arg\min_{\mathbf{u}\in\mathcal{U}} \mathcal{H}\left[\boldsymbol{\lambda}(t), \hat{\mathbf{x}}(t), \mathbf{u}\right]$$

almost everywhere on $[t_0, t_f]$. In addition, it guarantees the existence of nontrivial multipliers

$$\boldsymbol{\lambda} \neq \mathbf{0}$$

satisfying the costate differential equation with the respective boundary condition. It is important to mention that the variational principles, which were used in this section to illustrate a derivation of first-order necessary conditions are local in nature. For the minimum principle, it is, however, possible to prove a global version expressing that the optimal control $\hat{\mathbf{u}}(t)$ is the admissible control *globally* minimizing the Hamiltonian; that is,

$$\mathcal{H}\left[\boldsymbol{\lambda}(t), \hat{\mathbf{x}}(t), \hat{\mathbf{u}}(t)\right] \leqslant \mathcal{H}\left[\boldsymbol{\lambda}(t), \hat{\mathbf{x}}(t), \mathbf{u}\right], \; \forall \mathbf{u} \in \mathcal{U}$$

almost everywhere on $[t_0, t_f]$. Note that a local minimum is also global if the Hamiltonian is convex in the control. This holds, for example, if the system is control-affine. Because of its importance in practical applications, and in particular in the clearance context, let us consider such control-affine systems featuring a scalar control $u \in \mathbb{R}$ of the form

$$\dot{\mathbf{x}}(t) = \mathbf{f}_u\left[\mathbf{x}(t)\right] u(t) + \mathbf{f}_x\left[\mathbf{x}(t)\right] \tag{3.5}$$

with state dependent functions $\mathbf{f}_x, \mathbf{f}_u : \mathbb{R}^{n_x} \to \mathbb{R}^{n_x}$. Using the definition (3.3), the Hamiltonian in the case of Eq. (3.5) is expressed as

$$\mathcal{H}[\boldsymbol{\lambda}(t), \mathbf{x}(t), u(t)] = \boldsymbol{\lambda}^T(t)\mathbf{f}_u\left[\mathbf{x}(t)\right] u(t) + \boldsymbol{\lambda}^T(t)\mathbf{f}_x\left[\mathbf{x}(t)\right]$$

Introducing the so-called *switching function*,

$$\mathcal{S}\left[\boldsymbol{\lambda}(t), \mathbf{x}(t)\right] = \frac{\partial \mathcal{H}}{\partial u} = \boldsymbol{\lambda}^T(t)\mathbf{f}_u\left[\mathbf{x}(t)\right]$$

the Hamiltonian becomes

$$\mathcal{H}[\boldsymbol{\lambda}(t), \mathbf{x}(t), u(t)] = \mathcal{S}\left[\boldsymbol{\lambda}(t), \mathbf{x}(t)\right] u(t) + \boldsymbol{\lambda}^T(t)\mathbf{f}_x\left[\mathbf{x}(t)\right]$$

Recall that, according to the minimum principle, the optimal control is an admissible control that minimizes the Hamiltonian. Therefore, in the control-affine case, the optimal control $\hat{u}$ minimizes the product of the switching function and the control:

$$\hat{u}(t) = \arg \min_{\underline{u} \leqslant u \leqslant \overline{u}} \mathcal{S}[\boldsymbol{\lambda}(t), \hat{\mathbf{x}}(t)]u$$

This implies that the sign of the switching function determines the optimal control (see Fig. 3.1):

$$\hat{u} = \begin{cases} \underline{u}, & \mathcal{S} > 0 \\ \overline{u}, & \mathcal{S} < 0 \\ u_s, & \text{otherwise} \end{cases}$$

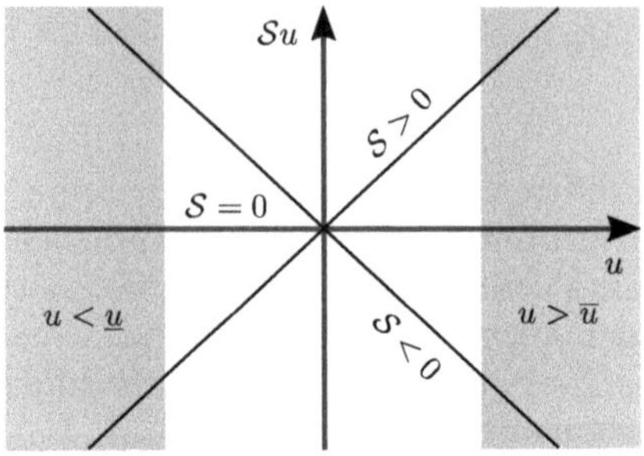

Fig. 3.1 Product of the switching function and control.

Observe that if $\mathcal{S}$ is zero on some interval $t \in [t_1, t_2]$, $t_1 < t_2$, the control is not uniquely determined by the value of the switching function alone. In this case, additional conditions are required to find the so-called *singular* control u_s. The key insight for deriving these additional conditions is that if the switching function remains zero, this directly implies that its time derivatives are also zero. The time derivatives of the switching function can be defined recursively as

$$\mathcal{S}^{(0)} = \mathcal{S}[\boldsymbol{\lambda}(t), \mathbf{x}(t)]$$

$$\mathcal{S}^{(i)} = \frac{d^i\mathcal{S}}{dt^i} = \frac{\partial \mathcal{S}^{(i-1)}}{\partial \mathbf{x}}\frac{d\mathbf{x}}{dt} + \frac{\partial \mathcal{S}^{(i-1)}}{\partial \boldsymbol{\lambda}}\frac{d\boldsymbol{\lambda}}{dt}, \qquad i = 1, \ldots, q$$

where the qth derivative satisfies

$$\frac{\partial \mathcal{S}^{(q-1)}}{\partial u} = 0, \qquad \frac{\partial \mathcal{S}^{(q)}}{\partial u} \neq 0$$

for which it can be shown that q is even [48]. Then, $\mathcal{S}^{(q)} = 0$ can be decomposed as

$$\underbrace{\mathcal{S}_x[\mathbf{x}(t)]}_{\mathcal{S}^{(q)}|_{u=0}} + \underbrace{\mathcal{S}_u[\mathbf{x}(t)]}_{\frac{\partial \mathcal{S}^{(q)}}{\partial u}} u(t) = 0$$

which can be solved for the singular control at an optimal solution:

$$u_s(t) = -\frac{\mathcal{S}_x[\boldsymbol{\lambda}(t), \hat{\mathbf{x}}(t)]}{\mathcal{S}_u[\boldsymbol{\lambda}(t), \hat{\mathbf{x}}(t)]}$$

Let us extend our discussion regarding the optimality conditions to the case in which the optimal control problem also features a purely state-dependent path constraint of the form

$$c[\mathbf{x}(t)] \leqslant 0 \tag{3.6}$$

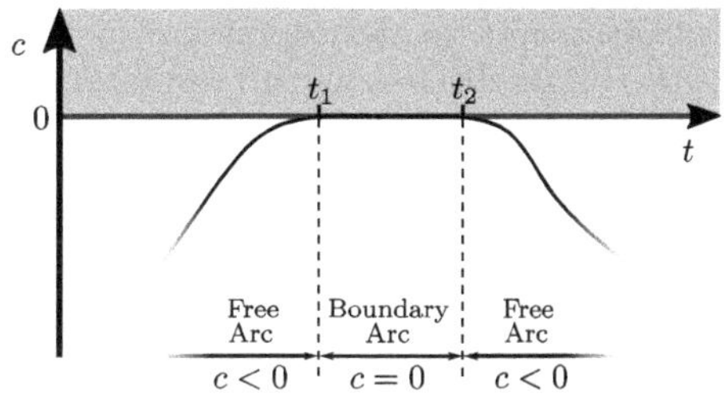

Fig. 3.2 Boundary and free arcs.

with $c : \mathbb{R}^{n_x} \to \mathbb{R}$. Observe that as our clearance problem considers box-bounded states, a path constraint of the general form (3.6) can be used to model both, either the constraint for the upper bound $x(t) - \overline{x} \leqslant 0$ or the lower bound $\underline{x} - x(t) \leqslant 0$.

Let us further consider the case where the state constraint is active, and the system rides the constraint on a so-called *boundary arc*, meaning $c\left[\mathbf{x}(t)\right] = 0$ for $t \in [t_1, t_2]$, $t_1 < t_2$ (see Fig. 3.2).

As for the discussion regarding the switching function, let us consider a control-affine system of the form (3.5). The control $u_b(t)$ holding the system at the boundary of the state constraint $c\left[\mathbf{x}(t)\right] = 0$ is termed the *boundary control.* Similar to the case for the singular control, the equation $c\left[\mathbf{x}(t)\right] = 0$ does not directly provide us with a condition revealing $u_b(t)$. However, the same insight as for the switching function regarding the stationarity of the time derivatives can be used to arrive at a condition for the boundary control. The time derivatives of the constraint function can be defined recursively as

$$
\begin{aligned}
c^{(0)} &= c\left[\mathbf{x}(t)\right] \\
c^{(i)} &= \frac{d^i c}{dt^i} = \frac{\partial c^{(i-1)}}{\partial \mathbf{x}} \frac{d\mathbf{x}}{dt}, \qquad i = 1, \ldots, p
\end{aligned}
$$

where the pth derivative satisfies

$$
\frac{\partial c^{(p-1)}}{\partial u} = 0, \qquad \frac{\partial c^{(p)}}{\partial u} \neq 0 \tag{3.7}
$$

Then, p is termed the *order of the constraint*, and the condition $c^{(p)} = 0$ can be decomposed as

$$
\underbrace{c_x\left[\mathbf{x}(t)\right]}_{c^{(p)}|_{u=0}} + \underbrace{c_u\left[\mathbf{x}(t)\right]}_{\frac{\partial c^{(p)}}{\partial u}} u(t) = 0
$$

allowing us to directly express the boundary control at an optimal solution:

$$
u_b(t) = -\frac{c_x\left[\hat{\mathbf{x}}(t)\right]}{c_u\left[\hat{\mathbf{x}}(t)\right]}
$$

The theoretical analysis and the development of optimality conditions for problems featuring purely state-dependent path constraints is quite involved (see Hartl et al. [49] and Gerdts [21]). Two approaches for the treatment of this type of constraint in optimal control theory are the indirect and the direct adjoining approach [49]. Under the indirect adjoining approach [50], an index reduction technique is applied to the constraint; the constraint is derived p times until the direct dependence of the constraint on the control is revealed [see Eq. (3.7)]. Under the direct adjoining approach [51], the Hamiltonian is augmented

$$\mathcal{H}[\boldsymbol{\lambda}(t), \mathbf{x}(t), \mathbf{u}(t), \mu(t)] = \boldsymbol{\lambda}^T(t)\mathbf{f}[\mathbf{x}(t), \mathbf{u}(t)] + \mu(t)c[\mathbf{x}(t)]$$

by directly adjoining the state constraint using a piecewise continuous multiplier satisfying the complementarity condition:

$$\mu(t) \geqslant 0, \qquad \mu(t)c[\mathbf{x}(t)] = 0$$

Moreover, it is worth mentioning that for problems with purely state-dependent path constraints, the costates are not necessarily continuous. At time points where a constraint becomes active/inactive, the costates can exhibit "jumps." Such a jump at a point of discontinuity t_* is of the form

$$\boldsymbol{\lambda}^T\Big|_{t=t_*^-} = \boldsymbol{\lambda}^T\Big|_{t=t_*^+} + \eta \left.\frac{\partial c}{\partial \mathbf{x}}\right|_{t=t_*}$$

where $\eta > 0$ defines the jump height and t_*^- (t_*^+) denote the left-(right-)sided limit of t_*. In particular, the fact that the sequence of free arcs and boundary arcs as well as the exact entry and exit points to these arcs are not known a priori typically poses challenges under indirect solution approaches. As pointed out earlier in this chapter, we do not focus on the development of indirect solution methods in this book but rather put the direct approach for the numerical solution of the clearance optimal control problem at the center of our attention. Nevertheless, the optimality conditions highlighted in this section prove valuable in Part II for the analysis of certain problem classes and to provide sanity checks of numerical solutions obtained under the direct solution approach.

PART II

Methods

The main subsystems of an aircraft closed-loop system are the controller, the plant, and sensors. Pilot commands, as well as other external disturbances such as wind, constitute relevant inputs to this type of system in the context of flight control law clearance (see Fig. 1).

Varga [52] provides an overview of criteria for civil aircraft, including the general model types commonly employed for this task:

- linear models
- linear-parametric models
- nonlinear-parametric models

Of these types, the last can be viewed as the most general (and challenging, from a clearance perspective), whereas the first two are usually defined based on linearizations of the nonlinear model at equilibrium (trim) points. Clearly, the validity of such a linearized model decreases the farther away this model is used from the respective reference point. Clearance criteria in the context of worst-case input analysis are typically related to [52]

- aerodynamic angles (angle of attack, angle of sideslip)
- load factors
- attitude/angular rates (pitch, roll, yaw)

In some cases, it is required to consider these criteria not only in isolation but jointly in the form of a multicriteria/set-based analysis.

The organization of chapters in this part of the book reflects the aforementioned categorizations. First, methods for linear and linear-parametric models are considered in Chapter 4 and Chapter 5. Moreover, Chapter 6 discusses methods for nonlinear-parametric models. For both linear and nonlinear models, set-based analysis methods for the optimal control-based clearance approach are presented in Chapter 7, allowing the joint investigation of multiple criteria. The methods under the optimal control approach share common modeling philosophies

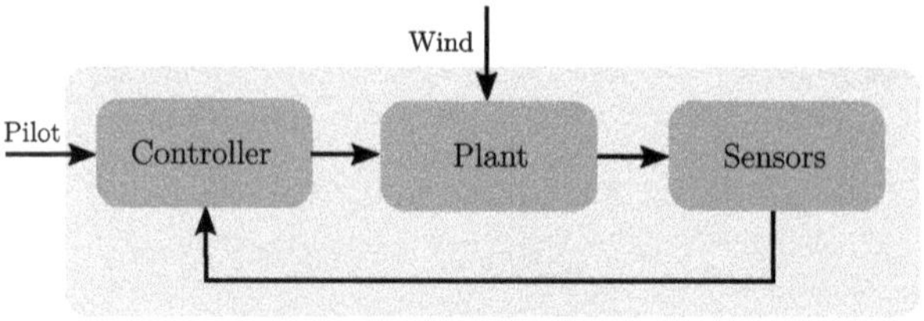

Fig. 1 Aircraft closed-loop system.

regarding the introduction of the criterion in the clearance problem as well as the consideration of internal limits in the servomechanism, which we discuss in the following.

Criterion (Cost): Under the optimal control approach, we choose to model the criterion as a *state-dependent cost J minimized at the final point of a fixed time interval* $[t_0, t_f]$. For this type of formulation, let us observe that the optimal cost is *nonincreasing for an increasing length of the time interval* if the system starts from an initial condition at which it can be held for an arbitrary amount of time. This can be shown as follows: Suppose that at t_0, the system originates from the initial condition, and the criterion evaluated at t_f is $\hat{J}$. Extending the interval by some $\Delta t > 0$, an admissible continuation strategy is to keep the system at the initial condition for Δt seconds and then apply the previous inputs, obtaining the previous cost $\hat{J}$. This argument implies that the set of attainable values of the criterion for the extended interval *is a superset of or equal to* the set of attainable values without the extension (see Fig. 2).

As a consequence, the maximal reasonable (fixed) final time can be considered for the clearance problem. In practice, the length of the time interval is often based

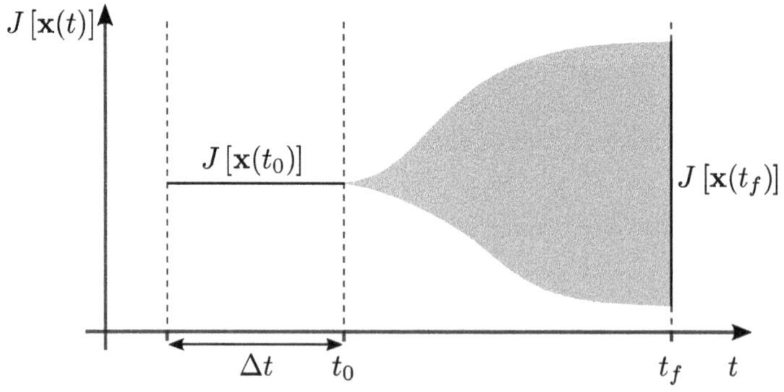

Fig. 2 Attainable values under the continuation strategy to keep the system at the initial condition for Δt seconds.

on characteristic time constants of the closed-loop system (e.g., 5–10 times the short period timescale [52]).

Let us discuss an alternative formulation regarding the modeling of the cost. Instead of minimizing J at t_f, it is also possible to consider the *minimization of the minimum of* J over the fixed interval $[t_0, t_f]$ (i.e., a *minimization of* $\tilde{J}$) defined as

$$\tilde{J} = \min_{t \in [t_0, t_f]} J[\mathbf{x}(t)]$$

We note that the attainable values of $J[\mathbf{x}(t)]$ at the end of an interval of length $t_f - t_0$ are *equal to* the attainable values of $\tilde{J}$. After all, the same value as $\tilde{J}$ occurring at some

$$t_* = \arg \min_{t \in [t_0, t_f]} J[\mathbf{x}(t)]$$

can always be obtained at the final point of a fixed interval by keeping the system in the initial condition on $\left[t_0 - (t_f - t_*), t_0\right]$ and then applying the same control on $[t_0, t_*]$. Naturally, the question arises: Is there some strong reason to prefer one formulation over the other?

Suppose that an optimization algorithm with some form of local search behavior is employed and that the minimum occurs at $t_* < t_f$ during the search. In this situation, all control parameters that influence the control function only for $t > t_*$ exhibit locally no effect on the cost. After all, we are dealing with a causal system in which future control actions cannot affect the past. For instance, if a local gradient-based algorithm was used, the derivatives of the cost with respect to these control parameters would be zero. This would imply that there would be no direct motivation for the local search algorithm to change these control parameters as, ultimately, no apparent effect of such a change on the cost can be inferred from the available information. This aspect, resulting from the local characteristics of these methods, effectively limits the search space because these parameters do not appear to be useful for advancing the worst-case search. In essence, the problem behaves in such a situation very much like a reduced problem solved on the narrower interval $[t_0, t_*]$ *with the cost evaluated at the final point* t_*. Therefore, under the optimal control approach, we prefer to directly model the criterion as a cost function evaluated at the final point t_f. This setting is deemed favorable because *past actions affect the future*, and we do not promote an unintended/undesired reduction of the search space due to the inherently local nature of the majority of optimization methods used in practice.

Servo Limits (Constraints): The servomechanism can be considered a critical path within an aircraft closed-loop system. After all, pilot commands translated by the controller need to pass through this channel to realize control surface deflections (see plant in Fig. 3). The criticality of this channel is contributed to the fact that physical actuators are subject to saturations (e.g., related to rate and position limits). These limits are, therefore, inherent characteristics of the flight

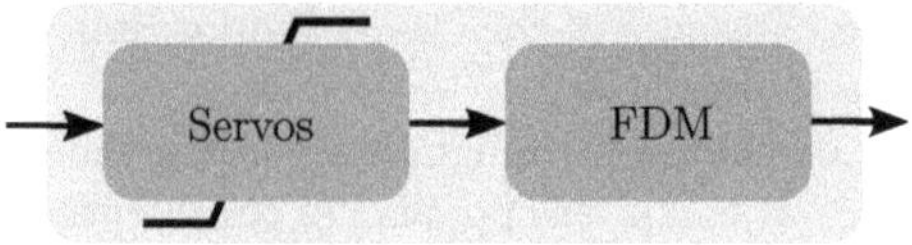

Fig. 3 The plant [servomechanism with limits, flight dynamic model (FDM)].

system. Essentially, a complete physical model of the servomechanism, including all mechanical and electrical components, reaction forces, and so forth, would, by itself, keep the system within the respective physical limits. Unfortunately, in practice, we often need to rely on simplifications/idealizations to represent complex model behavior. This is also true for the servomechanism, which is often represented by a low-order system featuring a logic to map limits directly onto states.

From a control perspective, it is highly undesirable to drive the actuators into saturation. Hence, the consideration of servo limits constitutes an important aspect to be addressed by methods concerned with the safety assessment of flight control systems. The importance of these limits within the servomechanism is also reflected in other areas, such as the investigation of pilot-induced oscillations (PIO). For example, the PIO categorization suggested by McRuer [53] considers not only the nonlinearity of the model but also surface rate and position limiting:

I Essentially Linear Pilot–Vehicle System Oscillations

II Quasi-Linear Pilot–Vehicle System Oscillations with Surface Rate or Position Limiting

III Essentially Nonlinear Pilot–Vehicle System Oscillations with Transitions

Even though PIO effects are not the focus of the methods presented in this book, this categorization appears to align well with the modeling intent behind optimal control approaches for worst-case input analysis.* Therefore, we adopt these modeling categories with slight modifications in the context of this book: In category I (Cat I), the dynamic system is represented by a linear(-parametric) model and is only subject to limits for external pilot commands/disturbances. Category II (Cat II) extends the first category by considering limits inherent to the servomechanism. All other cases are collected into category III (Cat III).

Within the optimal control framework of this book, the internal limits of the servomechanism are modeled as purely state-dependent path constraints. Under

*Interestingly, the excitement of oscillatory modes due to pilot---vehicle interactions can be observed as optimal (worst-case) solutions in certain clearance contexts.

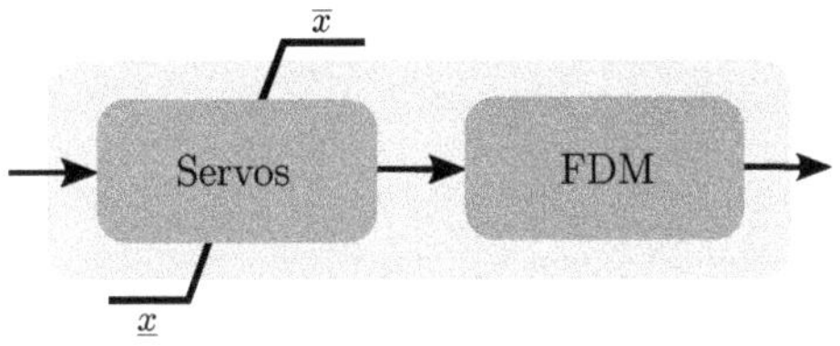

Fig. 4 Externalized limits of the servos.

such a modeling approach, the internal limits are in effect "externalized" and, therefore, exposed to the optimization problem (see Fig. 4).

Explicitly, the lower limit $\underline{x}$ and upper limit $\overline{x}$ on some state $x(t)$ are directly unveiled to the optimization problem through the introduction of path constraints of the form:

$$\underline{x} \leqslant x(t) \leqslant \overline{x}$$

Notably, under this approach, it is the responsibility of the control to ensure that the trajectory of the system does not breach the state limits. Therefore, the responsibility for not exceeding these limits is essentially transferred from the system itself to the control. Stated differently, under this type of modeling, the *optimal control* keeps the system inside the bounds and steers the system to a bound (if it is optimal to go there), holds the system on a bound (if it is optimal to remain there), or steers the system away from a bound (if it is optimal to leave it). Here, the optimal choice for the control holding the system on the bound is termed the *boundary control* in the context of optimal control theory (see Sec. 3.3). An important observation in this context is that if the limit would be modeled internally and the servomechanism was driven into saturation, there could be many possible control choices having the same effect on the criterion. This situation can make the worst-case input not uniquely defined in the context of an optimization problem. However, if the limits are imposed as constraints, the optimal control is in this situation defined by the condition to hold the system on the boundary.

CHAPTER 4

Linear Nominal Clearance

In this chapter, optimal control approaches for linear closed-loop aircraft models are introduced. Two types of optimal control formulations for worst-case input analysis are discussed in this context. The first type constitutes a linear optimal control problem with bounded controls (Cat I), and the second type additionally considers purely state-dependent constraints (Cat II). For both types of optimal control formulations, the worst-case input is characterized based on the minimum principle.

Moreover, to enable efficient solutions for clearance problems of realistic sizes, particularly those featuring a large number of states and inputs, we demonstrate how the respective optimal control problem in continuous time can be transformed into a linear program (LP) using the direct optimal control approach. Notably, a global solution to the discretized problem can be obtained for LPs by well-established and performant methods.

Furthermore, it is worth mentioning that the nominal direct optimal control formulation introduced in this chapter constitutes the basis for several optimal control-based clearance approaches presented throughout this book. This includes not only the parametric form as well as combined worst-case input analysis (static parameters and controls) but also reachability analysis of linear systems and the hybrid Cat III method for nonlinear systems.

This chapter is organized as follows: First, the nominal linear clearance problem is defined in Sec. 4.1 for control-constrained (Cat I) as well as control- and state-constrained (Cat II) models. Following, Sec. 4.2 illustrates the transcription to LPs using a selection of practical direct optimal control methods. For both problem types (Cat I in Sec. 4.3 and Cat II in Sec. 4.4), optimal control theoretical, analyses based on the minimum principle reveal the expected types of worst-case inputs. Moreover, for Cat II problems, an explicit expression for the boundary control is derived. Several numerical examples illustrate the application of the Cat I and II clearance methods for longitudinal and lateral aircraft closed-loop models.

4.1 PROBLEM FORMULATION

Clearance Optimal Control Problem (Cat I and II Nominal)

The nominal aircraft worst-case input problem (Cat I and II) including limits for the inputs and actuators can be modeled as a Mayer-type optimal control problem starting at $\mathbf{x}(t_0) = \mathbf{0}$:

$$\begin{aligned}
\underset{\mathbf{u}(t)}{\text{minimize}} \quad & \mathbf{c}^T \mathbf{x}(t_f) \\
\text{subject to} \quad & \dot{\mathbf{x}}(t) = \mathbf{A}\mathbf{x}(t) + \mathbf{B}\mathbf{u}(t) \\
& \underline{\mathbf{x}} \leqslant \mathbf{x}(t) \leqslant \overline{\mathbf{x}} \\
& \underline{\mathbf{u}} \leqslant \mathbf{u}(t) \leqslant \overline{\mathbf{u}} \\
& t \in \left[t_0, t_f\right], \ t_0, t_f \text{ fixed}
\end{aligned} \tag{4.1}$$

with the fixed initial time t_0 and final time t_f, the state vector $\mathbf{x}(t) \in \mathbb{R}^{n_x}$ bounded by $\underline{\mathbf{x}}, \overline{\mathbf{x}} \in \mathbb{R}^{n_x}$, the control vector $\mathbf{u}(t) \in \mathbb{R}^{n_u}$ bounded by $\underline{\mathbf{u}}, \overline{\mathbf{u}}, \in \mathbb{R}^{n_u}$, the system matrix $\mathbf{A} \in \mathbb{R}^{n_x \times n_x}$, the input matrix $\mathbf{B} \in \mathbb{R}^{n_x \times n_u}$, as well as the state output vector $\mathbf{c} \in \mathbb{R}^{n_x}$ modeling the criterion under investigation.

In this chapter, an aircraft closed-loop model in linear, time-invariant form

$$\dot{\mathbf{x}}(t) = \mathbf{A}\mathbf{x}(t) + \mathbf{B}\mathbf{u}(t) \tag{4.2}$$

with multiple inputs is considered for the purpose of worst-case input analysis. The output equation for a scalar output $y(t) \in \mathbb{R}$ with the corresponding output vector $\mathbf{c}$ is given as

$$y(t) = \mathbf{c}^T \mathbf{x}(t)$$

This output represents the clearance criterion in the optimal control problem. All trajectories are assumed to originate from a zero initial reference state $\mathbf{x}(t_0) = \mathbf{0}$.

Regarding the control vector, we assume finite box-bounds of the form

$$\underline{\mathbf{u}} \leqslant \mathbf{u}(t) \leqslant \overline{\mathbf{u}} \tag{4.3}$$

which constitutes the predominant form in clearance applications. Similarly, we also allow box-bounds of the state vector

$$\underline{\mathbf{x}} \leqslant \mathbf{x}(t) \leqslant \overline{\mathbf{x}} \tag{4.4}$$

These purely state-dependent constraints can be used to model the actuator rate and position limits of the system. For all unbounded state vector components, the respective limits may be defined as $\pm\infty$.

Throughout this chapter, the clearance problem is formulated as a single-phase Mayer-type optimal control problem. This implies that the problem is formulated on a fixed time interval $[t_0, t_f]$, and the cost function $J : \mathbb{R}^{n_x} \to \mathbb{R}$ representing the clearance criterion depends on the state vector at the final time point t_f:

$$J\left[\mathbf{x}(t_f)\right] = y(t_f) = \mathbf{c}^T\mathbf{x}(t_f)$$

Recall that there exist several transformations allowing conversion from more general problem types to this type of problem (see Sec. 3.2).

4.2 DISCRETIZED FORM

Discretized Clearance Optimal Control Problem (Cat I and II Nominal)

The transcription of nominal worst-case input problems (Cat I and II) based on direct methods yields an LP of the form

$$\begin{aligned} \underset{\mathbf{z}\in\mathbb{R}^n}{\text{minimize}} \quad & \mathbf{v}^T\mathbf{z} \\ \text{subject to} \quad & \mathbf{M}\mathbf{z} = \mathbf{0} \\ & \underline{\mathbf{z}} \leqslant \mathbf{z} \leqslant \overline{\mathbf{z}} \end{aligned} \tag{4.5}$$

with the equality constraint matrix $\mathbf{M} \in \mathbb{R}^{n_h \times n}$, the decision variable vector $\mathbf{z} \in \mathbb{R}^n$ bounded by $\underline{\mathbf{z}}, \overline{\mathbf{z}} \in \mathbb{R}^n$, as well as the cost coefficient vector $\mathbf{v} \in \mathbb{R}^n$.

The Cat I and II clearance optimal control problem in continuous time can be transcribed to a finite-dimensional optimization problem through appropriate discretization schemes. This transcription process is illustrated in the following using the backward Euler and trapezoidal collocation methods. In both cases, the discretization considers N discretization steps (segments) with discretization step length $h = (t_f - t_0)/N$. The discretized problem yields an LP of the form presented as Eq. (4.5). Global solutions of LPs can be obtained efficiently using numerical algorithms, such as Dantzig's simplex method [54].

4.2.1 BACKWARD EULER TRANSCRIPTION

In the case of the backward Euler method, state variable vectors $\mathbf{x}_i \in \mathbb{R}^{n_x}, i = 1, \ldots, N$ and control variable vectors $\mathbf{u}_i \in \mathbb{R}^{n_u}, i = 1, \ldots, N$ are introduced, and the time-discretized dynamic equality constraints are expressed as

$$\mathbf{x}_i = \mathbf{x}_{i-1} + h\left(\mathbf{A}\mathbf{x}_i + \mathbf{B}\mathbf{u}_i\right), \ i = 1, \ldots, N$$

For a fixed initial condition $\mathbf{x}_0 = \mathbf{0}$, the discretized Cat II problem can be written as

$$\begin{aligned} \underset{\mathbf{x}_i, \mathbf{u}_i}{\text{minimize}} \quad & \mathbf{c}^T \mathbf{x}_N \\ \text{subject to} \quad & \mathbf{x}_i = \mathbf{x}_{i-1} + h\left(\mathbf{A}\mathbf{x}_i + \mathbf{B}\mathbf{u}_i\right) \\ & \underline{\mathbf{x}} \leqslant \mathbf{x}_i \leqslant \overline{\mathbf{x}} \\ & \underline{\mathbf{u}} \leqslant \mathbf{u}_i \leqslant \overline{\mathbf{u}} \\ & i = 1, \ldots, N \end{aligned} \tag{4.6}$$

Note that this discretized problem is of Cat II type and, in the absence of state constraints, reduces to a Cat I–type problem. Moreover, Eq. (4.6) features $n = N(n_x + n_u)$ decision variables and $n_h = Nn_x$ equality constraints. The cost coefficient vector $\mathbf{v} \in \mathbb{R}^n$ as well as the decision variable vector $\mathbf{z} \in \mathbb{R}^n$ with corresponding lower bound $\underline{\mathbf{z}} \in \mathbb{R}^n$ and upper bound $\overline{\mathbf{z}} \in \mathbb{R}^n$ are

$$\mathbf{v} = \begin{bmatrix} \mathbf{0} \\ \mathbf{0} \\ \vdots \\ \mathbf{c} \\ \mathbf{0} \end{bmatrix}, \quad \underline{\mathbf{z}} = \begin{bmatrix} \underline{\mathbf{x}} \\ \underline{\mathbf{u}} \\ \vdots \\ \underline{\mathbf{x}} \\ \underline{\mathbf{u}} \end{bmatrix}, \quad \mathbf{z} = \begin{bmatrix} \mathbf{x}_1 \\ \mathbf{u}_1 \\ \vdots \\ \mathbf{x}_N \\ \mathbf{u}_N \end{bmatrix}, \quad \overline{\mathbf{z}} = \begin{bmatrix} \overline{\mathbf{x}} \\ \overline{\mathbf{u}} \\ \vdots \\ \overline{\mathbf{x}} \\ \overline{\mathbf{u}} \end{bmatrix}$$

and the equality constraints can be expressed through the matrix $\mathbf{M} \in \mathbb{R}^{n_h \times n}$:

$$\mathbf{M} = \begin{bmatrix} \mathbf{I} - h\mathbf{A} & -h\mathbf{B} & \mathbf{0} & \mathbf{0} & \mathbf{0} & \mathbf{0} & \cdots \\ -\mathbf{I} & \mathbf{0} & \mathbf{I} - h\mathbf{A} & -h\mathbf{B} & \mathbf{0} & \mathbf{0} & \cdots \\ \mathbf{0} & \mathbf{0} & -\mathbf{I} & \mathbf{0} & \mathbf{I} - h\mathbf{A} & -h\mathbf{B} & \\ \vdots & \vdots & & & \ddots & & \end{bmatrix}$$

May be skipped (↴) at first read.

Trapezoidal Transcription For the trapezoidal collocation method, the dynamic equality constraints are defined as

$$\mathbf{x}_i = \mathbf{x}_{i-1} + h\left(\mathbf{A}\frac{\mathbf{x}_i + \mathbf{x}_{i-1}}{2} + \mathbf{B}\frac{\mathbf{u}_i + \mathbf{u}_{i-1}}{2}\right), \; i = 1, \ldots, N$$

Compared to the backward Euler method, there is an additional control variable $\mathbf{u}_0$ introduced in the discretized problem

$$
\begin{aligned}
\underset{\mathbf{x}_i,\mathbf{u}_i,\mathbf{u}_0}{\text{minimize}} \quad & \mathbf{c}^T\mathbf{x}_N \\
\text{subject to} \quad & \mathbf{x}_i = \mathbf{x}_{i-1} + h\left(\mathbf{A}\frac{\mathbf{x}_i + \mathbf{x}_{i-1}}{2} + \mathbf{B}\frac{\mathbf{u}_i + \mathbf{u}_{i-1}}{2}\right) \\
& \underline{\mathbf{x}} \leqslant \mathbf{x}_i \leqslant \overline{\mathbf{x}} \\
& \underline{\mathbf{u}} \leqslant \mathbf{u}_i \leqslant \overline{\mathbf{u}} \\
& \underline{\mathbf{u}} \leqslant \mathbf{u}_0 \leqslant \overline{\mathbf{u}} \\
& i = 1, \ldots, N
\end{aligned}
$$

which implies that the discretized problem now features $n = N(n_x + n_u) + n_u$ decision variables. Therefore, the cost coefficient vector $\mathbf{v} \in \mathbb{R}^n$ and decision variable vector $\mathbf{z} \in \mathbb{R}^n$ with corresponding lower bound $\underline{\mathbf{z}} \in \mathbb{R}^n$ and upper bound $\overline{\mathbf{z}} \in \mathbb{R}^n$ are

$$
\mathbf{v} = \begin{bmatrix} \mathbf{0} \\ \mathbf{0} \\ \mathbf{0} \\ \vdots \\ \mathbf{c} \\ \mathbf{0} \end{bmatrix}, \quad
\underline{\mathbf{z}} = \begin{bmatrix} \underline{\mathbf{u}} \\ \underline{\mathbf{x}} \\ \underline{\mathbf{u}} \\ \vdots \\ \underline{\mathbf{x}} \\ \underline{\mathbf{u}} \end{bmatrix}, \quad
\mathbf{z} = \begin{bmatrix} \mathbf{u}_0 \\ \mathbf{x}_1 \\ \mathbf{u}_1 \\ \vdots \\ \mathbf{x}_N \\ \mathbf{u}_N \end{bmatrix}, \quad
\overline{\mathbf{z}} = \begin{bmatrix} \overline{\mathbf{u}} \\ \overline{\mathbf{x}} \\ \overline{\mathbf{u}} \\ \vdots \\ \overline{\mathbf{x}} \\ \overline{\mathbf{u}} \end{bmatrix}
$$

and the equality constraint matrix $\mathbf{M} \in \mathbb{R}^{n_h \times n}$ is defined as

$$
\mathbf{M} = \begin{bmatrix}
-\frac{h}{2}\mathbf{B} & \mathbf{I} - \frac{h}{2}\mathbf{A} & -\frac{h}{2}\mathbf{B} & \mathbf{0} & \mathbf{0} & \mathbf{0} & \mathbf{0} & \cdots \\
\mathbf{0} & -\mathbf{I} - \frac{h}{2}\mathbf{A} & -\frac{h}{2}\mathbf{B} & \mathbf{I} - \frac{h}{2}\mathbf{A} & -\frac{h}{2}\mathbf{B} & \mathbf{0} & \mathbf{0} & \cdots \\
\mathbf{0} & \mathbf{0} & \mathbf{0} & -\mathbf{I} - \frac{h}{2}\mathbf{A} & -\frac{h}{2}\mathbf{B} & \mathbf{I} - \frac{h}{2}\mathbf{A} & -\frac{h}{2}\mathbf{B} & \\
\vdots & \vdots & \vdots & & \ddots & & &
\end{bmatrix}
$$

There exist other popular transcription methods that can be employed to discretize the continuous time problem. It is worth mentioning that all transcription schemes based on Runge–Kutta one-step methods, which include collocation methods as an important special case, lead to LPs. For reference, Appendix G additionally provides the transcribed problem for a higher-order method (Hermite–Simpson).

4.3 CONTROL CONSTRAINTS

Worst-Case Input (Cat I)

The optimal (worst-case) input $\hat{u}(t)$ for the linear clearance problem with bounded controls and unbounded states (Cat I) is bang-bang

$$\hat{u}(t) = \begin{cases} \overline{u}, & \mathcal{S}(t) < 0 \\ \underline{u}, & \mathcal{S}(t) > 0 \end{cases}$$

if the system is controllable with respect to u. The sign of the switching function $\mathcal{S}$ determines if the control is on the upper bound $\overline{u}$ or the lower bound $\underline{u}$.

The line of reasoning regarding the characterization of the worst-case input follows the arguments for showing the nonsingularity of controllable, linear time-invariant systems for minimum time problems. Notably, these arguments are valid not only for fixed and free final time problems but also for minimization and maximization problems. In the following, a scalar input u with corresponding column $\mathbf{b} \in \mathbb{R}^{n_x}$ from the input matrix $\mathbf{B}$ is assumed. In the case of more than one control, the analysis remains valid if the system is controllable with respect to each control [55].

May be skipped (↴) at first read.

Define the Hamiltonian for Cat I clearance problems with a single input u as

$$\mathcal{H}\left[\boldsymbol{\lambda}(t), \mathbf{x}(t), u(t)\right] = \boldsymbol{\lambda}^T(t)\left[\mathbf{A}\mathbf{x}(t) + \mathbf{b}u(t)\right]$$

It is assumed that the system is controllable with respect to u, implying that the controllability matrix

$$\mathbf{U} = \left[\mathbf{A}^0\mathbf{b} \quad \mathbf{A}^1\mathbf{b} \quad \ldots \quad \mathbf{A}^{n_x-1}\mathbf{b}\right]$$

is full rank. The equation governing the evolution of the costates $\boldsymbol{\lambda}$ is obtained from

$$\frac{\partial\mathcal{H}}{\partial\mathbf{x}} = -\dot{\boldsymbol{\lambda}}^T(t) = \boldsymbol{\lambda}^T(t)\mathbf{A}$$

Defining the switching function as

$$\mathcal{S}(t) = \frac{\partial\mathcal{H}}{\partial u} = \boldsymbol{\lambda}^T(t)\mathbf{b}$$

the following cases characterize the optimal control on a time interval $[t_1, t_2]$, $t_2 > t_1$:

$$\hat{u}(t) = \begin{cases} \overline{u}, & \mathcal{S}(t) < 0 \\ \underline{u}, & \mathcal{S}(t) > 0 \\ u_s(t), & \mathcal{S}(t) = 0 \end{cases} \tag{4.7}$$

In the first two cases, the control is regular, and the sign of the switching function determines if the optimal control is on the upper bound $\overline{u}$ or on the lower bound $\underline{u}$. In the third case, the control $u_s(t)$ is singular and cannot be directly obtained as the switching function is zero on the interval $[t_1, t_2]$. The kth time derivative of the equation $\mathcal{S}(t) = 0$ is

$$\boldsymbol{\lambda}^T(t)\mathbf{A}^k\mathbf{b} = 0$$

Collecting the time derivatives for $k = 0, \ldots, n_x - 1$,

$$\boldsymbol{\lambda}^T(t)\underbrace{\begin{bmatrix}\mathbf{A}^0\mathbf{b} & \mathbf{A}^1\mathbf{b} & \ldots & \mathbf{A}^{n_x-1}\mathbf{b}\end{bmatrix}}_{\mathbf{U}} = \mathbf{0}$$

we can identify the controllability matrix $\mathbf{U}$. Because of our controllability assumption, this matrix is full rank, and only the trivial solution $\boldsymbol{\lambda} = \mathbf{0}$ satisfies this equation. This contradicts the nontriviality of the multipliers from the minimum principle. Thus, the switching function only exhibits isolated zeros, and the optimal control is purely bang-bang. In particular, the sign of the switching function determines if the control is on the lower or upper bound [see cases in Eq. (4.7)].

The results from the theoretical analysis are illustrated in the following by testing an experimental inner-loop controller of a fixed-wing aircraft model (see Steinert [56]) in a development stage. This inner-loop controller features three control inputs: the normal load factor command $n_{z,c}$, the lateral load factor command $n_{y,c}$, and the roll angle command Φ_c. Step responses as well as pole-zero plots for this model are provided in Appendix F. The models for the longitudinal and lateral plane are obtained from the linearization around a steady-state, straight and level flight condition at a reference height $h_{ref} \approx 2500\ m$ and reference velocity $V_{ref} \approx 67\ m/s$. All quantities (states, controls, criteria) are presented in the following relative to the trim (reference) values. The fixed final time t_f for the analysis is chosen as approximately five times the closed-loop short period timescale, which is $T_{SP} \approx 1.19$ s (i.e., $t_f \approx 6$ s for this particular system). A backward Euler discretization with a constant discretization step-size $h = 0.005$ s is used for all examples. Moreover, the switching functions (see Sec. 3.3) are computed using costate estimates based on the Lagrange multipliers of the discretized problems (see Sec. 5.4 in Gerdts [21]). Here and in what follows, the LP solver employed for the numerical solution of all studies is CPLEX 12.9.

Example 4.3.1: Worst-Case Normal Load Factor Input (Cat I)

The optimal (worst-case) normal load factor command $n_{z,c}(t)$ maximizing the load factor n_z (6 s) [or the angle of attack α_K (6 s)] starting from the initial condition $\left[x_{e,p}(0), x_{e,v}(0), \tilde{\mathbf{x}}_{lon}^T(0)\right] = \mathbf{0}$ in the longitudinal plane can be found as a solution to an optimal control problem of the form

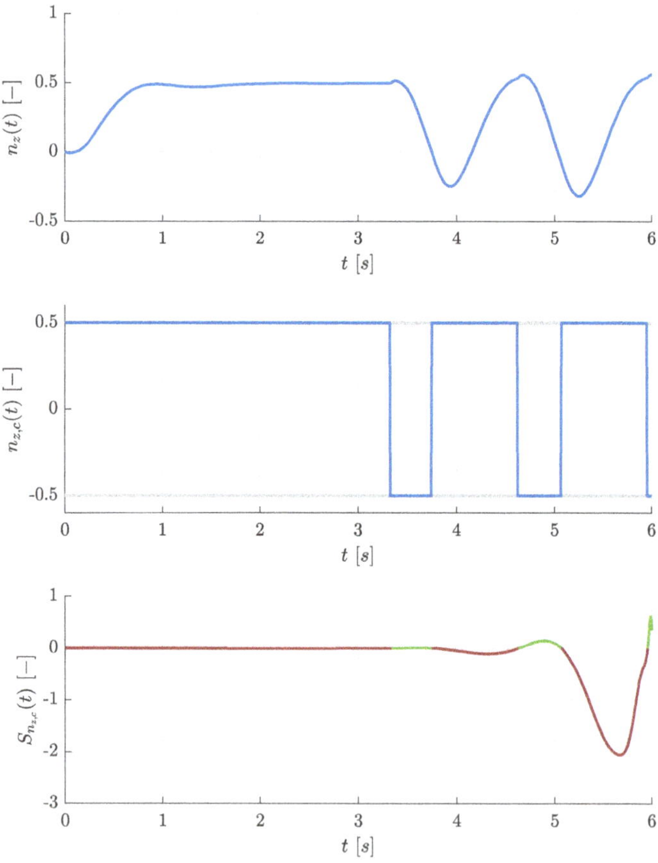

Fig. 4.1 Maximization of n_z (6 s), Cat I.

$$\begin{aligned}&\underset{-0.5\leqslant n_{z,c}(t)\leqslant 0.5}{\text{minimize}} \quad -n_z(6\,\text{s}) \text{ or } -\alpha_K(6\,\text{s})\\ &\text{subject to} \quad \frac{d}{dt}\begin{bmatrix} x_{e,p}(t) \\ x_{e,v}(t) \\ \tilde{\mathbf{x}}_{lon}(t) \end{bmatrix} = \begin{bmatrix} \mathbf{a}_{e,p} \\ \mathbf{a}_{e,v} \\ \tilde{\mathbf{A}}_{lon} \end{bmatrix} \begin{bmatrix} x_{e,p}(t) \\ x_{e,v}(t) \\ \tilde{\mathbf{x}}_{lon}(t) \end{bmatrix} + \begin{bmatrix} 0 \\ b_{e,n_{z,c}} \\ \tilde{\mathbf{b}}_{lon,n_{z,c}} \end{bmatrix} n_{z,c}(t) \\ &\qquad\qquad t \in [0, 6\,\text{s}]\end{aligned} \tag{4.8}$$

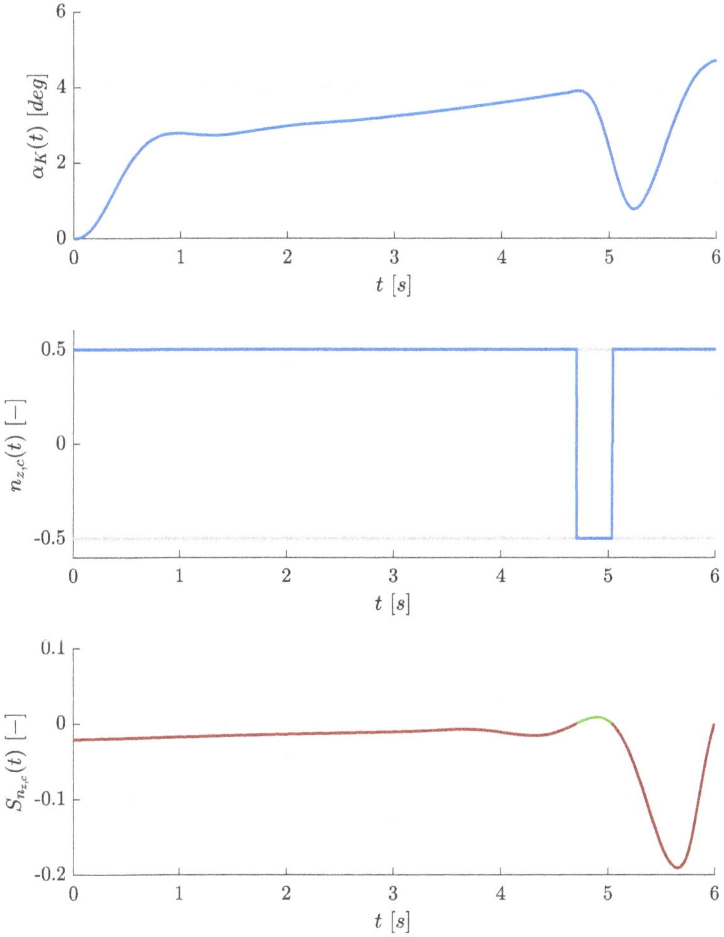

Fig. 4.2 Maximization of α_K (6 s), Cat I.

No limits are imposed on the internal position state $x_{e,p}$ and velocity state $x_{e,v}$ of the actuator related to the elevator. The optimal solutions, including the worst-case normal load factor input $n_{z,c}$ and corresponding switching function $\mathcal{S}_{n_{z,c}}$ (red/green, strictly negative/positive), are depicted in Fig. 4.1 for maximizing $n_z\,(6\,\mathrm{s})$ and Fig. 4.2 for maximizing $\alpha_K\,(6\,\mathrm{s})$.

Recall that, in the preceding example, the normal load factor command is bounded. However, no limits are imposed on the actuator states. Therefore, the clearance problem falls into Cat I. In the results, it can be observed that, in accordance with the theoretical analysis, the worst-case command is completely regular (i.e., purely bang-bang) in both examples. Note that the switching function only exhibits isolated zeros, and the time points where it switches signs correspond to the time points at which the optimal control jumps from one bound to the other. These switching time points for the maximization of $n_z(6\,\mathrm{s})$ are

$$t_1 \approx 3.32\,\mathrm{s}, \qquad t_2 \approx 3.74\,\mathrm{s}, \qquad t_3 \approx 4.62\,\mathrm{s}, \qquad t_4 \approx 5.06\,\mathrm{s}, \qquad t_5 \approx 5.95\,\mathrm{s}$$

The lengths of the intervals $[t_1, t_3]$ and $[t_3, t_5]$ both approximately match the short period timescale of the model. The short pulses where the control is at the lower bound on $[t_1, t_2]$ and $[t_3, t_4]$ occupy approximately one-third of these intervals. Note that at the very end of the time interval ($t \geqslant t_5$), the optimal control exploits the transient response in the normal load factor by switching from the upper bound to the lower bound just before the final time point.

This behavior is not observed for the maximization of $\alpha_K(6\,\mathrm{s})$, and the control is exclusively at the upper bound except for a short pulse during $[4.70\,\mathrm{s}, 5.04\,\mathrm{s}]$. Note that, also in this example, the control is purely bang-bang, and the switching time points of the optimal control match the isolated zeros of the corresponding switching function.

Example 4.3.2: Worst-Case Roll Angle Input (Cat I)

The optimal (worst-case) roll angle command $\Phi_c(t)$ maximizing the load factor $n_y\,(6\,\mathrm{s})$ [or the angle of sideslip $\beta_K\,(6\,\mathrm{s})$] starting from the initial condition $\left[x_{a,p}(0), x_{r,p}(0), x_{a,v}(0), x_{r,v}(0), \tilde{\mathbf{x}}_{lat}^T(0)\right] = \mathbf{0}$ in the lateral plane can be found as a solution to an optimal control problem of the form

$$\underset{-45\,\mathrm{deg} \leqslant \Phi_c(t) \leqslant 45\,\mathrm{deg}}{\text{minimize}} \quad -n_y\,(6\,\mathrm{s}) \;\; \text{or} \;\; -\beta_K\,(6\,\mathrm{s})$$

$$\text{subject to} \quad \frac{d}{dt}\begin{bmatrix} x_{a,p}(t) \\ x_{r,p}(t) \\ x_{a,v}(t) \\ x_{r,v}(t) \\ \tilde{\mathbf{x}}_{lat} \end{bmatrix} = \begin{bmatrix} \mathbf{a}_{a,p} \\ \mathbf{a}_{r,p} \\ \mathbf{a}_{a,v} \\ \mathbf{a}_{r,v} \\ \tilde{\mathbf{A}}_{lat} \end{bmatrix} \begin{bmatrix} x_{a,p}(t) \\ x_{r,p}(t) \\ x_{a,v}(t) \\ x_{r,v}(t) \\ \tilde{\mathbf{x}}_{lat} \end{bmatrix} + \begin{bmatrix} 0 \\ 0 \\ b_{a,\Phi_c} \\ b_{r,\Phi_c} \\ \tilde{\mathbf{b}}_{lat,\Phi_c} \end{bmatrix} \Phi_c(t) \quad (4.9)$$

$$t \in [0, 6\,\text{s}]$$

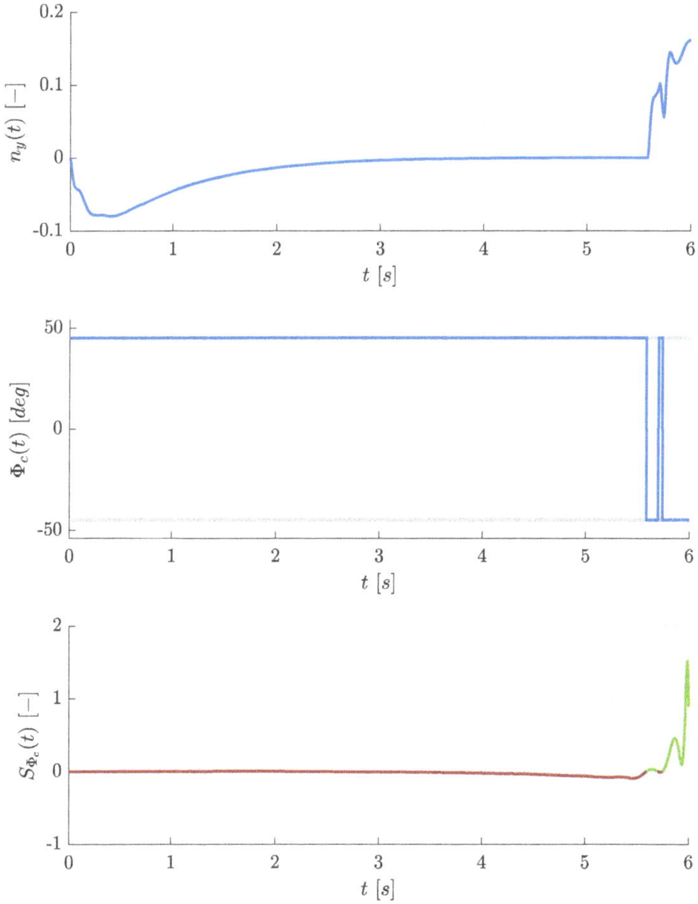

Fig. 4.3 Maximization of n_y (6 s), Cat I.

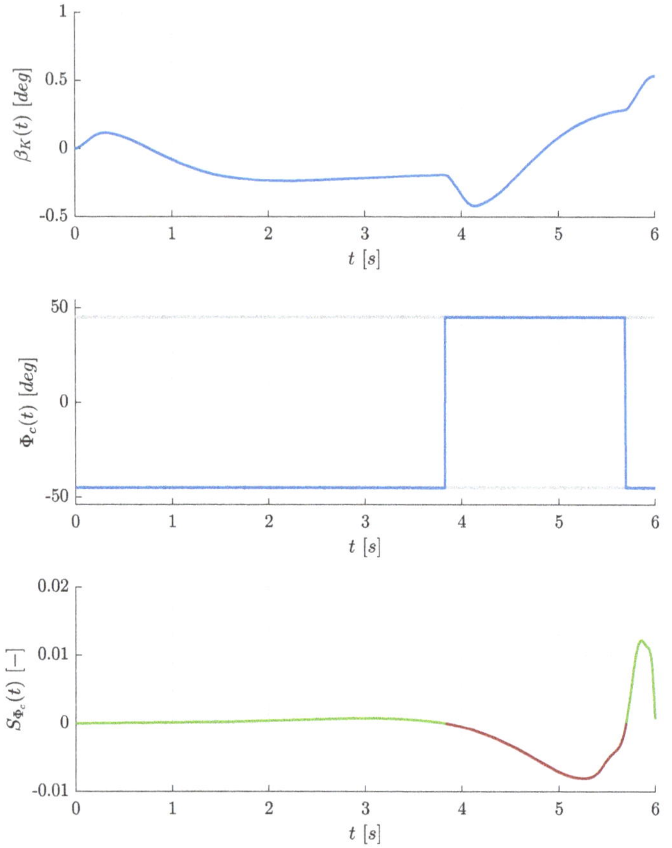

Fig. 4.4 Maximization of β_K (6 s), Cat I.

As for the longitudinal example, only a single control input (the roll angle command) is used; combined controls are discussed in the following chapter. No limits are imposed on the internal position states $x_{a,p}$ and $x_{r,p}$ as well as the velocity states $x_{a,v}$ and $x_{r,v}$ of the actuators related to the aileron and rudder. The optimal solutions, including the worst-case roll angle input Φ_c and corresponding switching function S_{Φ_c} (red/green, strictly negative/positive), are depicted in Fig. 4.3 for maximizing n_y (6 s) and Fig. 4.4 for maximizing β_K (6 s).

In the example for the lateral plane, it can be observed that the roll angle command Φ_c is effective for the maximization of the lateral load factor but rather ineffective for the maximization of the angle of sideslip (compare Fig. 4.3 and

Fig. 4.4). It is noteworthy that, similar to the longitudinal examples, the control is purely bang-bang in both cases, which aligns with the theoretical analysis. Even though the roll angle command is rather ineffective regarding the maximization of the angle of sideslip, this example may be exploited to illustrate a numerical effect arising in optimal control problems with a fixed time interval. In some cases, it is observed that a weak dependence of the cost function on the control actions at the beginning of the time interval results in high-frequency switches of the control ("control chattering"). For this example, the chattering effect is observed if the length of the fixed time interval is increased. The optimal cost and control histories for maximizing β_K (7 s) are presented in Fig. 4.5. Observe the chattering effect of the control (blue line). A direct result of fixing the final time is that the cost function can become increasingly insensitive with respect to control actions at the beginning of the time interval if the length of the interval is increased. For the value of the cost function, this is not detrimental because the solution of the LP is global. Loosely speaking, it simply matters very little for the value of the cost function at the final time point what exactly the control is doing at the beginning of the time interval. If this effect is undesired, for example, if not solely the value of the clearance criterion itself but a clean control history is of interest, it is possible to artificially motivate a shift in the control function. Let us illustrate this strategy for a single input system with control $u(t)$. Note that for more than one control the same strategy can be applied to each individual control. The control can be artificially pushed to some reference condition (zero in this case) on an interval $[t_0, t_p]$ through the introduction of an integral penalty term added to the cost with weight $\epsilon > 0$:

$$
\begin{aligned}
\underset{u(t)}{\text{minimize}} \quad & \mathbf{c}^T\mathbf{x}(t_f) + \epsilon \int_{t_0}^{t_p} |u(t)| dt \\
\text{subject to} \quad & \dot{\mathbf{x}}(t) = \mathbf{A}\mathbf{x}(t) + \mathbf{b}u(t) \\
& \underline{\mathbf{x}} \leqslant \mathbf{x}(t) \leqslant \overline{\mathbf{x}} \\
& \underline{u} \leqslant u(t) \leqslant \overline{u} \\
& t \in \left[t_0, t_f\right], \ t_0, t_f \text{ fixed}
\end{aligned}
\tag{4.10}
$$

This penalty can be interpreted as an "incentive" for the optimal control to keep the system at the reference condition on $[t_0, t_p]$. It is stressed, once again, that this additional "push" for the control is not part of the original clearance problem. Therefore, it is obvious that the effect of this artificial incentive (i.e., the contribution to the optimal cost) needs to be kept negligible to keep the interference with the clearance criterion under investigation in the optimization problem to an acceptable level. After all, the goal of introducing this penalty is a purely practical one to avoid the control chattering effect and, ultimately, obtain a "clean" control history without high-frequency switches. Let $\overline{\epsilon}$ be an acceptable accuracy for the criterion. Then, the weight ϵ can be chosen conservatively as

$$\epsilon = \frac{\overline{\epsilon}}{\int_{t_0}^{t_p} \max\{|\underline{u}|, |\overline{u}|\} dt}$$

which ensures that the contribution to the overall cost is at most $\overline{\epsilon}$. It should be mentioned that, despite the absolute value function in the cost, the optimal control problem can still be formulated as a linear problem. Let us illustrate this formulation for the preceding example.

Note that the absolute value of the control can be expressed as $|u(t)| = u^+(t) + u^-(t)$ in terms of the positive and negative parts $u^+(t), u^-(t) \geqslant 0$ of the control:

$$u(t) = u^+(t) - u^-(t)$$

The transformed problem can thus be written as the following linear optimal control problem:

$$\begin{aligned}
\underset{u^+(t),u^-(t)}{\text{minimize}} \quad & \mathbf{c}^T\mathbf{x}(t_f) + \epsilon \int_{t_0}^{t_p} u^+(t) + u^-(t) dt \\
\text{subject to} \quad & \dot{\mathbf{x}}(t) = \mathbf{A}\mathbf{x}(t) + \mathbf{b}u(t) \\
& \underline{\mathbf{x}} \leqslant \mathbf{x}(t) \leqslant \overline{\mathbf{x}} \\
& u(t) = u^+(t) - u^-(t) \\
& \underline{u} \leqslant u(t) \leqslant \overline{u} \\
& u^+(t), u^-(t) \geqslant 0 \\
& t \in \left[t_0, t_f\right], \; t_0, t_f \text{ fixed}
\end{aligned}$$

A practical strategy regarding the selection of t_p is to increase it in multiples of characteristic timescales of the closed-loop system, such as the short period timescale. The results obtained from following this strategy with $t_p = 2T_{SP} \approx 2.38$ s are presented in Fig. 4.5 (black lines). Note that the control is zero on the initial interval $t \in [0, \Delta t]$, $\Delta t < t_p$ and regular on the remaining interval $t \in [\Delta t, t_f]$. Note further that the control histories in this remaining interval essentially match. This result also solidifies the interval extension argument presented in the introduction of this part of the book regarding the attainable values of the cost when increasing the length of the interval. Note that by employing the penalty approach in this example the solver is slightly pushed toward a particular solution, which is to hold the system in the trim condition for the first Δt seconds and then apply the worst-case control to minimize the cost.

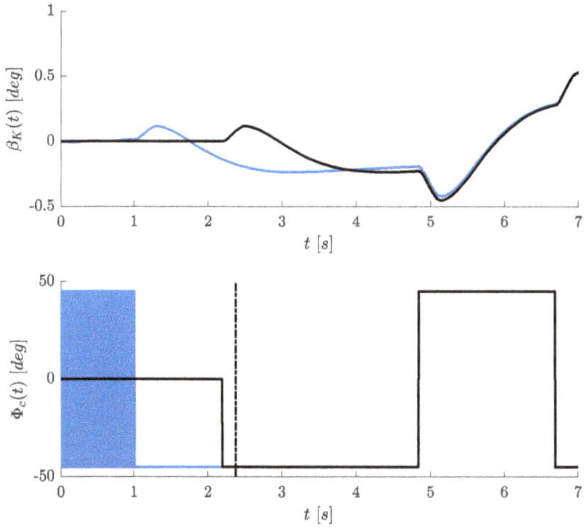

Fig. 4.5 Maximization of β_K (7 s) (blue without penalty, black with penalty).

4.4 CONTROL AND STATE CONSTRAINTS

Worst-Case Input (Cat II)

The worst-case input $\hat{u}(t)$ for the linear clearance problem with bounded states and bounded controls (Cat II) is bang-bang or takes the values of the boundary control $u_b(t)$:

$$\hat{u}(t) = \begin{cases} \overline{u}, & \mathcal{S}(t) < 0 \\ \underline{u}, & \mathcal{S}(t) > 0 \\ u_h(t), & \mathcal{S}(t) = 0 \end{cases}$$

if the system is controllable with respect to u. The boundary control for a linear state constraint related to the ith state vector component of order $p > 0$ is expressed as

$$u_b(t) = -\frac{\mathbf{a}_i \mathbf{A}^{p-1} \mathbf{x}(t)}{\mathbf{e}_i \mathbf{A}^{p-1} \mathbf{b}} \tag{4.11}$$

with $\mathbf{a}_i = \mathbf{e}_i \mathbf{A}$.

For simplicity, the arguments in the sequel regarding the characterization of the worst-case input are shown for a scalar state constraint $c\,[x_i(t)] \leqslant 0$ and control $u(t)$. For the particular clearance formulation presented in this chapter, this constraint is an affine function modeling one of the state constraints related to the upper bound

$$x_i - \overline{x}_i \leqslant 0$$

or lower bound

$$\underline{x}_i - x_i \leqslant 0$$

for some $i \in 1, \ldots, n_x$. As a direct consequence, the Jacobian of such a state constraint depending on the ith state component is of the form

$$\frac{\partial c}{\partial \mathbf{x}} = \pm \mathbf{e}_i$$

where the sign is determined by the type of bound (positive for an upper bound and negative for a lower bound). Clearly, for a linear constraint, the second derivatives of the constraint function vanish. This fact can be exploited in the following optimal control analysis regarding the characterization of the worst-case input of Cat II clearance problems.

May be skipped (↴) at first read.

Define the augmented Hamiltonian as

$$\mathcal{H}\left[\boldsymbol{\lambda}(t), \mu(t), \mathbf{x}(t), u(t)\right] = \boldsymbol{\lambda}^T(t)\left[\mathbf{A}\mathbf{x}(t) + \mathbf{b}u(t)\right] + \mu(t)c\left[x_i(t)\right]$$

where the purely state-dependent constraint $c\left[x_i(t)\right] \leqslant 0$ is adjoined using the multiplier

$$\mu(t) \begin{cases} = 0, & c\left[x_i(t)\right] < 0 \\ \geqslant 0, & c\left[x_i(t)\right] = 0 \end{cases}$$

The costate equations and the switching function are

$$\frac{\partial \mathcal{H}}{\partial \mathbf{x}} = -\dot{\boldsymbol{\lambda}}^T(t) = \boldsymbol{\lambda}^T(t)\mathbf{A} + \mu(t)\frac{\partial c}{\partial \mathbf{x}}$$

$$\mathcal{S}(t) = \frac{\partial \mathcal{H}}{\partial u} = \boldsymbol{\lambda}^T(t)\mathbf{b}$$

with the following cases regarding the optimal control:

$$\hat{u}(t) = \begin{cases} \overline{u}, & \mathcal{S}(t) < 0 \\ \underline{u}, & \mathcal{S}(t) > 0 \\ u_b(t), & \mathcal{S}(t) = 0 \end{cases}$$

On a state-constrained arc for $t \in [t_1, t_2]$, $t_2 > t_1$ we have $c[x_i(t)] = 0$ as well as $\mathcal{S}(t) = 0$, and the control takes the values of the boundary control $u_b(t)$. At the beginning and in the interior of this arc, all time derivatives of both the state constraint and the switching function must be zero.

Recall that the order p of the constraint is defined as the time derivative of the constraint in which the control appears explicitly for the first time. Consequently, for a state constraint of order p, we are required to take the time derivative of $c[x_i(t)] = 0$ until the control appears explicitly

$$\frac{\partial c}{\partial \mathbf{x}}\mathbf{A}^p\mathbf{x}(t) + \frac{\partial c}{\partial \mathbf{x}}\mathbf{A}^{p-1}\mathbf{b}u(t) = 0$$

using $\dot{\mathbf{x}} = \mathbf{A}\mathbf{x}(t) + \mathbf{b}u(t)$ and $\frac{\partial^2 c}{\partial \mathbf{x}^2} = \mathbf{0}$. Observe that in the process of taking the time derivative the product $\frac{\partial c}{\partial \mathbf{x}}\mathbf{A}^q\mathbf{b}$ appearing before the control is (by definition) zero for $q = 0, \ldots, p-2$ and must be nonzero for $q = p-1$. In addition, the pth time derivative of the condition $\mathcal{S}(t) = 0$ for the switching function yields

$$\boldsymbol{\lambda}^T(t)\mathbf{A}^p\mathbf{b} + \frac{\partial c}{\partial \mathbf{x}}\mathbf{A}^{p-1}\mathbf{b}\mu(t) = 0$$

using $\dot{\boldsymbol{\lambda}}^T(t) = -\boldsymbol{\lambda}^T(t)\mathbf{A} - \mu(t)\frac{\partial c}{\partial \mathbf{x}}$. We may now rearrange the expressions for the pth time derivative of both conditions and obtain the boundary control and constraint multiplier as

$$u_b(t) = -\frac{\frac{\partial c}{\partial \mathbf{x}}\mathbf{A}^p\mathbf{x}(t)}{\frac{\partial c}{\partial \mathbf{x}}\mathbf{A}^{p-1}\mathbf{b}}$$

$$\mu(t) = -\frac{\boldsymbol{\lambda}^T(t)\mathbf{A}^p\mathbf{b}}{\frac{\partial c}{\partial \mathbf{x}}\mathbf{A}^{p-1}\mathbf{b}}$$

Inserting $\frac{\partial c}{\partial \mathbf{x}} = \pm\mathbf{e}_i$ in the equation for the boundary control, we obtain

$$\begin{aligned} u_b(t) &= -\frac{\pm\mathbf{e}_i\mathbf{A}^p\mathbf{x}(t)}{\pm\mathbf{e}_i\mathbf{A}^{p-1}\mathbf{b}} \\ &= -\frac{\mathbf{e}_i\mathbf{A}^p\mathbf{x}(t)}{\mathbf{e}_i\mathbf{A}^{p-1}\mathbf{b}} \\ &= -\frac{\mathbf{a}_i\mathbf{A}^{p-1}\mathbf{x}(t)}{\mathbf{e}_i\mathbf{A}^{p-1}\mathbf{b}} \end{aligned}$$

Example 4.4.1: Worst-Case Normal Load Factor Input (Cat II)

In the Cat II problem for the longitudinal model, limits on the internal position state $x_{e,p}$ and velocity state $x_{e,v}$ of the actuator states related to the elevator are added to the Cat I problem (4.8):

$$\underline{x}_{e,p} \leqslant x_{e,p}(t) \leqslant \overline{x}_{e,p} \tag{4.12}$$

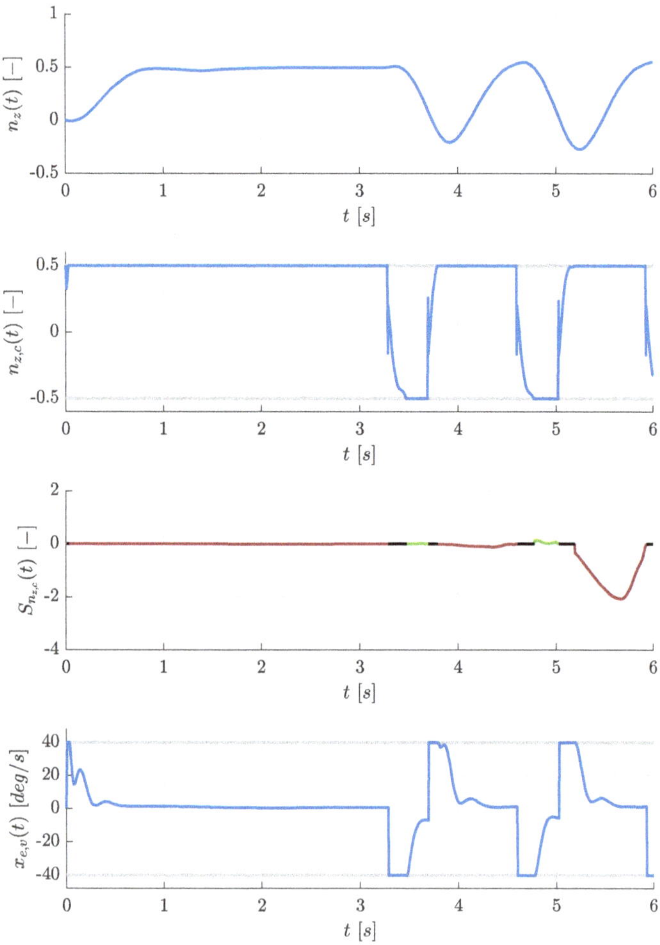

Fig. 4.6 Maximization of $n_z(6\ \mathrm{s})$, Cat II.

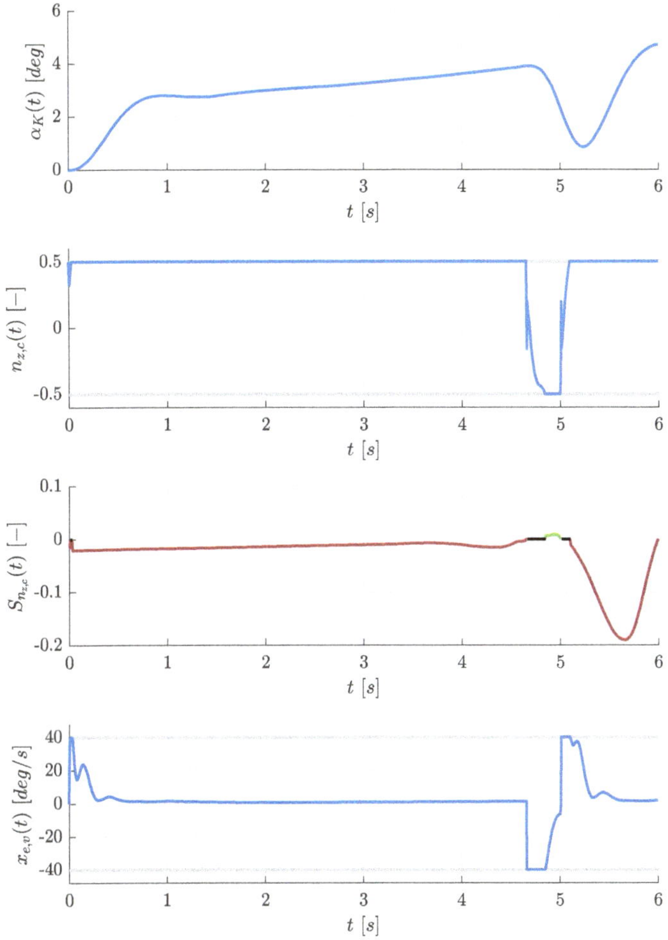

Fig. 4.7 Maximization of $\alpha_K(6\,\mathrm{s})$, Cat II.

$$\underline{x}_{e,v} \leqslant x_{e,v}(t) \leqslant \overline{x}_{e,v} \tag{4.13}$$

The optimal solutions, including the worst-case normal load factor input $n_{z,c}$ and corresponding switching function $\mathcal{S}_{n_{z,c}}$ (red/green, strictly negative/positive; otherwise black), are depicted in Fig. 4.6 for maximizing $n_z(6\,\mathrm{s})$ and Fig. 4.7 for maximizing $\alpha_K(6\,\mathrm{s})$.

It can be observed that the maximum values obtained for the angle of attack and normal load factor are slightly lower for Cat II compared to Cat I; compare Fig. 4.1 with Fig. 4.6 and Fig. 4.2 with Fig. 4.7. This is expected as the feasible region for Cat II – type problems is reduced by the addition of state constraints corresponding to the limits of the servomechanism. Note that in both cases presented in Fig. 4.6 and Fig. 4.7 the constraints on the actuator's velocity state for the lower and upper bound become active. Depending on the switching function $S_{n_{z,c}}$ for the normal load factor command, the following cases can be observed for the load factor command in the numerical results

$$n_{z,c}(t) = \begin{cases} 0.5, & S_{n_{z,b}}(t) < 0 \\ -0.5, & S_{n_{z,b}}(t) > 0 \\ n_{z,b}(t), & S_{n_{z,b}}(t) = 0 \end{cases}$$

that is, the control is on the upper bound (0.5) if the switching function is negative, on the lower bound (-0.5) if the switching function is positive, and takes intermediate values $n_{z,b}(t)$ where the switching function is zero (within numerical tolerances). On the intervals where the switching function is zero, the state constraints for the lower or upper limit on the actuator velocity state related to the elevator are active, confirming the theoretical analysis. This state constraint is first-order ($p = 1$). Thus, the boundary control $n_{z,b}(t)$ of the normal load factor command can be expressed based on Eq. (4.11) and $\mathbf{x}_{lon}^T = \left[x_{e,p}, x_{e,v}, \tilde{\mathbf{x}}_{lon}^T\right]$ as

$$n_{z,b}(t) = -\frac{\mathbf{a}_{e,v}\mathbf{x}_{lon}(t)}{b_{e,n_{z,c}}}$$

The comparisons of the boundary control obtained from this expression and the numerical results are depicted in Fig. 4.8. The numerical results depicted in this figure appear to confirm the predictions from the theoretical analysis for Cat II problems.

It can be observed that in both Cat II results (see Fig. 4.6 and Fig. 4.7) overshoots are visible in the control at the entry time points of the state-constrained arcs. This is an inherent effect originating from the discretization of the problem as the entry time points of the state-constrained arcs at which the points where the control jumps are not exactly resolved by the time grid. Practical experience suggests that, just like the backward Euler method employed for all numerical examples throughout this section, essentially all discretization methods can produce numerical artifacts of a similar form. Let us illustrate a related effect using the trapezoidal collocation method. For this method, control chattering (high-frequency switches of the optimal control) can be observed on state-constrained arcs. This effect is illustrated in Fig. 4.9.

In this figure, a detailed view of the optimal solution to the Cat II clearance problem for maximizing n_z (6 s) using the normal load factor command $n_{z,c}(t)$ is

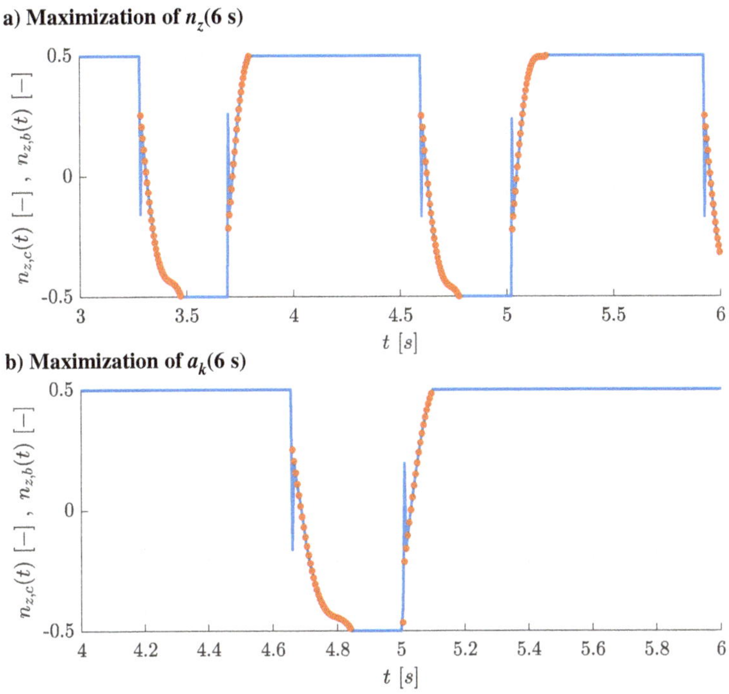

Fig. 4.8 Boundary controls (red) for maximization.

presented. This detailed view shows the entry to a first-order state-constrained arc coming from a regular part of the control. Because of the finite discretization step length, the entry point is not exactly resolved and, for both the backward Euler method (blue line) and trapezoidal method (black dashed line), the controls appear to overshoot at the entry of the arc. This may be interpreted as a compensatory action in order to enter the arc caused by the restricted input space of the discretized control. Note that the control obtained from the backward Euler method (blue line) exhibits only a single spike at the entry and produces a clean discretized version of the boundary control on the state-constrained arc (red dots, estimated based on the theoretical results using values from the backward Euler solution). However, the control history corresponding to the trapezoidal method shows heavy oscillations. A possible explanation for this behavior is that for the trapezoidal method, effectively, the mean of the neighboring discretized variables enters the discretized dynamics:

$$\mathbf{x}_i = \mathbf{x}_{i-1} + h\left(\mathbf{A}\frac{\mathbf{x}_i + \mathbf{x}_{i-1}}{2} + \mathbf{B}\frac{\mathbf{u}_i + \mathbf{u}_{i-1}}{2}\right)$$

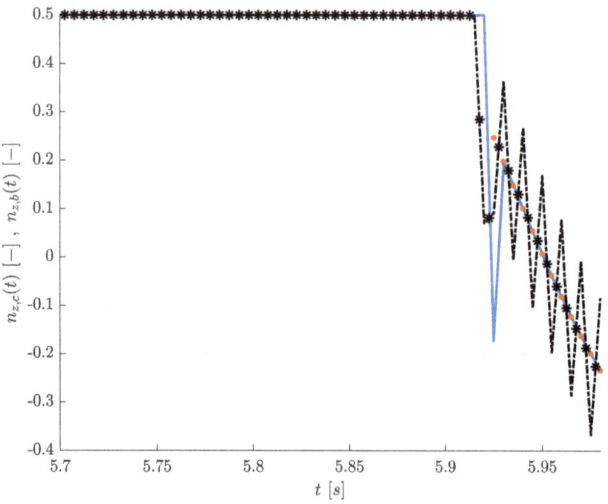

Fig. 4.9 Comparison of the optimal solutions (Cat II) for maximizing $n_z(6\ \mathrm{s})$ using the backward Euler method (blue line) and the trapezoidal method (black dashed line). Black star markers represent the mean control values between two control discretization points of the trapezoidal method, and red dots show the values of the boundary control based on the theoretical analysis.

The mean control values of neighboring time points are depicted in Fig. 4.9 for the trapezoidal method by black star markers. Note that these values appear to be very close to the boundary control found from the backward Euler method (blue line) and the boundary control estimated by the theoretical results using the values of the backward Euler solution (red dots).

Example 4.4.2: Worst-Case Roll Angle Input (Cat II)

In the Cat II problem for the lateral model, limits on the internal position states $x_{a,p}$ and $x_{r,p}$ as well as velocity states $x_{a,v}$ and $x_{r,v}$ of the actuator states related to the aileron and rudder are added to the Cat I problem (4.9):

$$\underline{x}_{a,p} \leqslant x_{a,p}(t) \leqslant \overline{x}_{a,p} \tag{4.14}$$

$$\underline{x}_{a,v} \leqslant x_{a,v}(t) \leqslant \overline{x}_{a,v} \tag{4.15}$$

$$\underline{x}_{r,p} \leqslant x_{r,p}(t) \leqslant \overline{x}_{r,p} \tag{4.16}$$

$$\underline{x}_{r,v} \leqslant x_{r,v}(t) \leqslant \overline{x}_{r,v} \tag{4.17}$$

The optimal solutions, including the worst-case roll angle input Φ_c and corresponding switching function $\mathcal{S}_{\Phi_c}$ (red/green, strictly negative/positive; otherwise black), are depicted in Fig. 4.10 for maximizing $n_y(6\,\mathrm{s})$ and Fig. 4.11 for maximizing $\beta_K(6\,\mathrm{s})$.

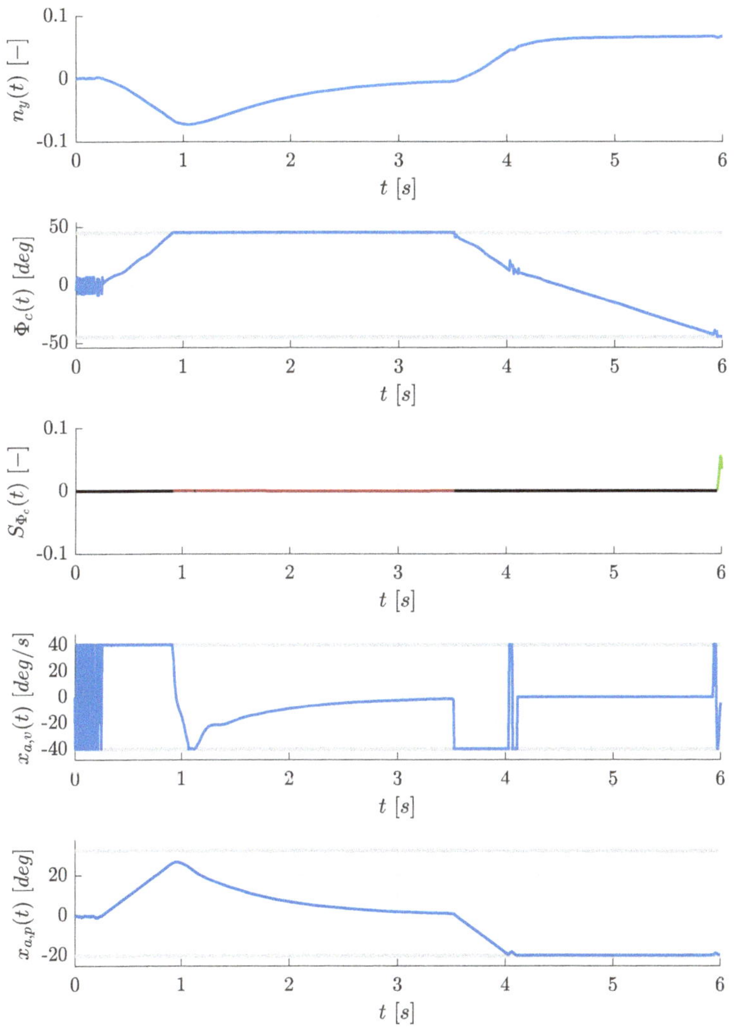

Fig. 4.10 Maximization of $n_y(6\,\mathrm{s})$, Cat II.

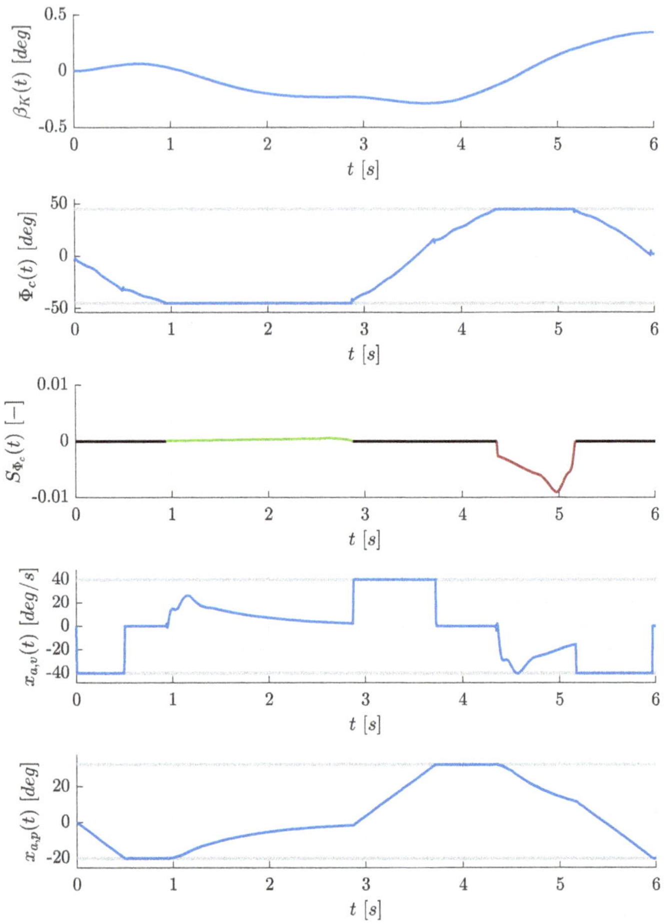

Fig. 4.11 Maximization of $\beta_K(6\,\mathrm{s})$, Cat II.

In the Cat II example for the lateral model, the position and rate limits for the actuator states related to the aileron and rudder are introduced as state constraints. The command is, as expected, less effective for both cases, that is, when maximizing $n_y(6\,\mathrm{s})$ (see Fig. 4.10) or $\beta_K(6\,\mathrm{s})$ (see Fig. 4.11). In particular, the considerable control effort required to attain a negligible value for the angle of sideslip at the final time point, as illustrated in Fig. 4.11, forces the servomechanism of the aileron in both the rate and position limits. This case is particularly interesting as the solution exhibits a combination of bang-bang and state-constrained arcs (first- and second-order) regarding the aileron's actuator states.

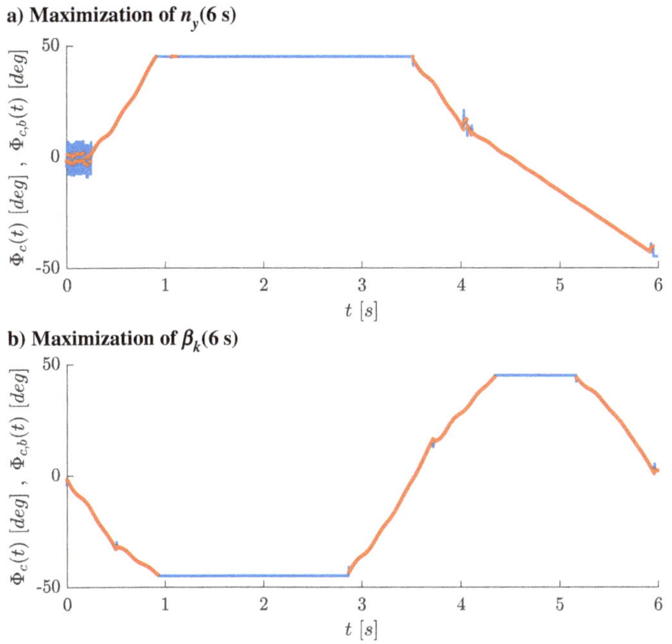

Fig. 4.12 Boundary controls (red) for maximization.

This actuator is modeled using a second-order system in which the time derivate of the position state $x_{a,p}$ is equal to the velocity $x_{a,v}$ (the third state vector component):

$$\dot{x}_{a,p}(t) = x_{a,v}(t) \tag{4.18}$$

Note that as soon as the position limit is reached, the rate rapidly drops to zero and remains at zero while the position bound is active. Clearly, this result expresses the implied condition $x_{a,v}(t) = 0$ for the boundary control, meaning that the rate must be kept zero to remain on the bound of the position state.

The boundary control for state constraints of first-order ($p = 1$) and second-order ($p = 2$) for the actuator states related to the aileron can be expressed as

$$\Phi_{c,b}(t) = -\frac{\mathbf{a}_{a,p}\mathbf{A}_{lat}\mathbf{x}_{lat}(t)}{\mathbf{e}_1\mathbf{A}_{lat}\mathbf{b}_{lat}} = -\frac{\mathbf{a}_{a,v}\mathbf{x}_{lat}(t)}{b_{a,\Phi_c}}$$

with $\mathbf{A}_{lat}^T = \left[\mathbf{a}_{a,p}^T, \mathbf{a}_{r,p}^T, \mathbf{a}_{a,v}^T, \mathbf{a}_{r,v}^T, \tilde{\mathbf{A}}_{lat}^T\right]$ and $\mathbf{b}_{lat}^T = \left[0, 0, b_{a,\Phi_c}, b_{r,\Phi_c}, \tilde{\mathbf{b}}_{lat,\Phi_c}^T\right]$, as well as $\mathbf{x}_{lat}^T = \left[x_{a,p}(t), x_{r,p}(t), x_{a,v}(t), x_{r,v}(t), \tilde{\mathbf{x}}_{lat}^T\right]$. This relationship is a direct consequence of Eq. (4.18): Because of $\mathbf{a}_{a,p} = \mathbf{e}_3$, we have $\mathbf{e}_3\mathbf{A}_{lat} = \mathbf{a}_{a,v}$ in the numerator as well as $\mathbf{e}_1\mathbf{A}_{lat}\mathbf{b}_{lat} = \mathbf{a}_{a,p}\mathbf{b}_{lat} = \mathbf{e}_3\mathbf{b}_{lat} = b_{a,\Phi_c}$ in the denominator. The comparisons of the boundary controls for maximizing the load factor and angle of sideslip between the solution of the discretized problem and the control predicted from the theoretical analysis are depicted in Fig. 4.12.

CHAPTER 5

Linear-Parametric Clearance

Aircraft closed-loop systems are often modeled in parametric form. In the context of the clearance problems explored in this chapter, the term *parametric* should be understood in the sense that parameters are exposed in the formulation but not necessarily changed in combination with the control inputs within the same problem. This mere exposure enables us to obtain sensitivity information for the worst-case solution regarding the parameters, which can provide valuable insights for flight control law clearance tasks. For example, it can aid root cause analyses by identifying influential parameters in the closed-loop model or provide sanity checks from a flight mechanical perspective. The particular parameterization depends on the type of model and, most importantly, the effects to be investigated. A reasonable selection of parameters is, in most cases, guided by flight mechanical considerations. Some parameters can be excluded a priori (such as the color of the aircraft), while others are expected to exhibit considerable potential for influencing the worst-case solution. Typical examples of such parameters include aerodynamic coefficients or parameters related to weight and weight distribution, such as mass, moments of inertia, and the center of gravity position. Similarly, gain values in the flight control law or quantities affecting the trim conditions can influence the worst-case solution.

This chapter provides a framework in which the influence of such parameters can be quantified and studied in the context of the optimal control-based clearance approach in a fairly general setting. First, the parametric form of the Cat I and II types of clearance problems is presented in Sec. 5.1. This formulation extends the nominal clearance problem [Eq. (4.1)] to include parameters in the system matrix, input matrix, and state output vector and considers parameter-dependent reference values for the states, controls, and outputs obtained from parameter-dependent trim conditions. Moreover, it is shown in Sec. 5.2 that this type of clearance problem can be transcribed to a parametric LP using the direct optimal control approach. At an optimal solution of this LP, first-order (Sec. 5.3) and second-order (Sec. 5.4) postoptimal sensitivity analysis allows us to study the influence of parameters on the worst-case solution of the discretized problem. Finally, as shown in Sec. 5.5, postoptimal sensitivities do not only represent a useful tool to quantify the influence of parameters on the worst-case solution but can also be exploited to perform a combined worst-case analysis in which both the inputs and parameters are varied simultaneously.

5.1 PROBLEM FORMULATION

Clearance Optimal Control Problem (Cat I and II Parametric)

Similar to the nominal clearance problem [Eq. (4.1)] the parametric aircraft worst-case input problem including limits for the inputs and actuators can be modeled as a parameter-dependent linear optimal control problem:

$$\begin{aligned} \underset{\mathbf{u}(t)}{\text{minimize}} \quad & \mathbf{c}^T(\mathbf{p})\,\mathbf{x}(t_f) \\ \text{subject to} \quad & \dot{\mathbf{x}}(t) = \mathbf{A}(\mathbf{p})\left[\mathbf{x}(t) - \mathbf{x}_r(\mathbf{p})\right] + \mathbf{B}(\mathbf{p})\left[\mathbf{u}(t) - \mathbf{u}_r(\mathbf{p})\right] \\ & \mathbf{x}(t_0) = \mathbf{x}_r(\mathbf{p}) \\ & \underline{\mathbf{x}} \leqslant \mathbf{x}(t) \leqslant \overline{\mathbf{x}} \\ & \underline{\mathbf{u}} \leqslant \mathbf{u}(t) \leqslant \overline{\mathbf{u}} \\ & t \in \left[t_0, t_f\right],\ t_0, t_f \text{ fixed} \end{aligned}$$

with the parameter-dependent reference state vector $\mathbf{x}_r(\mathbf{p}) : \mathbb{R}^{n_p} \rightarrow \mathbb{R}^{n_x}$ and reference (trimming) control vector $\mathbf{u}_r(\mathbf{p}) : \mathbb{R}^{n_p} \rightarrow \mathbb{R}^{n_x}$.

In linear-parametric form, the aircraft closed-loop model is represented as a linear dynamic system with the system matrix $\mathbf{A}(\mathbf{p}) : \mathbb{R}^{n_p} \rightarrow \mathbb{R}^{n_x \times n_x}$ and the input matrix $\mathbf{B}(\mathbf{p}) : \mathbb{R}^{n_p} \rightarrow \mathbb{R}^{n_x \times n_u}$ depending on the parameter vector $\mathbf{p} \in \mathbb{R}^{n_p}$:

$$\dot{\mathbf{x}}(t) = \mathbf{A}(\mathbf{p})\left[\mathbf{x}(t) - \mathbf{x}_r(\mathbf{p})\right] + \mathbf{B}(\mathbf{p})\left[\mathbf{u}(t) - \mathbf{u}_r(\mathbf{p})\right] \tag{5.1}$$

Note that this model also depends on the parametric reference state vector $\mathbf{x}_r(\mathbf{p})$ and reference control vector $\mathbf{u}_r(\mathbf{p})$ to consider parameter dependencies in the trim condition. Thus, trajectories of the system (5.1) originate from the parameter-dependent reference state vector

$$\mathbf{x}(t_0) = \mathbf{x}_r(\mathbf{p})$$

and $\mathbf{u}_r(\mathbf{p})$ is the input required for trimming at this point.

The controls and states are assumed box-bounded as in the nominal formulation [see Eq. (4.3) and Eq. (4.4)]. Note that because the control vector $\mathbf{u}(t)$ and state vector $\mathbf{x}(t)$ in the parametric linear optimal control problem are understood as absolute (and not relative to the reference condition as for the nominal case), the associated bounds do not change under parameter variations. This form is particularly convenient from a modeling perspective as the respective limits remain the same, even if the reference condition changes under parameter variations.

Moreover, not only the state output vector $\mathbf{c}(\mathbf{p}) : \mathbb{R}^{n_p} \to \mathbb{R}^{n_x}$ but also the reference output $y_r(\mathbf{p}) : \mathbb{R}^{n_p} \to \mathbb{R}$ for the output

$$y(t) = y_r(\mathbf{p}) + \mathbf{c}^T(\mathbf{p})\left[\mathbf{x}(t) - \mathbf{x}_r(\mathbf{p})\right]$$

representing the criterion is allowed to depend on the parameters. Note that this allows the criterion at the final time point t_f to be decomposed into a purely parameter-dependent part and another term depending on both the parameter vector and the final state vector:

$$y(t_f) = \underbrace{y_r(\mathbf{p}) - \mathbf{c}^T(\mathbf{p})\,\mathbf{x}_r(\mathbf{p})}_{d(\mathbf{p})} + \underbrace{\mathbf{c}^T(\mathbf{p})\,\mathbf{x}(t_f)}_{J[\mathbf{x}(t_f);\mathbf{p}]}$$

Clearly, the first part $d(\mathbf{p})$ has no influence on the optimal state and control histories for a fixed parameter value as it would merely offset the cost if it was introduced in the cost. Therefore, we treat this part purely in a postprocessing step and only consider the second term $J\left[\mathbf{x}(t_f);\mathbf{p}\right]$ in the cost.

5.2 DISCRETIZED FORM

Discretized Clearance Optimal Control Problem (Cat I and II Parametric)

The transcription of the linear-parametric clearance problem using direct methods yields a parametric LP of the form

$$\begin{aligned} \underset{\mathbf{z}\in\mathbb{R}^n}{\text{minimize}} \quad & \mathbf{v}^T(\mathbf{p})\,\mathbf{z} \\ \text{subject to} \quad & \mathbf{M}(\mathbf{p})\,\mathbf{z} = \mathbf{b}(\mathbf{p}) \\ & \underline{\mathbf{z}} \leqslant \mathbf{z} \leqslant \overline{\mathbf{z}} \end{aligned}$$

Given the minimizer $\hat{\mathbf{z}}$ of this LP, the optimal (worst-case) value of the criterion $\hat{y}_N(\mathbf{p})$ is expressed as

$$\hat{y}_N(\mathbf{p}) = y_r(\mathbf{p}) - \mathbf{c}^T(\mathbf{p})\,\mathbf{x}_r(\mathbf{p}) + \mathbf{v}^T(\mathbf{p})\,\hat{\mathbf{z}}$$

Similar to the nominal clearance problem (see Sec. 4.2), the transcription process of the linear optimal control problem to a finite-dimensional parameter optimization problem is illustrated using the backward Euler and trapezoidal collocation method.

It is noteworthy that compared to the nominal case the initial condition is embedded in the optimization problem as an additional equality constraint. This requires the introduction of an additional decision variable vector $\mathbf{x}_0 \in \mathbb{R}^{n_x}$ fixed to the reference state vector $\mathbf{x}_r(\mathbf{p})$ through the equality constraint:

$$\mathbf{x}_r(\mathbf{p}) - \mathbf{x}_0 = \mathbf{0} \tag{5.2}$$

It is shown in the sequel of this chapter that such an embedding constraint for the initial condition allows us to also recover sensitivity information regarding the influence of the parameters on the optimal (worst-case) solution through the parametric reference state vector $\mathbf{x}_r(\mathbf{p})$. As discussed in the preceeding section, the reference output vector $y_r(\mathbf{p})$ and the term $\mathbf{c}^T(\mathbf{p})\,\mathbf{x}_r(\mathbf{p})$ are not included in the cost as they merely offset the cost and are independent of the optimal state and control histories. After a minimizer $\hat{\mathbf{z}}$ of the LP is obtained, the parameter-dependent worst-case value of the criterion under investigation can then be recovered as

$$\hat{y}_N(\mathbf{p}) = y_r(\mathbf{p}) - \mathbf{c}^T(\mathbf{p})\,\mathbf{x}_r(\mathbf{p}) + \underbrace{\mathbf{v}^T(\mathbf{p})\,\hat{\mathbf{z}}}_{\mathbf{c}^T(\mathbf{p})\hat{\mathbf{x}}_N} \tag{5.3}$$

where $\hat{\mathbf{x}}_N$ represents the optimal state variable vector corresponding to the final time point.

BACKWARD EULER TRANSCRIPTION

For the backward Euler method state variable vectors $\mathbf{x}_i \in \mathbb{R}^{n_x}, i = 0, \ldots, N$ as well as control variable vectors $\mathbf{u}_i \in \mathbb{R}^{n_u}, i = 1, \ldots, N$ are introduced, and the dynamic constraints $\mathbf{d}_i = \mathbf{0}$, $i = 1, \ldots, N$ are expressed as

$$\mathbf{d}_i = \mathbf{x}_i - \mathbf{x}_{i-1} - h\left\{\mathbf{A}(\mathbf{p})\left[\mathbf{x}_i - \mathbf{x}_r(\mathbf{p})\right] + \mathbf{B}(\mathbf{p})\left[\mathbf{u}_i - \mathbf{u}_r(\mathbf{p})\right]\right\}.$$

Together with the embedding constraint (5.2) for the initial boundary condition, the parametric version of the discretized problem reads

$$\begin{aligned}
\underset{\mathbf{x}_0,\mathbf{x}_i,\mathbf{u}_i}{\text{minimize}} \quad & \mathbf{c}^T(\mathbf{p})\,\mathbf{x}_N \\
\text{subject to} \quad & \mathbf{x}_i = \mathbf{x}_{i-1} + h\left\{\mathbf{A}(\mathbf{p})\left[\mathbf{x}_i - \mathbf{x}_r(\mathbf{p})\right] + \mathbf{B}(\mathbf{p})\left[\mathbf{u}_i - \mathbf{u}_r(\mathbf{p})\right]\right\} \\
& \mathbf{x}_0 = \mathbf{x}_r(\mathbf{p}) \\
& \underline{\mathbf{x}} \leqslant \mathbf{x}_i \leqslant \overline{\mathbf{x}} \\
& \underline{\mathbf{u}} \leqslant \mathbf{u}_i \leqslant \overline{\mathbf{u}} \\
& i = 1, \ldots, N
\end{aligned}$$

This discretized problem features $n = (N+1)n_x + Nn_u$ decision variables and $n_h = (N+1)n_x$ equality constraints. The cost vector $\mathbf{v}(\mathbf{p}) : \mathbb{R}^{n_p} \to \mathbb{R}^n$ and decision variable vector $\mathbf{z} \in \mathbb{R}^n$ with corresponding lower bound $\underline{\mathbf{z}} \in \mathbb{R}^n$ and upper bound $\overline{\mathbf{z}} \in \mathbb{R}^n$ are

$$\mathbf{v}(\mathbf{p}) = \begin{bmatrix} \mathbf{0} \\ \mathbf{0} \\ \mathbf{0} \\ \vdots \\ \mathbf{c}(\mathbf{p}) \\ \mathbf{0} \end{bmatrix}, \quad \underline{\mathbf{z}} = \begin{bmatrix} -\infty \\ \underline{\mathbf{x}} \\ \underline{\mathbf{u}} \\ \vdots \\ \underline{\mathbf{x}} \\ \underline{\mathbf{u}} \end{bmatrix}, \quad \mathbf{z} = \begin{bmatrix} \mathbf{x}_0 \\ \mathbf{x}_1 \\ \mathbf{u}_1 \\ \vdots \\ \mathbf{x}_N \\ \mathbf{u}_N \end{bmatrix}, \quad \overline{\mathbf{z}} = \begin{bmatrix} \infty \\ \overline{\mathbf{x}} \\ \overline{\mathbf{u}} \\ \vdots \\ \overline{\mathbf{x}} \\ \overline{\mathbf{u}} \end{bmatrix}$$

Moreover, the equality constraints for the initial condition as well as dynamic constraints are expressed by the matrix $\mathbf{M}(\mathbf{p}) : \mathbb{R}^{n_p} \to \mathbb{R}^{n_h \times n}$:

$$\mathbf{M}(\mathbf{p}) = \begin{bmatrix} \mathbf{I} & \mathbf{0} & \mathbf{0} & \mathbf{0} & \mathbf{0} & \mathbf{0} & \mathbf{0} & \cdots \\ -\mathbf{I} & \mathbf{I} - h\mathbf{A}(\mathbf{p}) & -h\mathbf{B}(\mathbf{p}) & \mathbf{0} & \mathbf{0} & \mathbf{0} & \mathbf{0} & \cdots \\ \mathbf{0} & -\mathbf{I} & \mathbf{0} & \mathbf{I} - h\mathbf{A}(\mathbf{p}) & -h\mathbf{B}(\mathbf{p}) & \mathbf{0} & \mathbf{0} & \cdots \\ \mathbf{0} & \mathbf{0} & \mathbf{0} & -\mathbf{I} & \mathbf{0} & \mathbf{I} - h\mathbf{A}(\mathbf{p}) & -h\mathbf{B}(\mathbf{p}) & \\ \vdots & \vdots & \vdots & & & \ddots & & \end{bmatrix}$$

and the corresponding right-side vector $\mathbf{b}(\mathbf{p}) : \mathbb{R}^{n_p} \to \mathbb{R}^{n_h}$:

$$\mathbf{b}(\mathbf{p}) = \begin{bmatrix} \mathbf{x}_r(\mathbf{p}) \\ h\mathbf{A}(\mathbf{p})\,\mathbf{x}_r(\mathbf{p}) + h\mathbf{B}(\mathbf{p})\,\mathbf{u}_r(\mathbf{p}) \\ \vdots \\ h\mathbf{A}(\mathbf{p})\,\mathbf{x}_r(\mathbf{p}) + h\mathbf{B}(\mathbf{p})\,\mathbf{u}_r(\mathbf{p}) \end{bmatrix}$$

via the equality $\mathbf{M}(\mathbf{p})\,\mathbf{z} = \mathbf{b}(\mathbf{p})$.

Trapezoidal Transcription For the trapezoidal collocation method, state variable vectors $\mathbf{x}_i \in \mathbb{R}^{n_x}$, $i = 0, \ldots, N$ as well as control variable vectors $\mathbf{u}_i \in \mathbb{R}^{n_u}$, $i = 0, \ldots, N$ are introduced, and the dynamic constraints $\mathbf{d}_i = \mathbf{0}$, $i =$

May be skipped (↴ at first read.

$1, \ldots, N$ are defined as follows:

$$\mathbf{d}_i = \mathbf{x}_i - \mathbf{x}_{i-1} - h\left(\mathbf{A}(\mathbf{p})\,\frac{\mathbf{x}_i + \mathbf{x}_{i-1} - 2\mathbf{x}_r(\mathbf{p})}{2} + \mathbf{B}(\mathbf{p})\,\frac{\mathbf{u}_i + \mathbf{u}_{i-1} - 2\mathbf{u}_r(\mathbf{p})}{2}\right)$$

Similar to the backward Euler discretization, the parametric version of the discretized problem statement reads

$$\begin{aligned}
\underset{\mathbf{x}_0,\mathbf{u}_0,\mathbf{x}_i,\mathbf{u}_i}{\text{minimize}} \quad & \mathbf{c}^T(\mathbf{p})\,\mathbf{x}_N \\
\text{subject to} \quad & \mathbf{x}_i = \mathbf{x}_{i-1} + h\left(\mathbf{A}(\mathbf{p})\,\frac{\mathbf{x}_i + \mathbf{x}_{i-1} - 2\mathbf{x}_r(\mathbf{p})}{2} + \mathbf{B}(\mathbf{p})\,\frac{\mathbf{u}_i + \mathbf{u}_{i-1} - 2\mathbf{u}_r(\mathbf{p})}{2}\right) \\
& \mathbf{x}_0 = \mathbf{x}_r(\mathbf{p}) \\
& \underline{\mathbf{x}} \leqslant \mathbf{x}_i \leqslant \overline{\mathbf{x}} \\
& \underline{\mathbf{u}} \leqslant \mathbf{u}_0 \leqslant \overline{\mathbf{u}} \\
& \underline{\mathbf{u}} \leqslant \mathbf{u}_i \leqslant \overline{\mathbf{u}} \\
& i = 1, \ldots, N
\end{aligned}$$

This discretized problem features $n = (N+1)(n_x + n_u)$ decision variables and $n_h = (N+1)n_x$ equality constraints. The cost vector $\mathbf{v}(\mathbf{p}) : \mathbb{R}^{n_p} \to \mathbb{R}^n$ and decision variable vector $\mathbf{z} \in \mathbb{R}^n$ with corresponding lower bound $\underline{\mathbf{z}} \in \mathbb{R}^n$ and upper bound $\overline{\mathbf{z}} \in \mathbb{R}^n$ are

$$\mathbf{v}(\mathbf{p}) = \begin{bmatrix} \mathbf{0} \\ \mathbf{0} \\ \mathbf{0} \\ \mathbf{0} \\ \vdots \\ \mathbf{c}(\mathbf{p}) \\ \mathbf{0} \end{bmatrix}, \quad \underline{\mathbf{z}} = \begin{bmatrix} -\infty \\ \underline{\mathbf{u}} \\ \underline{\mathbf{x}} \\ \underline{\mathbf{u}} \\ \vdots \\ \underline{\mathbf{x}} \\ \underline{\mathbf{u}} \end{bmatrix}, \quad \mathbf{z} = \begin{bmatrix} \mathbf{x}_0 \\ \mathbf{u}_0 \\ \mathbf{x}_1 \\ \mathbf{u}_1 \\ \vdots \\ \mathbf{x}_N \\ \mathbf{u}_N \end{bmatrix}, \quad \overline{\mathbf{z}} = \begin{bmatrix} \infty \\ \overline{\mathbf{u}} \\ \overline{\mathbf{x}} \\ \overline{\mathbf{u}} \\ \vdots \\ \overline{\mathbf{x}} \\ \overline{\mathbf{u}} \end{bmatrix}$$

Moreover, the equality constraints for the initial condition as well as dynamic constraints are expressed by the matrix $\mathbf{M}(\mathbf{p}) : \mathbb{R}^{n_p} \to \mathbb{R}^{n_h \times n}$:

$$\mathbf{M}(\mathbf{p}) = \begin{bmatrix} \mathbf{I} & \mathbf{0} & \mathbf{0} & \mathbf{0} & \mathbf{0} & \mathbf{0} & \cdots \\ -\mathbf{I} - \frac{h}{2}\mathbf{A}(\mathbf{p}) & -\frac{h}{2}\mathbf{B}(\mathbf{p}) & \mathbf{I} - \frac{h}{2}\mathbf{A}(\mathbf{p}) & -\frac{h}{2}\mathbf{B}(\mathbf{p}) & \mathbf{0} & \mathbf{0} & \cdots \\ \mathbf{0} & \mathbf{0} & -\mathbf{I} - \frac{h}{2}\mathbf{A}(\mathbf{p}) & -\frac{h}{2}\mathbf{B}(\mathbf{p}) & \mathbf{I} - \frac{h}{2}\mathbf{A}(\mathbf{p}) & -\frac{h}{2}\mathbf{B}(\mathbf{p}) & \\ \vdots & \vdots & & & \ddots & & \end{bmatrix}$$

and the corresponding right-side vector $\mathbf{b}(\mathbf{p}) : \mathbb{R}^{n_p} \to \mathbb{R}^{n_h}$:

$$\mathbf{b}(\mathbf{p}) = \begin{bmatrix} \mathbf{x}_r(\mathbf{p}) \\ h\mathbf{A}(\mathbf{p})\,\mathbf{x}_r(\mathbf{p}) + h\mathbf{B}(\mathbf{p})\,\mathbf{u}_r(\mathbf{p}) \\ \vdots \\ h\mathbf{A}(\mathbf{p})\,\mathbf{x}_r(\mathbf{p}) + h\mathbf{B}(\mathbf{p})\,\mathbf{u}_r(\mathbf{p}) \end{bmatrix}$$

via the equality $\mathbf{M}(\mathbf{p})\,\mathbf{z} = \mathbf{b}(\mathbf{p})$.

Similar to the nominal clearance problem in the preceding chapter, the reader may refer to Appendix H regarding a description of the Hermite–Simpson method transcription method for linear-parametric clearance problems.

5.3 FIRST-ORDER POSTOPTIMAL SENSITIVITIES

First-Order Postoptimal Sensitivity (Cat I and II)

The first-order sensitivity of the optimal (worst-case) criterion $\hat{y}_N$ for the linear clearance problem (Cat I and II) in discretized form is

$$\frac{d\hat{y}_N}{d\mathbf{p}} = \frac{\partial y_r}{\partial \mathbf{p}} + \left[\hat{\mathbf{x}}_N - \mathbf{x}_r(\mathbf{p})\right]^T \frac{\partial \mathbf{c}}{\partial \mathbf{p}} + \left[\boldsymbol{\lambda}_0 - \mathbf{c}(\mathbf{p})\right]^T \frac{\partial \mathbf{x}_r}{\partial \mathbf{p}} + \sum_{i=1}^{N} \boldsymbol{\lambda}_i^T \frac{\partial \mathbf{d}_i}{\partial \mathbf{p}} \tag{5.4}$$

where $\hat{\mathbf{x}}_N$ is the optimal final state vector, $\boldsymbol{\lambda}_0 \in \mathbb{R}^{n_x}$ the Lagrange multiplier vector corresponding to the initial boundary condition, and $\boldsymbol{\lambda}_i \in \mathbb{R}^{n_x}$, $i = 1, \ldots, N$ are the Lagrange multiplier vectors corresponding to the dynamic equality constraints $\mathbf{d}_i(\hat{\mathbf{z}};\mathbf{p}) = \mathbf{0}$ depending on the optimal decision vector $\hat{\mathbf{z}}$ and the parameters $\mathbf{p}$. All quantities need to be evaluated at a KKT point for a nominal parameter value.

Postoptimal sensitivity analysis is well established for LPs [57–60]. In most references, the sensitivities of the optimal solution are investigated separately depending on the type of influence in the different parts (constraint matrix/right-hand side, cost coefficients) of the LP. The theorem from Castillo et al. [61], however, provides the sensitivities of the optimal cost $\hat{J}$ of optimization problems regarding the parameters $\mathbf{p}$ in a very general setting. In fact, it can be shown that for a general optimization problem, these sensitivities are equal to the *partial derivatives* of the Lagrangian $\mathcal{L}$ with respect to the parameters:

$$\frac{d\hat{J}}{d\mathbf{p}} = \frac{\partial \mathcal{L}}{\partial \mathbf{p}} \tag{5.5}$$

This relationship permits a parameter dependency within the LP of essentially arbitrary type. Based on Eq. (5.5), the first-order postoptimal sensitivity of Cat I and II types of clearance problems can directly be established.

May be skipped (↴) at first read.

The parameter-dependent terms of the Lagrangian $\mathcal{L}$ of the LP for the linear-parametric clearance problem are

$$\mathcal{L} = \mathbf{c}^T(\mathbf{p})\hat{\mathbf{x}}_N + \boldsymbol{\lambda}_0^T\left[\mathbf{x}_r(\mathbf{p}) - \hat{\mathbf{x}}_0\right] + \sum_{i=1}^{N} \boldsymbol{\lambda}_i^T \mathbf{d}_i(\hat{\mathbf{z}};\mathbf{p}) + \ldots$$

with the Lagrange multipliers $\boldsymbol{\lambda}_0, \ldots, \boldsymbol{\lambda}_N \in \mathbb{R}^{n_x}$. Using Eq. (5.5), the sensitivities of the optimal cost $\hat{J}$ with respect to the parameters are

$$\frac{d\hat{J}}{d\mathbf{p}} = \frac{\partial \mathcal{L}}{\partial \mathbf{p}} = \hat{\mathbf{x}}_N^T \frac{\partial \mathbf{c}}{\partial \mathbf{p}} + \boldsymbol{\lambda}_0^T \frac{\partial \mathbf{x}_r}{\partial \mathbf{p}} + \sum_{i=1}^{N} \boldsymbol{\lambda}_i^T \frac{\partial \mathbf{d}_i}{\partial \mathbf{p}}$$

Based on this result, the sensitivities of the optimal (worst-case) criterion $\hat{y}_N$ [see Eq. (5.3)] are expressed as

$$\frac{d\hat{y}_N}{d\mathbf{p}} = \frac{\partial y_r}{\partial \mathbf{p}} - \mathbf{x}_r^T(\mathbf{p}) \frac{\partial \mathbf{c}}{\partial \mathbf{p}} - \mathbf{c}^T(\mathbf{p}) \frac{\partial \mathbf{x}_r}{\partial \mathbf{p}} + \frac{d\hat{J}}{d\mathbf{p}}$$

Let us illustrate the application of the formula for the postoptimal sensitivities using parametric versions of the Cat I and Cat II clearance problems from Sec. 4.3 and Sec. 4.4. These parametric versions are based on the nominal solutions for the Cat I and Cat II problems with combined commands and final time $t_f = 6\,\text{s}$. The combined commands include inputs modeling wind disturbances. Here, the wind velocities are not directly controlled but indirectly, through artificial inputs fed into linear second-order systems. For the vertical wind velocity w_W, this system is defined as

$$\begin{bmatrix} \dot{w}_W(t) \\ \dot{w}_{W,a}(t) \end{bmatrix} = \begin{bmatrix} 0 & 1 \\ -\omega_W^2 & -2\zeta_W\omega_W \end{bmatrix} \begin{bmatrix} w_W(t) \\ w_{W,a}(t) \end{bmatrix} + \begin{bmatrix} 0 \\ \omega_W^2 \end{bmatrix} w_{W,c}(t) \tag{5.6}$$

with $\zeta_W, \omega_W = 1$. The step responses for this model are provided in Appendix F. The same form is used for the horizontal wind velocities u_W and v_W. Note that this implies that the wind inputs used to test the system are not able to produce instantaneous changes in the wind velocities or instantaneous changes in the rotational wind components approximated as

$$q_W(t) \approx \frac{w_{W,a}(t)}{V_{ref}}$$

$$r_W(t) \approx \frac{v_{W,a}(t)}{V_{ref}}$$

In addition to parameters belonging to the longitudinal and lateral plane, the trim conditions are considered in parametric form depending on the reference altitude h_{ref} and velocity V_{ref}.

Example 5.3.1: Postoptimal Sensitivities Longitudinal (Cat I and II)

Let us first illustrate the nominal solutions of the Cat I and II examples for the longitudinal model from Sec. 4.3 and Sec. 4.4 with combined control inputs including wind disturbances. Thus, the wind inputs

$$-7.62\,\text{m}\,\text{s}^{-1} \leqslant u_{W,c}(t) \leqslant 7.62\,\text{m}\,\text{s}^{-1}$$

$$-7.62\,\text{m}\,\text{s}^{-1} \leqslant w_{W,c}(t) \leqslant 7.62\,\text{m}\,\text{s}^{-1}$$

as well as the corresponding linear systems of type (5.6) are added to both clearance problems. This implies that each problem now features two additional inputs and four additional states.

As before, no limits are imposed on the internal position state $x_{e,p}$ and velocity state $x_{e,v}$ of the actuator related to the elevator in the Cat I version of the problem.

The nominal Cat I results are depicted in Fig. 5.1 for maximizing the normal load factor $n_z(6\text{ s})$ and Fig. 5.2 for maximizing the aerodynamic angle of attack $\alpha_A(6\text{ s})$.

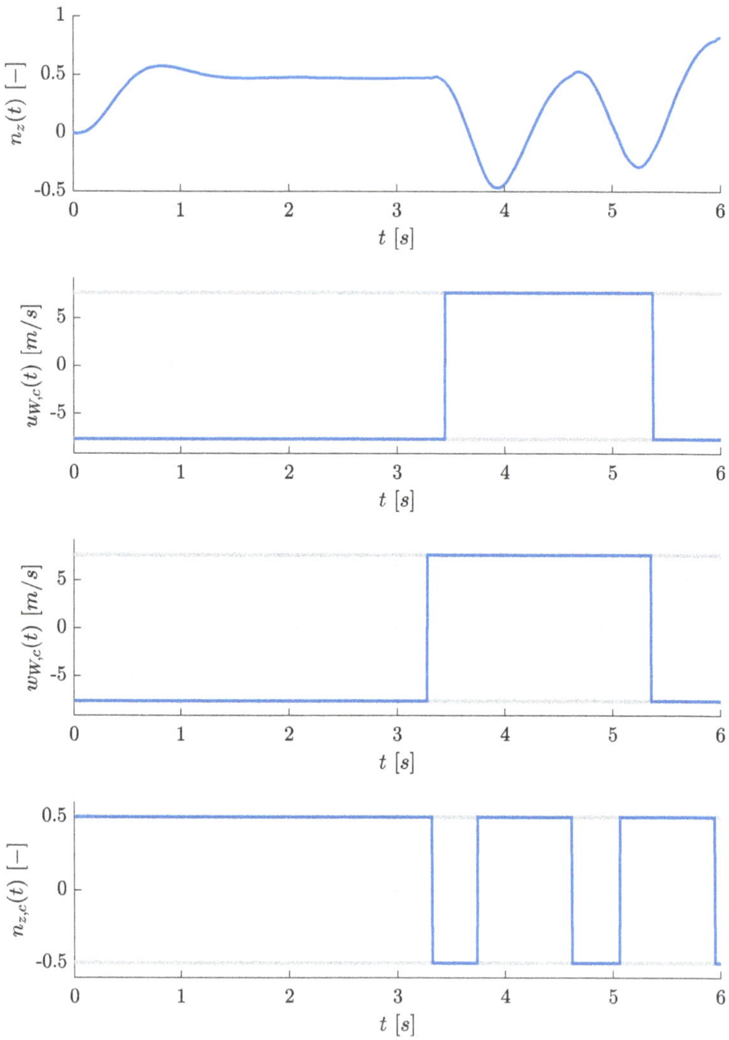

Fig. 5.1 Maximization of $n_z(6\text{ s})$, Cat I.

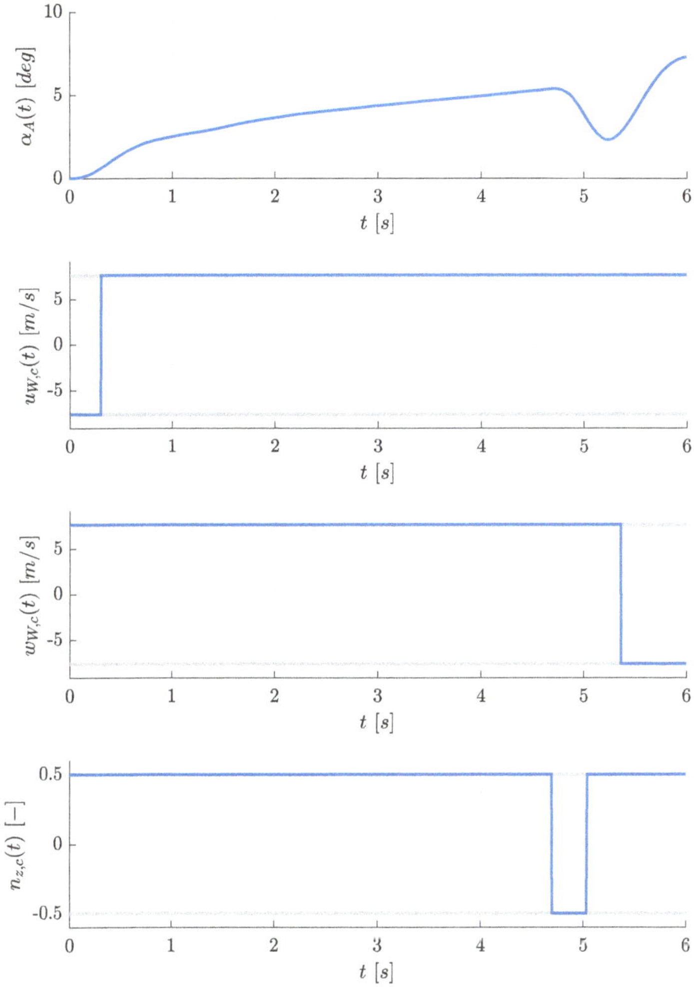

Fig. 5.2 Maximization of $\alpha_K(6\ \mathrm{s})$, Cat I.

In the case of the Cat II type of problems, the state constraints (4.12) and (4.13) on the position and velocity states of the actuator related to the elevator are added to the Cat I problems. The optimal solutions are depicted in Fig. 5.3 for maximizing the normal load factor $n_z(6\ \mathrm{s})$ and Fig. 5.4 for maximizing the aerodynamic angle of attack $\alpha_A(6\ \mathrm{s})$.

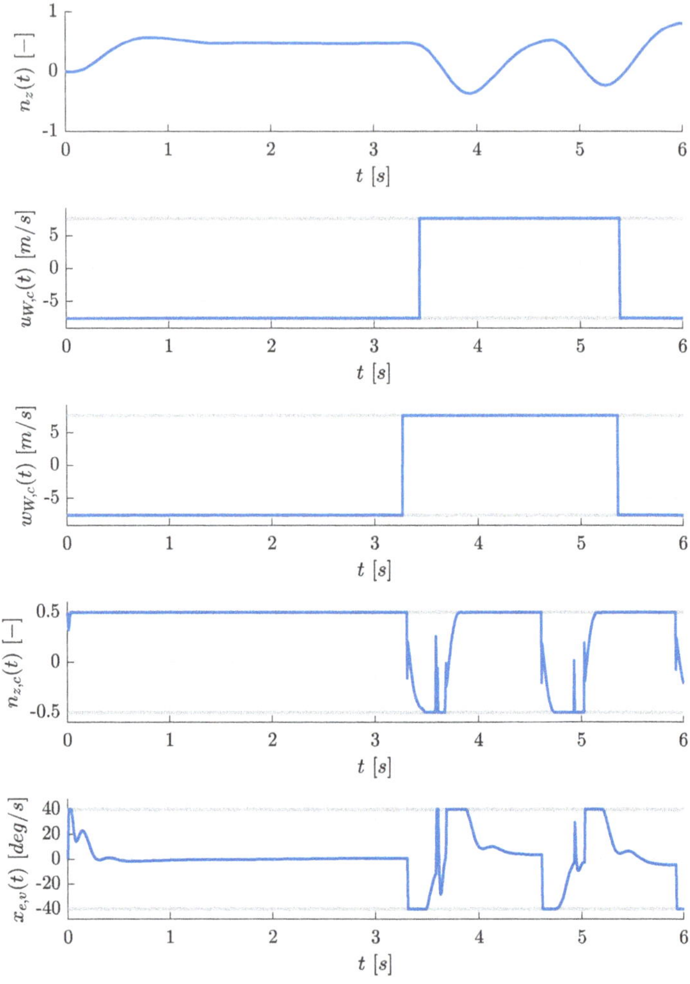

Fig. 5.3 Maximization of n_z (6 s), Cat II.

Next, the application of the first-order postoptimal sensitivity analysis is presented. For this purpose, parameters related to pitch stiffness ΔM_α, pitch damping ΔM_q, elevator effectiveness ΔM_η, mass Δm, and moment of inertia about the body's y-axis ΔI_{yy} are introduced. Moreover, the parameter Δx^{RG} is considered, where for $\Delta x^{RG} = -1$ the center of gravity coincides with the neutral point. All of these parameters are introduced as multiplicative

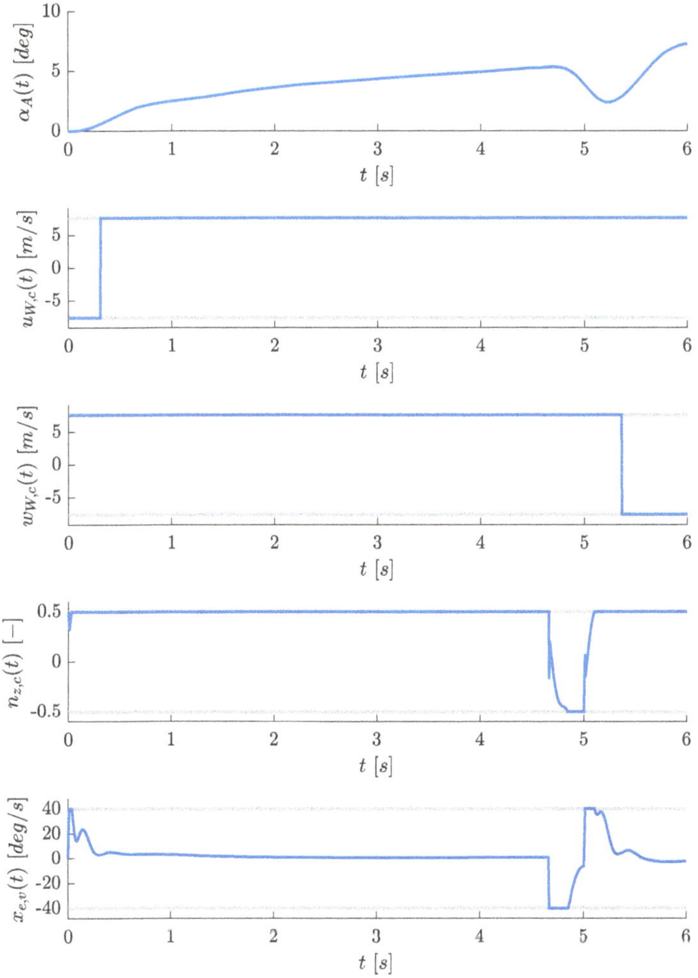

Fig. 5.4 Maximization of α_A (6 s), Cat II.

parameters centered at zero. Additionally, the influence of trim parameters h_{ref} and V_{ref} is investigated. The sensitivities corresponding to the maximum normal load factor $\hat{n}_z(6\,\text{s})$ and the aerodynamic angle of attack $\hat{\alpha}_A(6\,\text{s})$ using the normal load factor command $n_{z,c}(t)$ in combination with the wind commands $u_{W,c}(t)$ and $w_{W,c}(t)$ are depicted in Fig. 5.5 and Fig. 5.6. In each chart, the sensitivities (absolute values, normalized to the highest value) are shown for the optimal cost of the Cat I (blue) and Cat II (black) types of clearance problems.

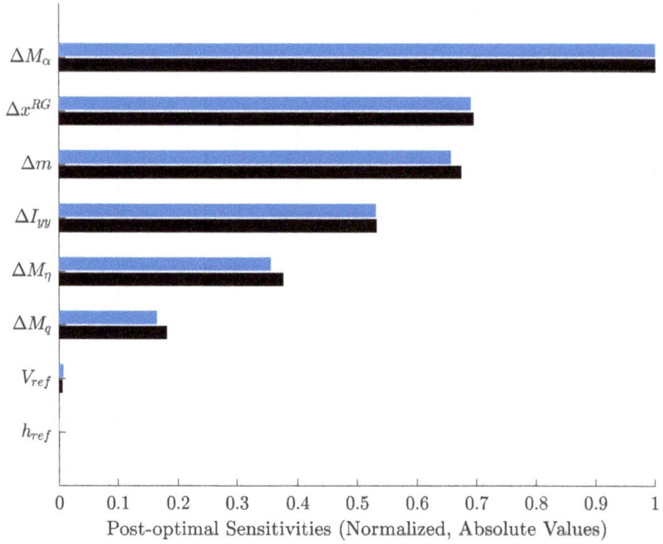

Fig. 5.5 Postoptimal sensitivities $\hat{n}_z(6\text{ s})$, Cat I/II (blue/black).

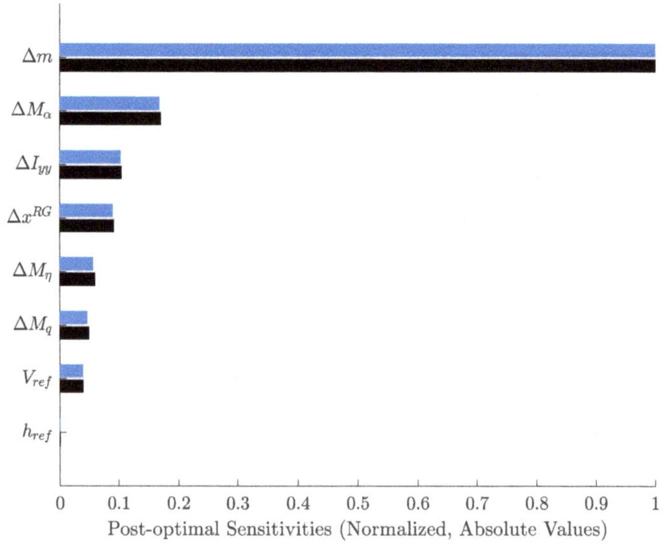

Fig. 5.6 Postoptimal sensitivities $\hat{\alpha}_A(6\text{ s})$, Cat I/II (blue/black).

Observe that the indications regarding the parameter sensitivities for both Cat I and II types of clearance problems appear to be similar for the maximization of n_z (6 s) and α_A (6 s). This is most likely contributed to the fact that even though the actuators saturate in the Cat II cases, the worst-case inputs exhibit a comparable structure for the optimal control actions under Cat I and Cat II models (compare Fig. 5.1 with Fig. 5.3 and Fig. 5.2 with Fig. 5.4). For both the maximization of the load factor (Fig. 5.5) and angle of attack (Fig. 5.6), high influences are identified for the parameters Δx^{RG} and ΔM_α. This result is not unexpected as both quantities are indicators for static stability and, in particular, robustness with respect to disturbances in the longitudinal plane. Additionally, the parameters associated with the mass Δm and the moment of inertia ΔI_{yy} show high sensitivities. The reference velocity V_{ref} and height h_{ref}, however, appear to have little to no influence relative to the other parameters on the optimal (worst-case) cost function values.

Example 5.3.2: Postoptimal Sensitivities Lateral (Cat I and II)

As for the longitudinal model, first, the nominal solutions of the Cat I and II examples for the lateral model from Sec. 4.3 and Sec. 4.4 with combined control

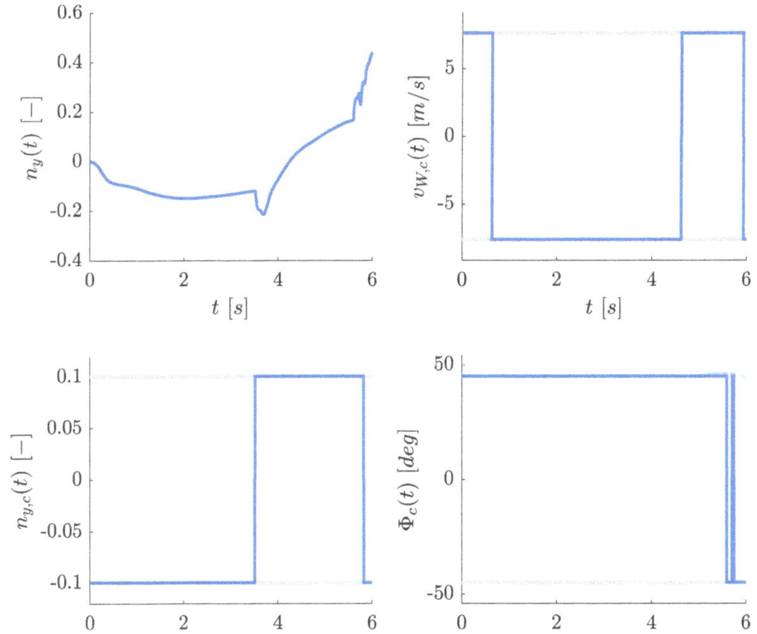

Fig. 5.7 Maximization of n_y (6 s), Cat I.

inputs including wind disturbances are illustrated. For this model, the lateral load factor command

$$-0.1 \leqslant n_{y,c}(t) \leqslant 0.1$$

and the lateral wind input

$$-7.62\,\mathrm{m\,s^{-1}} \leqslant v_{W,c}(t) \leqslant 7.62\,\mathrm{m\,s^{-1}}$$

together with the corresponding linear system of type (5.6) are added to the clearance problems. Thus, each problem now features two additional inputs and two additional states.

For the Cat I version of the problem, no limits are imposed on the internal position states $x_{a,p}$ and $x_{r,p}$, as well as velocity states $x_{a,v}$ and $x_{r,v}$ of the actuators related to the aileron and rudder. The nominal Cat I results are presented in Fig. 5.7 for maximizing the lateral load factor $n_y(6\,\mathrm{s})$ and Fig. 5.8 for maximizing the aerodynamic angle of sideslip $\beta_A(6\,\mathrm{s})$.

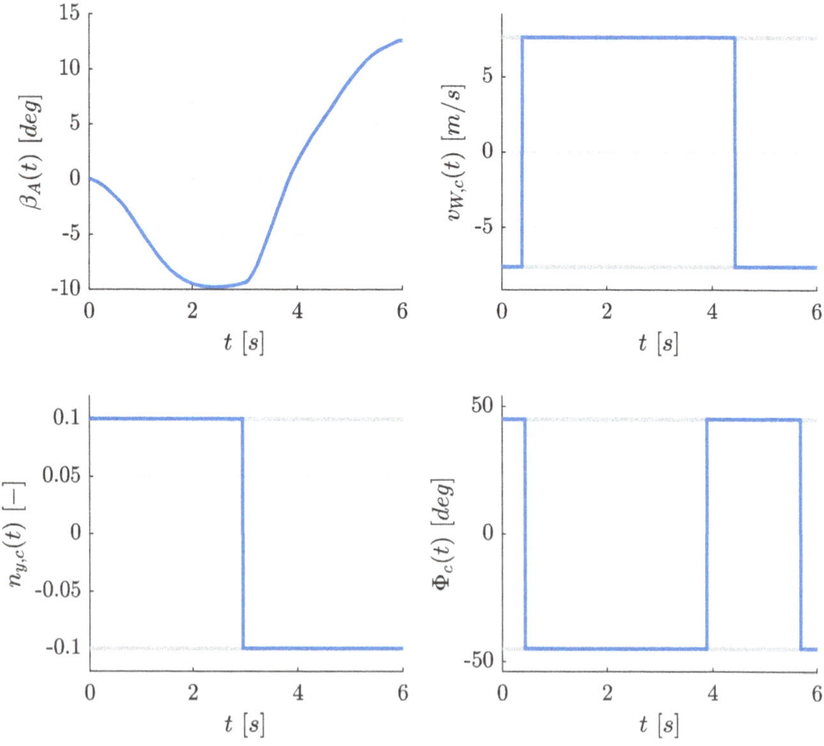

Fig. 5.8 Maximization of $\beta_A(6\,\mathrm{s})$, Cat I.

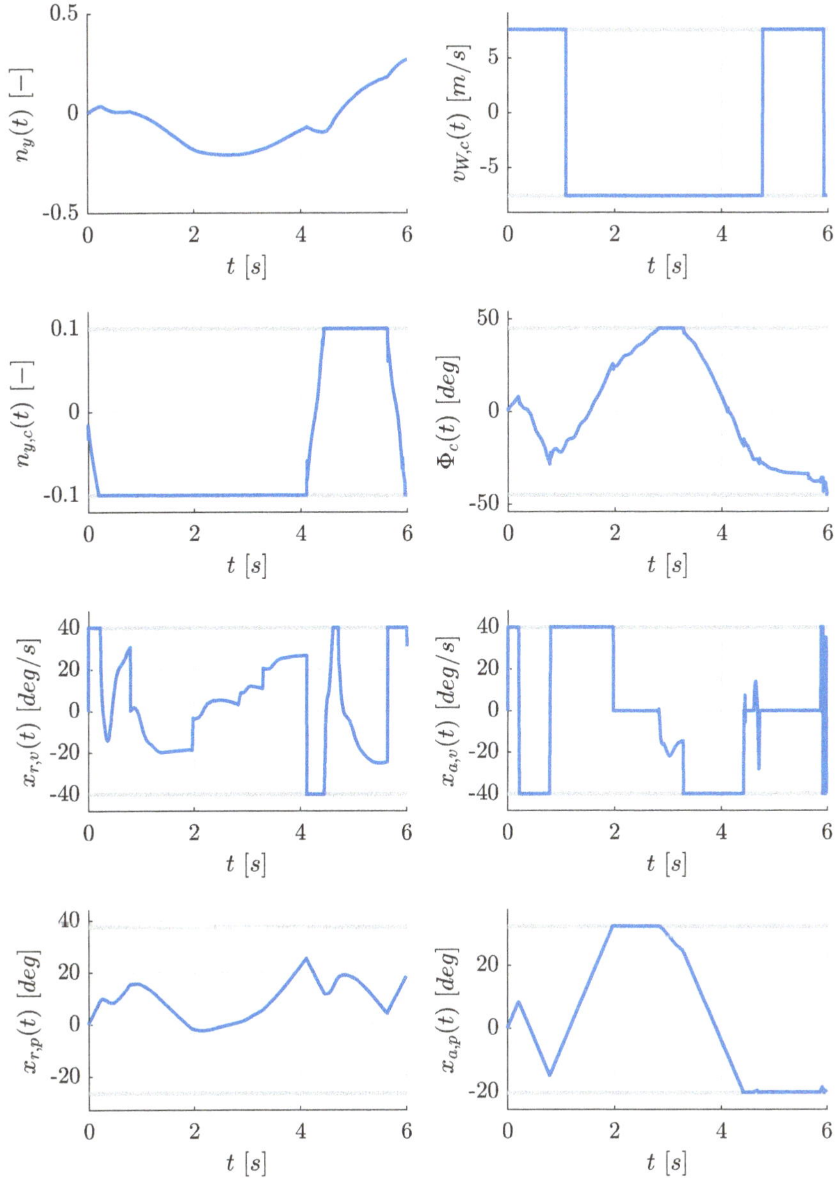

Fig. 5.9 Maximization of n_y(6 s), Cat II.

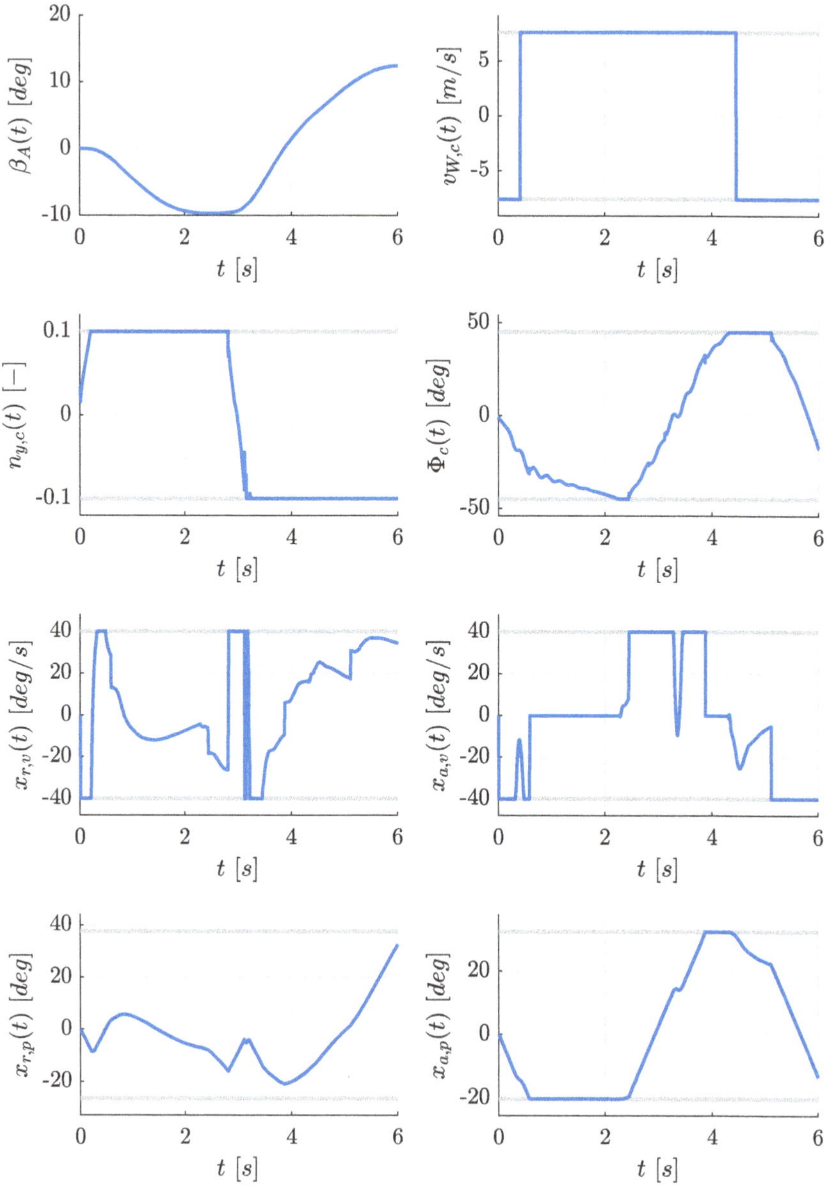

Fig. 5.10 Maximization of β_A(6 s), Cat II.

Compared to the Cat I problems, state constraints (4.14), (4.15), (4.16), and (4.17) on the position and velocity states of the actuators related to the aileron and rudder are imposed in the Cat II problems. The optimal solutions for these cases are shown in Fig. 5.9 for maximizing $n_y(6\,\text{s})$ and Fig. 5.10 for maximizing $\beta_A(6\,\text{s})$.

For the illustration of the postoptimal sensitivity analysis, parameters related to the dihedral stability (ΔL_β), weathercock stability (ΔN_β), roll and yaw damping (ΔL_p, ΔL_r, ΔN_r), aileron and rudder effectiveness (ΔL_ξ, ΔL_ζ, ΔN_ζ), as well as the moments of inertia about the body's x-axis (ΔI_{xx}) and z-axis (ΔI_{zz}) are considered. Similar to the longitudinal model, these parameters are introduced as multiplicative parameters centered at zero. Additionally, the trim parameters h_{ref} and V_{ref} are added. The sensitivities corresponding to the maximum lateral load factor $\hat{n}_y$ (6 s) and the aerodynamic angle of sideslip $\hat{\beta}_A$ (6 s) using the lateral load factor command $n_{y,c}(t)$ and roll angle command $\Phi_c(t)$ in combination with the wind command $v_{W,c}(t)$ are depicted in Fig. 5.11 and Fig. 5.12. In each chart, the sensitivities (absolute values, normalized to the largest absolute value) are shown both for the Cat I (blue) and Cat II (black) types of models.

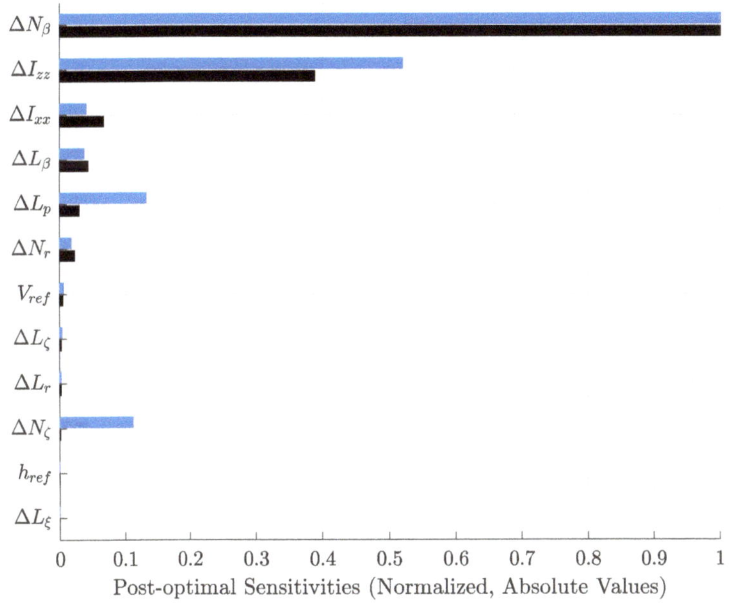

Fig. 5.11 Postoptimal sensitivities $\hat{n}_y(6\,\text{s})$, Cat I/II (blue/black).

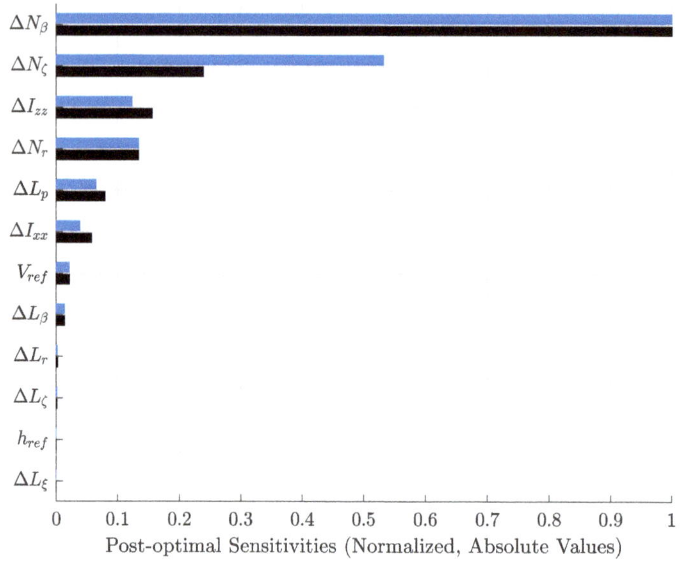

Fig. 5.12 Postoptimal sensitivities $\hat{\beta}_A$(6 s), Cat I/II (blue/black).

For the lateral plane, the parameters corresponding to the weathercock stability ΔN_β and the moment of inertia ΔI_{zz} around the yaw axis appear to be influential for both the maximization of the lateral load factor (see Fig. 5.11) and the aerodynamic angle of sideslip (see Fig. 5.12). Moreover, the parameter related to the rudder effectiveness ΔN_ζ appears to play a crucial role in the maximization of the angle of sideslip (see Fig. 5.12). It is noteworthy, that the optimal control actions in the nominal results primarily excite the yaw motion in the worst-case solutions. As such, it is reasonable that the quantities associated with the yaw axis exhibit high sensitivities. Conversely, the quantities corresponding to the roll axis (roll moment coefficients and aileron effectiveness) only seem to play a minor role. Similar to the results for the longitudinal model, the reference velocity and height have a negligible effect on the optimal values of the criteria relative to the other parameters.

5.4 SECOND-ORDER POSTOPTIMAL SENSITIVITIES

Second-Order Postoptimal Sensitivity (Cat I and II)

The second-order sensitivity of the optimal (worst-case) criterion $\hat{y}_N$ for the linear clearance problem (Cat I and II) in discretized form is

$$\frac{d^2\hat{y}_N}{d\mathbf{p}^2} = \frac{\partial^2\left[y_r(\mathbf{p}) - \mathbf{c}^T(\mathbf{p})\,\mathbf{x}_r(\mathbf{p})\right]}{\partial\mathbf{p}^2} + \frac{\partial^2\mathcal{L}_s}{\partial\mathbf{p}^2} - \left(\mathbf{D} + \mathbf{D}^T\right)$$

with

$$\mathbf{D} = \frac{\partial^2\mathcal{L}_s}{\partial\mathbf{p}\partial\mathbf{z}_b}\mathbf{M}_B^{-1}\frac{\partial\left[\mathbf{M}_s(\mathbf{p})\,\hat{\mathbf{z}}_s - \mathbf{b}_s(\mathbf{p})\right]}{\partial\mathbf{p}} \tag{5.7}$$

All quantities need to be evaluated at a basic feasible optimal point of a parametric LP in standard form with Lagrangian $\mathcal{L}_s$, basis $\mathbf{M}_B$, and basic variables $\mathbf{z}_b > 0$ for a nominal parameter value. This LP in standard form features only equality and nonnegativity constraints $\mathbf{M}_s(\mathbf{p})\,\mathbf{z}_s = \mathbf{b}_s(\mathbf{p})$, $\mathbf{z}_s \geqslant \mathbf{0}$. Additionally, strict complementarity needs to hold.

To derive the expression for the second-order postoptimal sensitivity, it is favorable to consider the LP of the discretized Cat I and II clearance problems in standard form:

$$\begin{aligned} \underset{\mathbf{z}_s\in\mathbb{R}^{n_s}}{\text{minimize}} \quad & \mathbf{v}_s^T(\mathbf{p})\mathbf{z}_s \\ \text{subject to} \quad & \mathbf{M}_s(\mathbf{p})\mathbf{z}_s = \mathbf{b}_s(\mathbf{p}) \\ & \mathbf{z}_s \geqslant \mathbf{0} \end{aligned} \tag{5.8}$$

with $\mathbf{v}_s : \mathbb{R}^{n_p} \to \mathbb{R}^{n_s}$, $\mathbf{M}_s : \mathbb{R}^{n_p} \to \mathbb{R}^{n_m\times n_s}$, and $\mathbf{b}_s : \mathbb{R}^{n_p} \to \mathbb{R}^{n_m}$. The parameter-dependent Lagrangian for this LP is defined as

$$\mathcal{L}_s\left(\hat{\mathbf{z}}_s, \boldsymbol{\lambda}_s, \boldsymbol{\mu}_s, \mathbf{p}\right) = \mathbf{v}_s^T(\mathbf{p})\hat{\mathbf{z}}_s + \boldsymbol{\lambda}_s^T\left[\mathbf{M}_s(\mathbf{p})\hat{\mathbf{z}}_s - \mathbf{b}_s(\mathbf{p})\right] - \boldsymbol{\mu}_s^T\hat{\mathbf{z}}_s \tag{5.9}$$

based on the optimal decision vector $\hat{\mathbf{z}}_s$ as well as the Lagrange multiplier vectors $\boldsymbol{\lambda}_s \in \mathbb{R}^{n_m}$ and $\boldsymbol{\mu}_s \in \mathbb{R}^{n_s}$. As discussed in Sec. 1.2, a conversion to this standard form can always be achieved by standard transformations. In the following, it is assumed that the sensitivity analysis is performed at a basic feasible optimal point for a nominal parameter value, and strict complementarity holds.

Basic Solutions and Strict Complementarity: A globally optimal solution of the LP [Eq. (5.8)] is guaranteed to be at a vertex of the convex polytope defining the feasible set. An effective method for finding such an optimal vertex is the Simplex method [54]. Essentially, this algorithm visits the vertices of the feasible set until an optimal solution is found. Geometrically, this can be interpreted as moving along the edges of the polytope in directions of nonincreasing cost until the cost cannot be improved further. All

May be skipped at first read.

vertices that are visited during the iterations of the Simplex algorithm are *basic feasible*. This means that they are vertices of the feasible polytope and at each point there exists an index set $\mathcal{B} \subseteq \{1, \ldots, n_s\}$ of cardinality n_c with $\hat{z}_{s,i} = 0, i \notin \mathcal{B}$. Moreover, the *basis matrix* $\mathbf{M}_\mathcal{B}$ composed of the columns $\mathbf{m}_i, i \in \mathcal{B}$ of $\mathbf{M}_s$ has full rank, implying invertibility. Furthermore, let us define the *nonbasic* index set $\mathcal{N} = \{1, \ldots, n_s\} \setminus \mathcal{B}$ of cardinality $n_s - n_c$. Using the definition of the basic and nonbasic index sets, the variables $\hat{\mathbf{z}}_s$ and $\boldsymbol{\mu}_s$ can be rearranged into the *basic variables* $\mathbf{z}_b \in \mathbb{R}^{n_c}$ and $\boldsymbol{\mu}_b \in \mathbb{R}^{n_c}$ that contain all $\hat{z}_{s,i}, i \in \mathcal{B}$ and $\mu_{s,i}, i \in \mathcal{B}$ as well as the *nonbasic variables* $\mathbf{z}_n \in \mathbb{R}^{n_s - n_c}$ and $\boldsymbol{\mu}_n \in \mathbb{R}^{n_s - n_c}$ collecting all $\hat{z}_{s,i}, i \in \mathcal{N}$ and $\mu_{s,i}, i \in \mathcal{N}$. Furthermore, the columns of the matrix $\mathbf{M}_s$ can be rearranged to $[\mathbf{M}_\mathcal{B}, \mathbf{M}_\mathcal{N}]$ with the invertible basis matrix $\mathbf{M}_\mathcal{B}$ and the matrix $\mathbf{M}_\mathcal{N}$ collecting all columns of $\mathbf{M}_s$ not indexed by $\mathcal{B}$. A basic feasible *optimal* point is a basic feasible point that is a solution for the LP [Eq. (5.8)]. A basic feasible optimal solution is guaranteed to exist under relatively mild assumptions (see Theorem 13.2 in Nocedal and Wright [23]):

- "If [Eq. (5.8)] has a nonempty feasible region, then there is at least one basic feasible point."
- "If [Eq. (5.8)] has solutions, then at least one such solution is a basic optimal point."
- "If [Eq. (5.8)] is feasible and bounded, then it has an optimal solution."

Moreover, if the strict complementarity condition

$$\hat{\mathbf{z}}_s + \boldsymbol{\mu}_s > \mathbf{0}$$

holds at an optimal solution, then the optimal solution is unique [59], and the basic and nonbasic variables satisfy $\mathbf{z}_b > \mathbf{0}, \mathbf{z}_n = \mathbf{0}$, respectively, $\boldsymbol{\mu}_b = \mathbf{0}, \boldsymbol{\mu}_n > \mathbf{0}$.

A formula allowing for an efficient evaluation of the second-order sensitivity of the optimal cost $\hat{J}_s$ for Eq. (5.8) can be derived as

$$\frac{d^2\hat{J}_s}{d\mathbf{p}^2} = \frac{\partial^2 \mathcal{L}_s}{\partial \mathbf{p}^2} - \left(\mathbf{D} + \mathbf{D}^T\right)$$

with $\mathbf{D}$ defined as in Eq. (5.7). It is noteworthy that the matrices in this formula typically exhibit high sparsity if the direct discretization approach is based on segmented collocation methods of low order, as is the case for the backward Euler and trapezoidal collocation methods.

Consider the following augmented form (see Castillo et al. [61]):

$$\begin{aligned}
\underset{\mathbf{z}_s,\tilde{\mathbf{p}}}{\text{minimize}} \quad & \mathbf{v}_s^T(\tilde{\mathbf{p}})\mathbf{z}_s \\
\text{subject to} \quad & \mathbf{M}_s(\tilde{\mathbf{p}})\mathbf{z}_s = \mathbf{b}_s(\tilde{\mathbf{p}}) \\
& \mathbf{z}_s \geqslant \mathbf{0} \\
& \tilde{\mathbf{p}} = \mathbf{p}
\end{aligned}$$

May be skipped at first read.

of the LP [Eq. (5.8)] featuring a parameter embedding constraint for a fixed nominal parameter $\mathbf{p}$. The Lagrangian of this augmented problem is expressed in terms of the Lagrangian Eq. (5.9) as

$$\tilde{\mathcal{L}}_s\left(\hat{\mathbf{z}}_s, \boldsymbol{\lambda}_s, \boldsymbol{\mu}_s, \tilde{\mathbf{p}}, \eta; \mathbf{p}\right) = \mathcal{L}_s\left(\hat{\mathbf{z}}_s, \boldsymbol{\lambda}_s, \boldsymbol{\mu}_s, \tilde{\mathbf{p}}\right) + \eta^T\left(\tilde{\mathbf{p}} - \mathbf{p}\right)$$

and the KKT conditions of the augmented problem are

$$\begin{aligned}
\left(\frac{\partial \mathcal{L}_s}{\partial \mathbf{z}_s}\right)^T &= \mathbf{0} \\
\left(\frac{\partial \mathcal{L}_s}{\partial \tilde{\mathbf{p}}}\right)^T + \eta &= \mathbf{0} \\
\mathbf{h}_s(\hat{\mathbf{z}}_s, \tilde{\mathbf{p}}) &= \mathbf{0} \\
\tilde{\mathbf{p}} - \mathbf{p} &= \mathbf{0} \\
\mathbf{Z}\boldsymbol{\mu}_s &= \mathbf{0} \\
\boldsymbol{\mu}_s &\geqslant \mathbf{0} \\
\hat{\mathbf{z}}_s &\geqslant \mathbf{0}
\end{aligned} \tag{5.10}$$

with $\left(\frac{\partial \mathcal{L}_s}{\partial \mathbf{z}_s}\right)^T = \mathbf{v}_s(\tilde{\mathbf{p}}) + \mathbf{M}_s^T(\tilde{\mathbf{p}})\boldsymbol{\lambda}_s - \boldsymbol{\mu}_s$, $\mathbf{h}_s(\hat{\mathbf{z}}_s, \tilde{\mathbf{p}}) = \mathbf{M}_s(\tilde{\mathbf{p}})\hat{\mathbf{z}}_s - \mathbf{b}_s(\tilde{\mathbf{p}})$, and $\mathbf{Z} = diag(\hat{\mathbf{z}}_s)$. Note that there are two additional conditions compared to the original problem due to the parameter embedding constraint, namely, $\left(\frac{\partial \mathcal{L}_s}{\partial \tilde{\mathbf{p}}}\right)^T = -\eta$ and $\tilde{\mathbf{p}} = \mathbf{p}$. Obviously, this augmented problem has the same optimal cost function value as the problem without the parameter embedding constraint. For the KKT conditions to remain valid in the neighborhood of the nominal solution, the total derivative $\frac{d\mathbf{F}}{d\mathbf{p}}$ of the first-order conditions [Eq. (5.10)] defined by

$$\mathbf{F}[\mathbf{w}(\mathbf{p});\mathbf{p}] = \begin{bmatrix} \left(\frac{\partial \mathcal{L}_s}{\partial \mathbf{z}_s}\right)^T \\ \left(\frac{\partial \mathcal{L}_s}{\partial \tilde{\mathbf{p}}}\right)^T + \boldsymbol{\eta} \\ \mathbf{h}_s(\hat{\mathbf{z}}_s, \tilde{\mathbf{p}}) \\ \tilde{\mathbf{p}} - \mathbf{p} \\ \mathbf{Z}\boldsymbol{\mu}_s \end{bmatrix} = \mathbf{0}$$

with $\mathbf{w}^T = \left[\hat{\mathbf{z}}_s^T, \tilde{\mathbf{p}}^T, \boldsymbol{\lambda}_s^T, \boldsymbol{\eta}^T, \boldsymbol{\mu}_s^T\right]$ needs to satisfy $\frac{d\mathbf{F}}{d\mathbf{p}} = \mathbf{0}$, which implies

$$\frac{\partial \mathbf{F}}{\partial \mathbf{w}}\frac{\partial \mathbf{w}}{\partial \mathbf{p}} = -\frac{\partial \mathbf{F}}{\partial \mathbf{p}}$$

or explicitly

$$\begin{bmatrix} \mathbf{0} & \frac{\partial^2 \mathcal{L}_s}{\partial \mathbf{z}_s \partial \tilde{\mathbf{p}}} & \mathbf{M}_s^T & \mathbf{0} & \mathbf{I} \\ \frac{\partial^2 \mathcal{L}_s}{\partial \tilde{\mathbf{p}} \partial \mathbf{z}_s} & \frac{\partial^2 \mathcal{L}_s}{\partial \tilde{\mathbf{p}}^2} & \left(\frac{\partial \mathbf{h}_s}{\partial \tilde{\mathbf{p}}}\right)^T & \mathbf{I} & \mathbf{0} \\ \mathbf{M}_s & \frac{\partial \mathbf{h}_s}{\partial \tilde{\mathbf{p}}} & \mathbf{0} & \mathbf{0} & \mathbf{0} \\ \mathbf{0} & \mathbf{I} & \mathbf{0} & \mathbf{0} & \mathbf{0} \\ \bar{\mathbf{M}} & \mathbf{0} & \mathbf{0} & \mathbf{0} & \mathbf{Z} \end{bmatrix} \begin{bmatrix} \frac{d\hat{\mathbf{z}}_s}{d\mathbf{p}} \\ \frac{d\tilde{\mathbf{p}}}{d\mathbf{p}} \\ \frac{d\boldsymbol{\lambda}_s}{d\mathbf{p}} \\ \frac{d\boldsymbol{\eta}}{d\mathbf{p}} \\ \frac{d\boldsymbol{\mu}_s}{d\mathbf{p}} \end{bmatrix} = \begin{bmatrix} \mathbf{0} \\ \mathbf{0} \\ \mathbf{0} \\ \mathbf{I} \\ \mathbf{0} \end{bmatrix} \tag{5.11}$$

with $\bar{\mathbf{M}} = diag(\boldsymbol{\mu}_s)$. If strict complementarity $\hat{\mathbf{z}}_s + \boldsymbol{\mu}_s > \mathbf{0}$ holds, the matrices $\bar{\mathbf{M}}$ and $\mathbf{Z}$ can be expressed as

$$\bar{\mathbf{M}} = \begin{bmatrix} \mathbf{0} & \mathbf{0} \\ \mathbf{0} & \mathbf{M}_n \end{bmatrix}, \qquad \mathbf{Z} = \begin{bmatrix} \mathbf{Z}_b & \mathbf{0} \\ \mathbf{0} & \mathbf{0} \end{bmatrix}$$

with $\mathbf{M}_n = diag\left(\boldsymbol{\mu}_n\right)$ and $\mathbf{Z}_b = diag\left(\mathbf{z}_b\right)$ as the variables $\boldsymbol{\mu}_b$ and $\mathbf{z}_n$ are zero at a basic feasible optimal point. Together with the partitioning of $\mathbf{M}_s$ in $\mathbf{M}_B$ and $\mathbf{M}_N$, as well as $\mathbf{z}_s$ in $\mathbf{z}_b$ and $\mathbf{z}_n$, and similarly $\boldsymbol{\mu}_s$ in $\boldsymbol{\mu}_b$ and $\boldsymbol{\mu}_n$,

Eq. (5.11) can be expanded to

$$\begin{bmatrix} \mathbf{0} & \mathbf{0} & \frac{\partial^2 \mathcal{L}_s}{\partial \mathbf{z}_b \partial \tilde{\mathbf{p}}} & \mathbf{M}_B^T & \mathbf{0} & \mathbf{I} & \mathbf{0} \\ \mathbf{0} & \mathbf{0} & \frac{\partial^2 \mathcal{L}_s}{\partial \mathbf{z}_n \partial \tilde{\mathbf{p}}} & \mathbf{M}_N^T & \mathbf{0} & \mathbf{0} & \mathbf{I} \\ \frac{\partial^2 \mathcal{L}_s}{\partial \tilde{\mathbf{p}} \partial \mathbf{z}_b} & \frac{\partial^2 \mathcal{L}_s}{\partial \tilde{\mathbf{p}} \partial \mathbf{z}_n} & \frac{\partial^2 \mathcal{L}_s}{\partial \tilde{\mathbf{p}}^2} & \left(\frac{\partial \mathbf{h}_s}{\partial \tilde{\mathbf{p}}}\right)^T & \mathbf{I} & \mathbf{0} & \mathbf{0} \\ \mathbf{M}_B & \mathbf{M}_N & \frac{\partial \mathbf{h}_s}{\partial \tilde{\mathbf{p}}} & \mathbf{0} & \mathbf{0} & \mathbf{0} & \mathbf{0} \\ \mathbf{0} & \mathbf{0} & \mathbf{I} & \mathbf{0} & \mathbf{0} & \mathbf{0} & \mathbf{0} \\ \mathbf{0} & \mathbf{0} & \mathbf{0} & \mathbf{0} & \mathbf{0} & \mathbf{Z}_b & \mathbf{0} \\ \mathbf{0} & \mathbf{M}_n & \mathbf{0} & \mathbf{0} & \mathbf{0} & \mathbf{0} & \mathbf{0} \end{bmatrix} \begin{bmatrix} \frac{d\mathbf{z}_b}{d\mathbf{p}} \\ \frac{d\mathbf{z}_n}{d\mathbf{p}} \\ \frac{d\tilde{\mathbf{p}}}{d\mathbf{p}} \\ \frac{d\boldsymbol{\lambda}_s}{d\mathbf{p}} \\ \frac{d\eta}{d\mathbf{p}} \\ \frac{d\boldsymbol{\mu}_b}{d\mathbf{p}} \\ \frac{d\boldsymbol{\mu}_n}{d\mathbf{p}} \end{bmatrix} = \begin{bmatrix} \mathbf{0} \\ \mathbf{0} \\ \mathbf{0} \\ \mathbf{0} \\ \mathbf{I} \\ \mathbf{0} \\ \mathbf{0} \end{bmatrix}$$

The fifth row block matrix equation simply expresses the consistent result $\frac{d\tilde{\mathbf{p}}}{d\mathbf{p}} = \mathbf{I}$ for the total differential of the parameters $\tilde{\mathbf{p}}$ with respect to the parameters $\mathbf{p}$. The last two row block matrix equations yield

$$\mathbf{Z}_b \frac{d\boldsymbol{\mu}_b}{d\mathbf{p}} = \mathbf{0} \rightarrow \frac{d\boldsymbol{\mu}_b}{d\mathbf{p}} = \mathbf{0}$$

$$\mathbf{M}_n \frac{d\mathbf{z}_n}{d\mathbf{p}} = \mathbf{0} \rightarrow \frac{d\mathbf{z}_n}{d\mathbf{p}} = \mathbf{0}$$

as the matrices $\mathbf{M}_n$ and $\mathbf{Z}_b$ are both diagonal matrices with strictly positive entries on the main diagonal. Note that this result reflects the requirement that the variables $\boldsymbol{\mu}_b$ and $\mathbf{z}_n$ remain zero and the basis matrix $\mathbf{M}_B$ is unchanged in a neighborhood of the solution around the nominal parameter. From the first and fourth row block matrix equations, the total derivatives of the Lagrange multipliers $\boldsymbol{\lambda}_s$ and the basic variables $\mathbf{z}_b$ are obtained as

$$\frac{\partial^2 \mathcal{L}_s}{\partial \mathbf{z}_b \partial \tilde{\mathbf{p}}} + \mathbf{M}_B^T \frac{d\boldsymbol{\lambda}_s}{d\mathbf{p}} = \mathbf{0} \rightarrow \frac{d\boldsymbol{\lambda}_s}{d\mathbf{p}} = -\left(\mathbf{M}_B^T\right)^{-1} \frac{\partial^2 \mathcal{L}_s}{\partial \mathbf{z}_b \partial \tilde{\mathbf{p}}}$$

$$\mathbf{M}_B \frac{d\mathbf{z}_b}{d\mathbf{p}} + \frac{\partial \mathbf{h}_s}{\partial \tilde{\mathbf{p}}} = \mathbf{0} \rightarrow \frac{d\mathbf{z}_b}{d\mathbf{p}} = -\mathbf{M}_B^{-1} \frac{\partial \mathbf{h}_s}{\partial \tilde{\mathbf{p}}}$$

Furthermore, for completeness, the second row block matrix equation yields

$$\frac{\partial^2 \mathcal{L}_s}{\partial \mathbf{z}_n \partial \tilde{\mathbf{p}}} + \mathbf{M}_N^T \frac{d\boldsymbol{\lambda}_\mathbf{s}}{d\mathbf{p}} + \frac{d\boldsymbol{\mu}_n}{d\mathbf{p}} = \mathbf{0}$$

$$\rightarrow \frac{d\boldsymbol{\mu}_n}{d\mathbf{p}} = -\frac{\partial^2 \mathcal{L}_s}{\partial \mathbf{z}_n \partial \tilde{\mathbf{p}}} + \mathbf{M}_N^T \left(\mathbf{M}_B^T\right)^{-1} \frac{\partial^2 \mathcal{L}_s}{\partial \mathbf{z}_b \partial \tilde{\mathbf{p}}}$$

Finally, note that the negative multiplier η corresponding to the parameter embedding constraint represents the first-order sensitivity of the optimal cost $\hat{J}_s$ with respect to the parameter $\mathbf{p}$ [see Eq. (5.4)]:

$$\frac{d\hat{J}_s}{d\mathbf{p}} = \frac{\partial \mathcal{L}_s}{\partial \tilde{\mathbf{p}}} = -\eta^T$$

Together with the third row block matrix equation

$$\frac{\partial^2 \mathcal{L}_s}{\partial \tilde{\mathbf{p}} \partial \mathbf{z}_b} \frac{d\mathbf{z}_b}{d\mathbf{p}} + \frac{\partial^2 \mathcal{L}_s}{\partial \tilde{\mathbf{p}}^2} + \left(\frac{\partial \mathbf{h}_s}{\partial \tilde{\mathbf{p}}}\right)^T \frac{d\boldsymbol{\lambda}_s}{d\mathbf{p}} + \frac{d\eta}{d\mathbf{p}} = \mathbf{0}$$

the relationship for the second-order cost sensitivity is readily derived

$$\begin{aligned}
\frac{d^2\hat{J}_s}{d\mathbf{p}^2} &= -\frac{d\eta}{d\mathbf{p}} \\
&= \frac{\partial^2 \mathcal{L}_s}{\partial \tilde{\mathbf{p}} \partial \mathbf{z}_b} \frac{d\mathbf{z}_b}{d\mathbf{p}} + \frac{\partial^2 \mathcal{L}_s}{\partial \tilde{\mathbf{p}}^2} + \left(\frac{\partial \mathbf{h}_s}{\partial \tilde{\mathbf{p}}}\right)^T \frac{d\boldsymbol{\lambda}_s}{d\mathbf{p}} \\
&= -\frac{\partial^2 \mathcal{L}_s}{\partial \tilde{\mathbf{p}} \partial \mathbf{z}_b} \mathbf{M}_B^{-1} \frac{\partial \mathbf{h}_s}{\partial \tilde{\mathbf{p}}} + \frac{\partial^2 \mathcal{L}_s}{\partial \tilde{\mathbf{p}}^2} - \left(\frac{\partial \mathbf{h}_s}{\partial \tilde{\mathbf{p}}}\right)^T \left(\mathbf{M}_B^T\right)^{-1} \frac{\partial^2 \mathcal{L}_s}{\partial \mathbf{z}_b \partial \tilde{\mathbf{p}}} \\
&= -\frac{\partial^2 \mathcal{L}_s}{\partial \tilde{\mathbf{p}} \partial \mathbf{z}_b} \mathbf{M}_B^{-1} \frac{\partial \mathbf{h}_s}{\partial \tilde{\mathbf{p}}} + \frac{\partial^2 \mathcal{L}_s}{\partial \tilde{\mathbf{p}}^2} - \left(\frac{\partial \mathbf{h}_s}{\partial \tilde{\mathbf{p}}}\right)^T \left(\mathbf{M}_B^{-1}\right)^T \left(\frac{\partial^2 \mathcal{L}_s}{\partial \tilde{\mathbf{p}} \partial \mathbf{z}_b}\right)^T \\
&= -\frac{\partial^2 \mathcal{L}_s}{\partial \tilde{\mathbf{p}} \partial \mathbf{z}_b} \mathbf{M}_B^{-1} \frac{\partial \mathbf{h}_s}{\partial \tilde{\mathbf{p}}} + \frac{\partial^2 \mathcal{L}_s}{\partial \tilde{\mathbf{p}}^2} - \left(\frac{\partial \mathbf{h}_s}{\partial \tilde{\mathbf{p}}}\right)^T \left(\frac{\partial^2 \mathcal{L}_s}{\partial \tilde{\mathbf{p}} \partial \mathbf{z}_b} \mathbf{M}_B^{-1}\right)^T \\
&= -\frac{\partial^2 \mathcal{L}_s}{\partial \tilde{\mathbf{p}} \partial \mathbf{z}_b} \mathbf{M}_B^{-1} \frac{\partial \mathbf{h}_s}{\partial \tilde{\mathbf{p}}} + \frac{\partial^2 \mathcal{L}_s}{\partial \tilde{\mathbf{p}}^2} - \left(\frac{\partial^2 \mathcal{L}_s}{\partial \tilde{\mathbf{p}} \partial \mathbf{z}_b} \mathbf{M}_B^{-1} \frac{\partial \mathbf{h}_s}{\partial \tilde{\mathbf{p}}}\right)^T
\end{aligned}$$

5.5 COMBINED ANALYSIS

Clearance Optimal Control Problem (Cat I and II Bilevel)

The bilevel decomposition approach separates the parameter search from the linear clearance optimal control problem for a combined search in the parameter and control space.

The parameter optimization problem for $\underline{\mathbf{p}} \leqslant \mathbf{p} \leqslant \overline{\mathbf{p}}$, $\underline{\mathbf{p}}, \overline{\mathbf{p}} \in \mathbb{R}^{n_p}$ in the upper level solves

$$\min_{\underline{\mathbf{p}} \leqslant \mathbf{p} \leqslant \overline{\mathbf{p}}} \hat{J}(\mathbf{p}) \tag{5.12}$$

Here, $\hat{J}$ represents the *optimal* cost of the lower level concerned with the solution to a linear clearance optimal control problem using fixed parameters provided by the upper level.

In this section, we extend the optimal control methods for linear-parametric Cat I and II types of problems to a combined optimization of parameters and controls. Unfortunately, this problem cannot be treated as linear due to nonlinearity introduced by the parameter dependency. Hence, a decomposition approach in the form of a bilevel setup is employed to tackle this type of problem in a modular setup. Under this approach, the problem is split into two levels (see Fig. 5.13) [62].

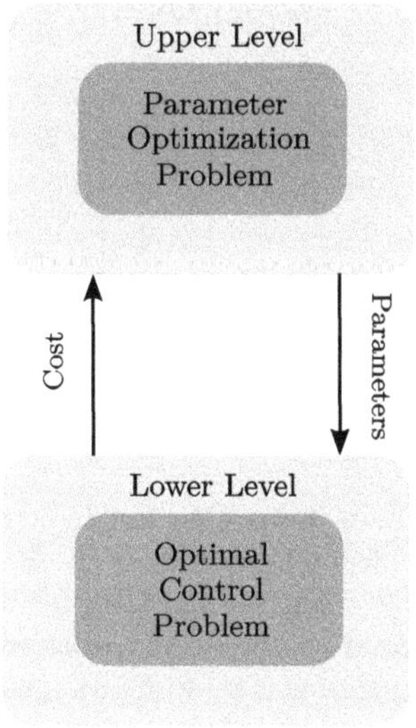

Fig. 5.13 Bilevel setup.

The higher level is concerned with the parameter search, while the lower level solves LPs with fixed parameters prescribed by the upper level. This approach leverages the efficiency of LP algorithms to obtain global solutions to the nominal, high-dimensional clearance optimal control problem for the lower level. Moreover, the upper-level problem is usually of lower dimension, allowing the efficient use of not only local (gradient-based) but also global optimization methods.

The bilevel method is illustrated in the following, based on the linear-parametric Cat II examples with combined control inputs from Sec. 5.3. Because a perturbation of the parameters also affects the trim condition, all worst-case values regarding the aerodynamic angles, as well as the load factors, are provided as absolute values and not relative to their respective reference values. Only a subset of influential parameters is considered for both the longitudinal and lateral models. For the longitudinal model, these parameters are

$$\mathbf{p}_{lon} = \begin{bmatrix} \Delta M_\alpha \\ \Delta M_q \\ \Delta M_\eta \\ \Delta m \\ \Delta I_{yy} \\ \Delta x^{RG} \end{bmatrix}$$

For the lateral model, the following parameters are considered:

$$\mathbf{p}_{lat} = \begin{bmatrix} \Delta L_\beta \\ \Delta L_p \\ \Delta N_\beta \\ \Delta N_\zeta \\ \Delta I_{xx} \\ \Delta I_{zz} \end{bmatrix}$$

All numerical results are obtained using the trapezoidal transcription method. An acceptable accuracy of the clearance criteria is chosen as $\overline{\epsilon} = 0.001$ for all criteria, which equates to $\overline{\epsilon} \cdot 180/\pi$ deg ≈ 0.0573 deg for the aerodynamic angles α_A and β_A. This accuracy is used to impose penalties on all controls according to Eq. (4.10). Moreover, the final time is set to $t_f = 8$ s, which is slightly higher compared to the results for the nominal solutions presented in Sec. 5.3, where a final time of $t_f = 6$ s is used.

The step length for all analyses is set at $h = 0.025$ s. In this context, it should be noted that the bilevel search at the upper level focuses exclusively on minimizing the optimal value of the cost returned from the lower level. Unlike the optimal control and state histories, this value is observed to converge rapidly and yields values within acceptable limits, even with a coarse discretization.

Example 5.5.1: Convergence of the Optimal Cost

Convergence studies regarding the optimal values (Cat II, nominal) of $\alpha_A(8\,\text{s})$ and $n_z(8\,\text{s})$ for the longitudinal model and $\beta_A(8\,\text{s})$ and $n_y(8\,\text{s})$ for the lateral model with the discretization step length h ranging from 0.005 s to 0.25 s are shown in Figs. 5.14–5.17 (top). The time required per LP solution (t_{LP}) is normalized to the maximum time ($t_{LP,max}$) of each study and provided in the corresponding figures (bottom). Moreover, the tolerance intervals are represented relative to the solution with the smallest discretization step length $h = 0.005$ s (green boxes).

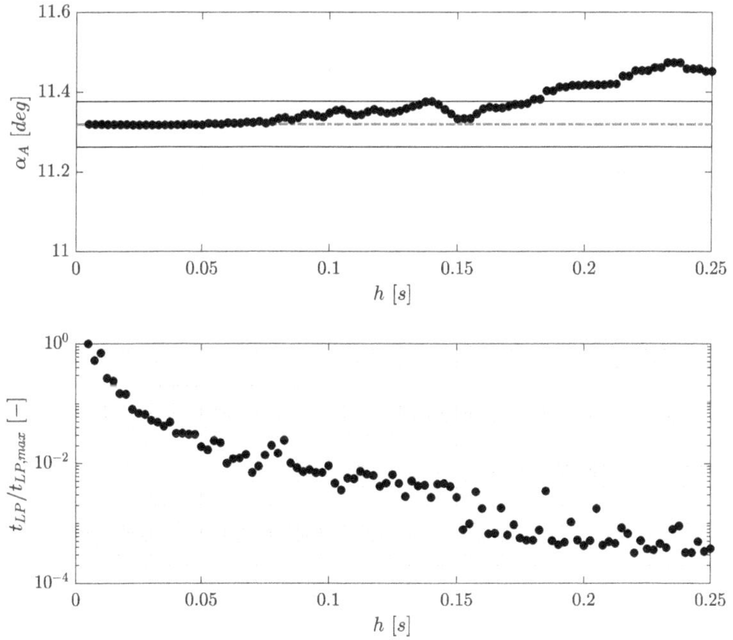

Fig. 5.14 Maximization of $\alpha_A(8\,\text{s})$.

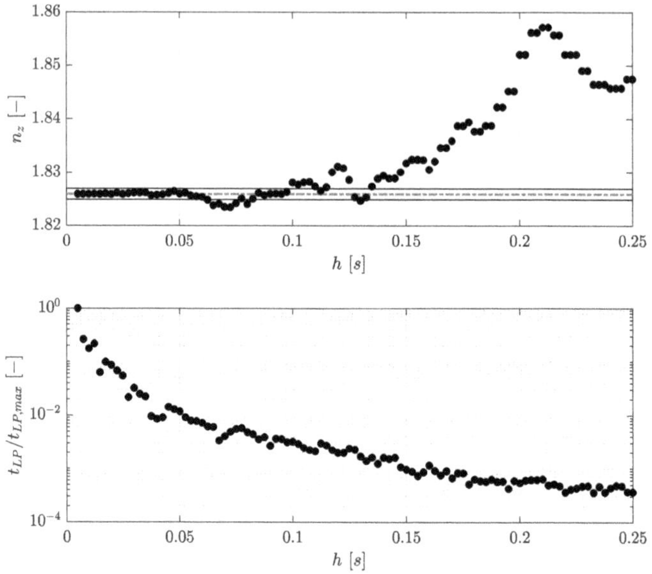

Fig. 5.15 Maximization of n_z (8 s).

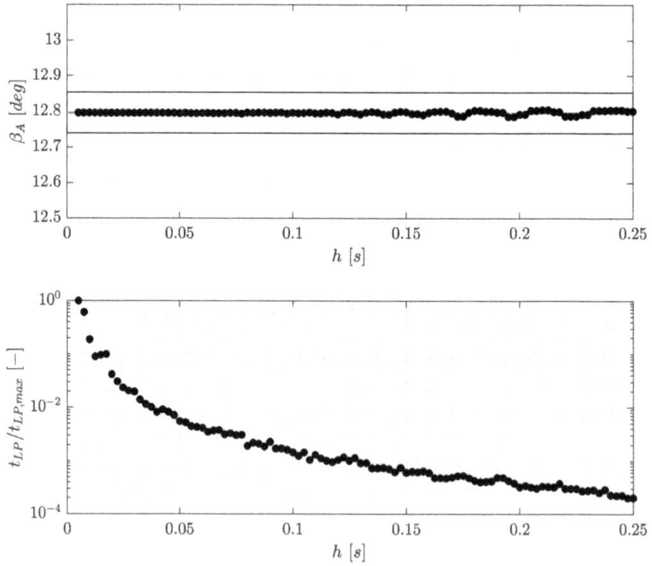

Fig. 5.16 Maximization of β_A (8 s).

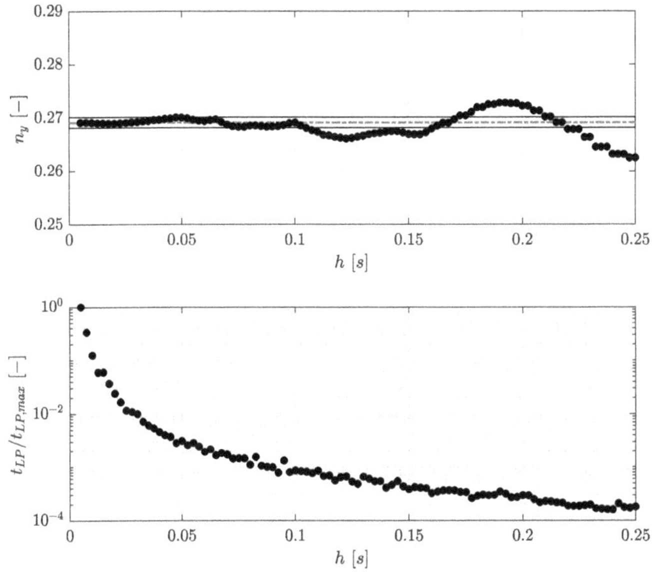

Fig. 5.17 Maximization of n_y(8 s).

Observe that the values of all criteria in the convergence studies exhibit values within the tolerance intervals (green boxes) for relatively coarse discretization step lengths. Additionally, for smaller step sizes ($h \lesssim 0.05$ s), the solutions of the LPs appear to require a rapidly increasing amount of computational work with little effect on the worst-case values. The convergence studies indicate that sufficiently accurate approximations of the optimal cost for all criteria below the acceptable accuracy can be expected even for coarse discretization step sizes.

Engineering Note 5.5.1: Efficient Linear Programming Solutions

Several of the clearance methods presented in this book rely on solving a large number of LPs. In particular, observe that if two parametric problems with the same discretization differ only in their specific parameter values, the discretized problem's structure remains unaffected. In such cases, the basis matrix produced from solving an LP can be used to warm-start other LPs.

Moreover, some solvers also offer an option to solve a single problem using several methods simultaneously, such as primal and dual Simplex as well as interior point methods. CPLEX's "concurrent" method is an example of such a strategy.

Numerical experience suggests that employing these two acceleration strategies can substantially reduce the required solution time.

The bilevel optimization of the parameter-dependent longitudinal and lateral Cat II models is illustrated in the following using two different types of limits regarding the admissible parameter region. The first type allows all parameters to vary between 10% and −10% (i.e., simple box-bounds are used).

The optimization algorithms for the upper-level problem employed in the following examples are as follows:

- **Gradient-based (Jacobian)**: The sequential quadratic programming (SQP) solver implemented in the MATLAB® function `fmincon` uses a quasi-Newton Hessian approximation and is able to handle nonlinear inequality constraints as well as linear equality and inequality constraints.
- **Gradient-based (Jacobian and Hessian)**: The trust-region-reflective (TR) solver implemented in the MATLAB® function `fmincon` uses the exact Hessian and can handle box-bounds as well as linear equality constraints.
- **Global (Stochastic)**: The surrogate optimization (SR) solver implemented in the MATLAB® function `surrogateopt`* represents a global optimization solver and allows box-bounds as well as nonlinear inequality constraints.
- **Global (Deterministic)**: The DIRECT algorithm (DR, see Jones et al. [63] and implementation see Bjorkman and Holmström [64]) divides the search space with a strategy balancing local and global search subject to box-bounds. This algorithm "*is guaranteed to converge to the globally optimal function value if the objective function is continuous—or at least continuous in the neighborhood of a global optimum*" [63].

Example 5.5.2: Combined Analysis (Box Constraints)

The results for the longitudinal Cat II model with box-bounded parameters are summarized in Table 5.1 and Table 5.2. Similarly, Table 5.3 and Table 5.4 show the respective results for the lateral Cat II model. Recall that the parameters represent multiplicative scaling factors centered at zero. Thus, the results represent changes in the absolute value of the corresponding physical quantities.

The solutions are obtained using the local gradient-based solvers (SQP, TR) as well as the global solvers (SG, DR). The optimality tolerance for the SQP and TR solvers is set to 10^{-4}. Regarding the global solvers, it is found that the SG solver converges quickly for the clearance problems under investigation. Therefore, the maximum number of function evaluations is set to 50 as a termination criterion. The same number is set for the DR solver for the maximum number of iterations.

*Introduced in MATLAB® R2018B.

TABLE 5.1 MAXIMIZATION OF $n_z(8\ \mathrm{s})$ (BOX-BOUNDED PARAMETERS)

Solver	$\hat{n}_z(8\ \mathrm{s})$	n_{LP}	$\Delta\hat{M}_\alpha$	$\Delta\hat{M}_q$	$\Delta\hat{M}_\eta$	$\Delta\hat{m}$	$\Delta\hat{I}_{yy}$	$\Delta\hat{x}^{RG}$
SQP	2.222	3	−0.100	−0.100	0.100	−0.100	0.100	−0.100
TR	2.222	31	−0.100	−0.100	0.100	−0.100	0.100	−0.100
SG	2.222	50	−0.100	−0.100	0.100	−0.100	0.100	−0.100
DR	2.222	1041	−0.100	−0.100	0.100	−0.100	0.100	−0.100

TABLE 5.2 MAXIMIZATION OF $\alpha_A(8\ \mathrm{s})$ (BOX-BOUNDED PARAMETERS)

Solver	$\hat{\alpha}_A(8\ \mathrm{s})$	n_{LP}	$\Delta\hat{M}_\alpha$	$\Delta\hat{M}_q$	$\Delta\hat{M}_\eta$	$\Delta\hat{m}$	$\Delta\hat{I}_{yy}$	$\Delta\hat{x}^{RG}$
SQP	14.166 deg	7	−0.100	−0.100	0.100	0.100	0.100	−0.100
TR	14.165 deg	6	−0.100	−0.100	0.100	0.100	0.100	−0.100
SG	14.164 deg	50	−0.100	−0.099	0.100	0.100	0.100	−0.100
DR	14.147 deg	1473	−0.100	−0.099	0.100	0.100	0.100	−0.099

TABLE 5.3 MAXIMIZATION OF $n_y(8\ \mathrm{s})$ (BOX-BOUNDED PARAMETERS)

Solver	$\hat{n}_y(8\ \mathrm{s})$	n_{LP}	$\Delta\hat{L}_\beta$	$\Delta\hat{L}_p$	$\Delta\hat{N}_\beta$	$\Delta\hat{N}_\zeta$	$\Delta\hat{I}_{xx}$	$\Delta\hat{I}_{zz}$
SQP	0.323	11	−0.100	−0.100	−0.100	0.100	0.100	0.100
TR	0.323	6	−0.100	−0.100	−0.100	0.100	0.100	0.100
SG	0.323	50	−0.100	−0.100	−0.100	0.100	0.100	0.100
DR	0.323	1115	−0.100	−0.100	−0.100	0.100	0.100	0.100

TABLE 5.4 MAXIMIZATION OF $\beta_A(8\ \mathrm{s})$ (BOX-BOUNDED PARAMETERS)

Solver	$\hat{\beta}_A(8\ \mathrm{s})$	n_{LP}	$\Delta\hat{L}_\beta$	$\Delta\hat{L}_p$	$\Delta\hat{N}_\beta$	$\Delta\hat{N}_\zeta$	$\Delta\hat{I}_{xx}$	$\Delta\hat{I}_{zz}$
SQP	16.325 deg	13	−0.100	0.007	−0.100	0.100	0.100	0.100
TR	16.332 deg	7	−0.100	−0.100	−0.100	0.100	0.100	0.100
SG	16.343 deg	50	−0.091	0.100	−0.100	0.100	0.100	0.100
DR	16.326 deg	1053	−0.100	−0.100	−0.100	0.100	0.100	0.100

First, let us observe that the worst-case values of the criteria for the optimal parameter combinations presented in Tables 5.1–5.4 are considerably higher than the worst-case values for the nominal parameter values: $\hat{n}_{z,\mathrm{nom}}(8\ \mathrm{s}) = 1.826$, $\hat{\alpha}_{A,\mathrm{nom}}(8\ \mathrm{s}) = 11.317, \mathrm{deg}$, $\hat{n}_{y,\mathrm{nom}}(8\ \mathrm{s}) = 0.269$, and $\hat{\beta}_{A,\mathrm{nom}}(8\ \mathrm{s}) = 12.796, \mathrm{deg}$.

Moreover, the solvers appear to yield consistent results regarding not only the optimal values of the criteria but also the worst-case parameter combinations. Interestingly, the worst-case values of the parameters are almost exclusively found at one of the bounds. In some cases, intermediate values are observed (see, for example, $\Delta\hat{L}_p$ in Table 5.4). However, for these cases, the values of the criterion seem to be of comparable magnitude. This can be seen as an indication that the criterion only weakly depends on these parameters, and the solvers converge to one particular point based on the respective termination criteria.

The gradient-based schemes appear to work very efficiently and converge within a few iterations. This is true for both the SQP solver with quasi-Newton Hessian update and the TR solver with the exact Hessian. For the maximization of $\alpha_A(8\ \mathrm{s})$, these solvers require the solution of a comparable number of LPs until convergence. For the maximization of $n_z(8\ \mathrm{s})$, the SQP method requires fewer solutions than the TR method, while for the maximization of $\beta_A(8\ \mathrm{s})$ and $n_y(8\ \mathrm{s})$, the TR method appears to be more efficient.

Note that for the longitudinal Cat II model, the pitch stiffness and pitch damping are decreased, and the center of gravity is moved toward the neutral point. These changes in the parameter values decrease the aircraft's resistance to stick commands, which seems a reasonable result from a flight mechanical perspective. Moreover, the elevator's effectiveness is increased, which essentially augments the effect of the worst-case control actions. Also, the moment of inertia about the body's y-axis is increased, whereas the mass is at the upper bound for the maximization of $\alpha_A(8\ \mathrm{s})$ and at the lower bound for the maximization of $n_z(8\ \mathrm{s})$. In the lateral cases [i.e., the maximization of $\beta_A(8\ \mathrm{s})$ and $n_y(8\ \mathrm{s})$], the weathercock and dihedral stability are lowered in absolute values, and the moments of inertia for both axes (roll and yaw) are increased. As mentioned earlier, the roll damping $\Delta\hat{L}_p$ appears to have little influence on the worst-case value for the maximization of $\beta_A(8\ \mathrm{s})$.

Note that the upper-level problem (5.12) only allows for box-bounded parameters. This form was chosen for notational simplicity as it represents a standard formulation in most practical applications. However, it should be mentioned that there is no inherent restriction in the bilevel setup that would prevent us from considering general inequality constraints to define the admissible parameter region.

The second example illustrates the use of a polytopal-shaped region, where a subset of uncertain parameters (aerodynamic coefficients) is constrained to a region that contains at least a $1 - \alpha$ portion of the probability mass. The motivation behind this type of constraint is as follows: If no worst-case value in the admissible region exceeds a certain critical limit and the solution is global, then the exceedance probability of this criterion is at most α.

Example 5.5.3: Combined Analysis (Polytopal Constraints)

For the second type of parameter region, the parameters Δx^{RG}, Δm, ΔI_{xx}, ΔI_{yy}, and ΔI_{zz} are allowed to vary within $\pm 10\%$, as in the box-bounded case. However, all aerodynamic parameters, i.e., ΔM_α, ΔM_q, and ΔM_η for the longitudinal model, and ΔL_β, ΔL_p, ΔN_β, and ΔN_ζ for the lateral model, are now subject to independent univariate normal distributions with a zero mean and a standard deviation of 0.05. For these distributions, the admissible parameter space is defined as the region containing a probability mass of $1-\alpha$. For a zero mean, this region can be defined, for example, as an ellipsoidal-shaped region of the form [65]

$$\underbrace{\mathbf{p}^T \boldsymbol{\Sigma}^{-1} \mathbf{p} - \chi^2_{n_p,1-\alpha}}_{g_c(\mathbf{p};\alpha)} \leqslant 0$$

with the diagonal positive definite covariance matrix $\boldsymbol{\Sigma} \in \mathbb{R}^{n_p \times n_p}$. Moreover, $\chi^2_{n_p,1-\alpha}$ represents the $(1-\alpha)$-quantile of the χ^2 distribution with n_p degrees of freedom.

To provide a general approach that can be easily applied to other distribution types, we use an overapproximation of the region. This approximation is defined as a convex polytope of the form

$$\mathbf{A}_c \mathbf{p} \leqslant \mathbf{b}_c$$

For the longitudinal plane, we construct a three-dimensional polytope based on the tangent planes at 14 support points on the surface defined by the level-set $g_c(\mathbf{p};\alpha) = 0$, where $\alpha = 10^{-3}$. Please refer to Fig. 5.18 for a visualization of the polytope.

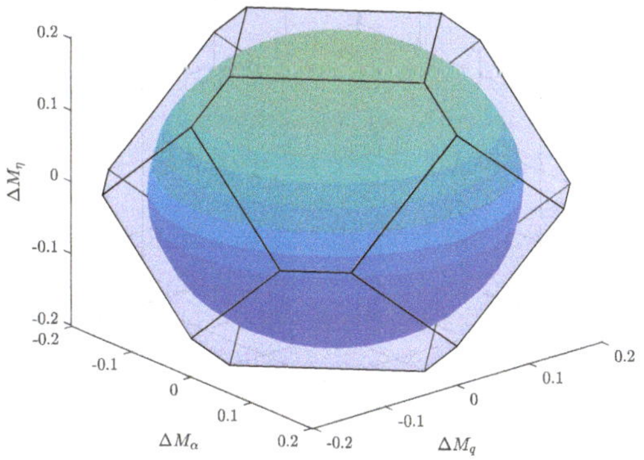

Fig. 5.18 Admissible parameter region for ΔM_α, ΔM_q, and ΔM_η (longitudinal model).

Similarly, for the lateral plane, we construct a four-dimensional polytope using 24 support points with the same level-set definition and value for α. These conservative regions contain approximately a probability mass of $1 - \alpha \approx 1 - 3.67 \cdot 10^{-4}$ for the longitudinal plane and $1 - \alpha \approx 1 - 2.01 \cdot 10^{-4}$ for the lateral plane. Both figures are estimated based on 10^7 random samples.

The combined search under the bilevel approach in the control space and these polytopal-shaped parameter regions are illustrated using the SQP method (optimality tolerance 10^{-4}). The results are summarized in Table 5.5 for the longitudinal and Table 5.6 for the lateral Cat II model.

TABLE 5.5 MAXIMIZATION OF α_A(8 s) AND n_z(8 s) (POLYTOPAL-SHAPED CONSTRAINTS)

Criterion	Optimal Value	n_{LP}	$\Delta\hat{M}_\alpha$	$\Delta\hat{M}_q$	$\Delta\hat{M}_\eta$	$\Delta\hat{m}$	$\Delta\hat{I}_{yy}$	$\Delta\hat{x}^{RG}$
$\alpha_A(8\,s)$	14.867 deg	15	−0.202	0.000	0.148	0.100	0.100	−0.100
$n_z(8\,s)$	2.402	7	−0.202	−0.148	0.000	−0.100	0.100	−0.100

TABLE 5.6 MAXIMIZATION OF β_A(8 s) AND n_y(8 s) (POLYTOPAL-SHAPED CONSTRAINTS)

Criterion	Optimal Value	n_{LP}	$\Delta\hat{L}_\beta$	$\Delta\hat{L}_p$	$\Delta\hat{N}_\beta$	$\Delta\hat{N}_\zeta$	$\Delta\hat{I}_{xx}$	$\Delta\hat{I}_{zz}$
$\beta_A(8\,s)$	20.949 deg	11	0.000	−0.000	−0.215	0.215	0.100	0.100
$n_y(8\,s)$	0.393	35	0.000	0.000	−0.215	0.215	−0.100	0.100

Note that, as with the box-bounded case, the solver only requires the evaluation of a few lower-level problems to converge for this type of parameter region. It is observed that the criteria exhibit higher (absolute) values compared to the box-bounded case. This can be attributed to the fact that the solutions are found at the vertices of the polytopal-shaped parameter regions and that the vertices for the particular parameter distributions and confidence level used in this example allow for higher (absolute) parameter values compared to those in the box-bounded case.

The numerical experiments from this section indicate that the bilevel decomposition approach is suitable for solving combined worst-case analyses for parameter-dependent clearance problems. In particular, gradient-based methods appear to not only be efficient as convergence is achieved within a few steps, but they also attain essentially the same optimal values as global optimization algorithms. Moreover, an important advantage of these methods is that they allow checking optimality using the KKT conditions (see Sec. 1.1) in the termination criteria. Conversely, defining meaningful termination criteria for global optimization algorithms is generally more challenging.

We conclude with the remark that, in a practical setting, a simple yet expectedly effective method for the bilevel approach is to start the local gradient-based solvers from several randomized initial points to increase the chance of identifying the global worst case.

CHAPTER 6

Nonlinear Clearance

In this chapter, we discuss two optimal control-based approaches for analyzing nonlinear systems. The first approach relies on the application of direct optimal control methods for parametric nonlinear models. This approach allows us to perform a combined search in both the parameter and control space for Cat III problems. The transcription process to a nonlinear program (NLP) and the computation of postoptimal sensitivities concerning the optimal cost with respect to fixed parameters are discussed.

We would like to emphasize that the direct optimal control approach to flight control law clearance only produces globally optimal (worst-case) solutions in practice if the numerical optimization method can solve the discretized problem to global optimality. For instance, this is the case for linear models (Cat I and II, cf. Chapter 4), where the direct optimal control approach leads to linear programs. Efficient algorithms exist in this case guaranteeing global solutions. However, in general, the current state of the art for NLP methods does not provide such a guarantee. In this context, it is noteworthy that there are other strategies that in fact converge toward globally optimal solutions for nonlinear optimal control problems. For example, dynamic programming (DP) can provide such a guarantee. Unfortunately, for nonlinear systems, DP requires a discretization of the problem not only in time but also in the state space. This leads to the so-called *curse of dimensionality*, where the number of grid points grows rapidly in the number of state dimensions. For instance, the discretization of a problem featuring six states on a Cartesian grid, using one hundred grid points for each dimension, requires a trillion state discretization points. In practice, aircraft closed-loop models typically feature several tens or sometimes more than a hundred states. Thus, the computational burden resulting from the curse of dimensionality is usually prohibitive in the case of full aircraft closed-loop models; for practical reasons, we thus prefer to focus our attention on direct optimal control methods.

The second approach presented in this chapter is referred to as the "hybrid" method. The idea behind this approach is to combine optimal control analyses based on reduced models. In particular, the solution obtained from a reduced model is used to define a suitable parameterization as well as a corresponding initial guess for the control function. Specifically, Cat I and II models are employed as reduced models, and the control parameterization is based on the switching structure of the optimal control. This parameterization and initial guess are then refined for the nonlinear simulation model using a black-box (single shooting) optimization method [5, 66].

This chapter is organized as follows: First, in Sec. 6.1, we present the direct optimal control approach for nonlinear-parametric models with control and state constraints. The transcription to an NLP is illustrated using the same direct optimal control methods as in the linear case (backward Euler, trapezoidal collocation). We then compare the results obtained from Cat II and Cat III models (lateral, longitudinal) in numerical examples regarding the nominal solution, postoptimal sensitivities as well as the combined search in the parameter and control spaces. Finally, in Sec. 6.2, we describe the hybrid method for Cat III–type problems.

6.1 DIRECT APPROACH

Clearance Optimal Control Analysis (Cat III Direct)

The transcription of the nonlinear-parametric worst-case input problem (3.1) based on direct methods yields an NLP of the form

$$\begin{aligned} \underset{\mathbf{z}\in\mathbb{R}^n}{\text{minimize}} \quad & J(\mathbf{z}) \\ \text{subject to} \quad & \mathbf{a}(\mathbf{z}) = \mathbf{0} \\ & \underline{\mathbf{z}} \leqslant \mathbf{z} \leqslant \overline{\mathbf{z}} \end{aligned}$$

with the nonlinear equality constraint function $\mathbf{a} \in \mathbb{R}^n \to \mathbb{R}^{n_a}$ and the decision variable vector $\mathbf{z} \in \mathbb{R}^n$ bounded by $\underline{\mathbf{z}}, \overline{\mathbf{z}} \in \mathbb{R}^n$.

Similar to the linear case, the nonlinear clearance problem in continuous time is transcribed to a finite-dimensional optimization problem. In this section, we illustrate the transcription process for the backward Euler and trapezoidal collocation method using a mesh with N segments [discretization step size $h = (t_f - t_0)/N$]. Moreover, Appendix H provides a description of the Hermite–Simpson transcription method.

For Cat III models, the discretized problem results in a high-dimensional NLP. However, the sparse structures emerging from segmented collocation methods enable an efficient solution using numerical solvers (e.g., IPOPT [24]) specifically tailored to this type of problem.

BACKWARD EULER TRANSCRIPTION

In the case of the backward Euler method, we introduce the parameter variable vector $\mathbf{p} \in \mathbb{R}^{n_p}$; the state variable vectors $\mathbf{x}_i \in \mathbb{R}^{n_x}$, $i = 0, \ldots, N$; and the control variable vectors $\mathbf{u}_i \in \mathbb{R}^{n_u}$, $i = 1, \ldots, N$. The time-discretized dynamic equality constraints are expressed as

$$\mathbf{x}_i = \mathbf{x}_{i-1} + h\mathbf{f}\left(\mathbf{x}_i, \mathbf{u}_i, \mathbf{p}\right), \; i = 1, \ldots, N$$

and the discretized Cat III problem (3.1) can be written as

$$\begin{aligned}
\underset{\mathbf{x}_0, \mathbf{x}_i, \mathbf{u}_i, \mathbf{p}}{\text{minimize}} \quad & J\left(\mathbf{x}_N, \mathbf{p}\right) \\
\text{subject to} \quad & \mathbf{x}_i = \mathbf{x}_{i-1} + h\mathbf{f}\left(\mathbf{x}_i, \mathbf{u}_i, \mathbf{p}\right) \\
& \boldsymbol{\gamma}\left(\mathbf{x}_0, \mathbf{p}\right) = \mathbf{0} \\
& \underline{\mathbf{p}} \leqslant \mathbf{p} \leqslant \overline{\mathbf{p}} \\
& \underline{\mathbf{x}} \leqslant \mathbf{x}_i \leqslant \overline{\mathbf{x}} \\
& \underline{\mathbf{u}} \leqslant \mathbf{u}_i \leqslant \overline{\mathbf{u}} \\
& i = 1, \ldots, N
\end{aligned}$$

This NLP features $n = N(n_x + n_u) + n_x + n_p$ decision variables and $n_h = Nn_x + n_\gamma$ equality constraints. The decision variable vector $\mathbf{z} \in \mathbb{R}^n$ and the corresponding lower bound $\underline{\mathbf{z}} \in \mathbb{R}^n$ as well as upper bound $\overline{\mathbf{z}} \in \mathbb{R}^n$ are

$$\underline{\mathbf{z}} = \begin{bmatrix} \underline{\mathbf{p}} \\ -\infty \\ \underline{\mathbf{x}} \\ \underline{\mathbf{u}} \\ \vdots \\ \underline{\mathbf{x}} \\ \underline{\mathbf{u}} \end{bmatrix}, \qquad \mathbf{z} = \begin{bmatrix} \mathbf{p} \\ \mathbf{x}_0 \\ \mathbf{x}_1 \\ \mathbf{u}_1 \\ \vdots \\ \mathbf{x}_N \\ \mathbf{u}_N \end{bmatrix}, \qquad \overline{\mathbf{z}} = \begin{bmatrix} \overline{\mathbf{p}} \\ \infty \\ \overline{\mathbf{x}} \\ \overline{\mathbf{u}} \\ \vdots \\ \overline{\mathbf{x}} \\ \overline{\mathbf{u}} \end{bmatrix}$$

The equality constraints can be collected in $\mathbf{a} : \mathbb{R}^n \to \mathbb{R}^{n_a}$:

$$\mathbf{a}\left(\mathbf{z}\right) = \begin{bmatrix} \boldsymbol{\gamma}\left(\mathbf{x}_0, \mathbf{p}\right) \\ \mathbf{x}_1 - \mathbf{x}_0 - h\mathbf{f}\left(\mathbf{x}_1, \mathbf{u}_1, \mathbf{p}\right) \\ \vdots \\ \mathbf{x}_N - \mathbf{x}_{N-1} - h\mathbf{f}\left(\mathbf{x}_N, \mathbf{u}_N, \mathbf{p}\right) \end{bmatrix}$$

Moreover, the constraint Jacobian $\mathbf{A} : \mathbb{R}^n \to \mathbb{R}^{n_a \times n}$ is expressed as

$$\mathbf{A}(\mathbf{z}) = \begin{bmatrix} \frac{\partial \boldsymbol{\gamma}}{\partial \mathbf{p}} & \frac{\partial \boldsymbol{\gamma}}{\partial \mathbf{x}_0} & \mathbf{0} & \mathbf{0} & \mathbf{0} & \mathbf{0} & \mathbf{0} & \mathbf{0} & \cdots \\ -h\frac{\partial \mathbf{f}_1}{\partial \mathbf{p}} & -\mathbf{I} & \mathbf{I} - h\frac{\partial \mathbf{f}_1}{\partial \mathbf{x}_1} & -h\frac{\partial \mathbf{f}_1}{\partial \mathbf{u}_1} & \mathbf{0} & \mathbf{0} & \mathbf{0} & \mathbf{0} & \cdots \\ -h\frac{\partial \mathbf{f}_2}{\partial \mathbf{p}} & \mathbf{0} & -\mathbf{I} & \mathbf{0} & \mathbf{I} - h\frac{\partial \mathbf{f}_2}{\partial \mathbf{x}_2} & -h\frac{\partial \mathbf{f}_2}{\partial \mathbf{u}_2} & \mathbf{0} & \mathbf{0} & \cdots \\ -h\frac{\partial \mathbf{f}_3}{\partial \mathbf{p}} & \mathbf{0} & \mathbf{0} & \mathbf{0} & -\mathbf{I} & \mathbf{0} & \mathbf{I} - h\frac{\partial \mathbf{f}_3}{\partial \mathbf{x}_3} & -h\frac{\partial \mathbf{f}_3}{\partial \mathbf{u}_3} & \\ \vdots & \vdots & \vdots & \vdots & & & \ddots & & \end{bmatrix}$$

using the short notation $\mathbf{f}_i = \mathbf{f}(\mathbf{x}_i, \mathbf{u}_i, \mathbf{p})$.

May be skipped (↴) at first read.

TRAPEZOIDAL TRANSCRIPTION

In the case of the trapezoidal method, the parameter variable vector $\mathbf{p} \in \mathbb{R}^{n_p}$, state variable vectors $\mathbf{x}_i \in \mathbb{R}^{n_x}, i = 0, \ldots, N$, and control variable vectors $\mathbf{u}_i \in \mathbb{R}^{n_u}, i = 0, \ldots, N$ are introduced. The time-discretized dynamic equality constraints are expressed as

$$\mathbf{x}_i = \mathbf{x}_{i-1} + h\left(\frac{1}{2}\mathbf{f}(\mathbf{x}_{i-1}, \mathbf{u}_{i-1}, \mathbf{p}) + \frac{1}{2}\mathbf{f}(\mathbf{x}_i, \mathbf{u}_i, \mathbf{p})\right), \quad i = 1, \ldots, N$$

and the discretized Cat III problem (3.1) can be written as

$$\begin{aligned} \underset{\mathbf{x}_0, \mathbf{x}_i, \mathbf{u}_0, \mathbf{u}_i, \mathbf{p}}{\text{minimize}} \quad & J(\mathbf{x}_N, \mathbf{p}) \\ \text{subject to} \quad & \mathbf{x}_i = \mathbf{x}_{i-1} + h\left(\frac{1}{2}\mathbf{f}(\mathbf{x}_{i-1}, \mathbf{u}_{i-1}, \mathbf{p}) + \frac{1}{2}\mathbf{f}(\mathbf{x}_i, \mathbf{u}_i, \mathbf{p})\right) \\ & \boldsymbol{\gamma}(\mathbf{x}_0, \mathbf{p}) = \mathbf{0} \\ & \underline{\mathbf{p}} \leqslant \mathbf{p} \leqslant \overline{\mathbf{p}} \\ & \underline{\mathbf{x}} \leqslant \mathbf{x}_i \leqslant \overline{\mathbf{x}} \\ & \underline{\mathbf{u}} \leqslant \mathbf{u}_i \leqslant \overline{\mathbf{u}} \\ & \underline{\mathbf{u}} \leqslant \mathbf{u}_0 \leqslant \overline{\mathbf{u}} \\ & i = 1, \ldots, N \end{aligned}$$

This NLP features $n = (N+1)(n_x + n_u) + n_p$ decision variables and $n_h = Nn_x + n_\gamma$ equality constraints. The decision variable vector $\mathbf{z} \in \mathbb{R}^n$ and the corresponding lower bound $\underline{\mathbf{z}} \in \mathbb{R}^n$ as well as upper bound $\overline{\mathbf{z}} \in \mathbb{R}^n$ are

$$
\underline{\mathbf{z}} = \begin{bmatrix} \underline{\mathbf{p}} \\ -\infty \\ \underline{\mathbf{u}} \\ \underline{\mathbf{x}} \\ \underline{\mathbf{u}} \\ \vdots \\ \underline{\mathbf{x}} \\ \underline{\mathbf{u}} \end{bmatrix}, \qquad \mathbf{z} = \begin{bmatrix} \mathbf{p} \\ \mathbf{x}_0 \\ \mathbf{u}_0 \\ \mathbf{x}_1 \\ \mathbf{u}_1 \\ \vdots \\ \mathbf{x}_N \\ \mathbf{u}_N \end{bmatrix}, \qquad \overline{\mathbf{z}} = \begin{bmatrix} \overline{\mathbf{p}} \\ \infty \\ \overline{\mathbf{u}} \\ \overline{\mathbf{x}} \\ \overline{\mathbf{u}} \\ \vdots \\ \overline{\mathbf{x}} \\ \overline{\mathbf{u}} \end{bmatrix}
$$

The equality constraints can be collected in $\mathbf{a} : \mathbb{R}^n \rightarrow \mathbb{R}^{n_a}$:

$$
\mathbf{a}(\mathbf{z}) = \begin{bmatrix} \gamma(\mathbf{x}_0, \mathbf{p}) \\ \mathbf{x}_1 - \mathbf{x}_0 - h\left[\frac{1}{2}\mathbf{f}(\mathbf{x}_0, \mathbf{u}_0, \mathbf{p}) + \frac{1}{2}\mathbf{f}(\mathbf{x}_1, \mathbf{u}_1, \mathbf{p})\right] \\ \vdots \\ \mathbf{x}_N - \mathbf{x}_{N-1} - h\left[\frac{1}{2}\mathbf{f}(\mathbf{x}_{N-1}, \mathbf{u}_{N-1}, \mathbf{p}) + \frac{1}{2}\mathbf{f}(\mathbf{x}_N, \mathbf{u}_N, \mathbf{p})\right] \end{bmatrix}
$$

Moreover, the constraint Jacobian $\mathbf{A} : \mathbb{R}^n \rightarrow \mathbb{R}^{n_a \times n}$ is expressed as

$$
\mathbf{A}(\mathbf{z}) = \begin{bmatrix} \frac{\partial \gamma}{\partial \mathbf{p}} & \frac{\partial \gamma}{\partial \mathbf{x}_0} & \mathbf{0} & \mathbf{0} & \mathbf{0} & \mathbf{0} & \mathbf{0} & \cdots \\ -\frac{h}{2}\frac{\partial \mathbf{f}_0}{\partial \mathbf{p}} - \frac{h}{2}\frac{\partial \mathbf{f}_1}{\partial \mathbf{p}} & -\mathbf{I} - \frac{h}{2}\frac{\partial \mathbf{f}_0}{\partial \mathbf{x}_0} & -\frac{h}{2}\frac{\partial \mathbf{f}_0}{\partial \mathbf{u}_0} & \mathbf{I} - \frac{h}{2}\frac{\partial \mathbf{f}_1}{\partial \mathbf{x}_1} & -\frac{h}{2}\frac{\partial \mathbf{f}_1}{\partial \mathbf{u}_1} & \mathbf{0} & \mathbf{0} & \cdots \\ -\frac{h}{2}\frac{\partial \mathbf{f}_1}{\partial \mathbf{p}} - \frac{h}{2}\frac{\partial \mathbf{f}_2}{\partial \mathbf{p}} & \mathbf{0} & \mathbf{0} & -\mathbf{I} - \frac{h}{2}\frac{\partial \mathbf{f}_1}{\partial \mathbf{x}_1} & -\frac{h}{2}\frac{\partial \mathbf{f}_1}{\partial \mathbf{u}_1} & \mathbf{I} - \frac{h}{2}\frac{\partial \mathbf{f}_2}{\partial \mathbf{x}_2} & -\frac{h}{2}\frac{\partial \mathbf{f}_2}{\partial \mathbf{u}_2} & \\ \vdots & \vdots & \vdots & & & \ddots & & \end{bmatrix}
$$

using the short notation $\mathbf{f}_i = \mathbf{f}(\mathbf{x}_i, \mathbf{u}_i, \mathbf{p})$.

As for linear Cat I and II types of problems, it can be desirable to obtain postoptimal cost sensitivities regarding fixed parameters. This can be achieved by embedding these parameters as free variables in the problem along with the introduction of equality constraints fixing these parameters to their respective values. The sensitivities of the optimal cost $\hat{J}$ with respect to a subset of n_e parameters $p_i, i \in \mathcal{E}$ fixed to $\overline{\underline{p}}_i$ can then be obtained from the Lagrange multipliers associated with the parameter embedding constraints:

$$p_i = \overline{\underline{p}}_i,\ i \in \mathcal{E}$$

These embedding constraints can be imposed in Eq. (3.1) by setting the lower and upper bounds of the respective elements in the parameter vector to the same value, that is, $\underline{p}_i, \overline{p}_i = \overline{\underline{p}}_i, i \in \mathcal{E}$ in

$$\underline{p}_i \leqslant p_i \leqslant \overline{p}_i,\ i \in \mathcal{E} \tag{6.1}$$

In this case, the terms of the Lagrangian $\mathcal{L}$ depending on $\overline{\underline{p}}_i, i \in \mathcal{E}$ from Eq. (6.1) are

$$\mathcal{L} = \ldots + \sum_{i\in\mathcal{E}} \underline{\mu}_i \left(\overline{\underline{p}}_i - p_i\right) + \sum_{i\in\mathcal{E}} \overline{\mu}_i \left(p_i - \overline{\underline{p}}_i\right)$$

where $\underline{\mu}_i \in \mathbb{R}$ and $\overline{\mu}_i \in \mathbb{R}$ are the multipliers associated with the lower bound and upper bound. Based on Eq. (5.5), the postoptimal sensitivities are thus directly obtained as

$$\frac{d\hat{J}}{dp_i} = \frac{\partial \mathcal{L}}{\partial \overline{\underline{p}}_i} = \underline{\mu}_i - \overline{\mu}_i,\ i \in \mathcal{E}$$

Engineering Note 6.1.1: IPOPT Postoptimal Cost Sensitivities

In addition to the optimal decision variables, IPOPT also returns Lagrange multipliers, including the multiplier vectors `z_L` and `z_U` for the lower and upper bounds of the decision variables.

To efficiently implement the parameter embedding constraints in IPOPT, one can set the lower and upper bounds of the respective decision variables to the same value. However, older versions of IPOPT (before version 3.14.0) detect that the decision variables are fixed, remove them from the problem by default, and compute no multipliers for these variables. This default behavior can be changed by altering the IPOPT setting `fixed_variable_treatment` to `make_constraint`. In more recent versions of IPOPT (starting with version 3.14.0), bound multipliers are computed by default and for all settings except `make_parameter_nodual` in `fixed_variable_treatment`.

A rather remarkable observation found from numerical solutions under the Cat III direct approach is that in several instances the solutions obtained from using a nonlinear model appear to be similar to the corresponding solutions obtained from using linearized models. Specifically, even though linearized models clearly constitute relatively simple approximations of the nonlinear dynamics, the optimal solutions for typical clearance criteria (e.g., load factors or aerodynamic angles) often tend to behave *qualitatively* similar for both model types. This means that despite the fact that the optimal values do not match quantitatively, at least the type of expected worst-case inputs from the optimal control analysis for linear systems (bang-bang/boundary controls) as well as the arc sequences defining the optimal solution structure are often found to be comparable. This observation is illustrated in the following by comparing the nominal Cat II results from Sec. 5.3 with Cat III results for both the maximization of the normal load factor in the longitudinal plane and the aerodynamic angle of sideslip in the lateral plane.

The nonlinear model for the Cat III direct approach used in the following analysis features 82 states and includes a nonlinear plant model with aeroelastic effects, sensor models, the gain scheduled controller, and the dynamic model for the servomechanisms of all primary control surfaces including rate and position limits. Furthermore, six control inputs are used to test the system, including the normal load factor command, the lateral load factor command, the roll angle command, and the wind commands for the longitudinal, normal, and lateral wind gusts. For the nonlinear closed-loop model, a reduced simulation model is used (the full model is used under the Cat III hybrid approach, see Sec. 6.2); in particular, the subsystems responsible for not exceeding the limits of the servomechanism are removed and replaced by state constraints for the rate and position states. The linear Cat II models for the longitudinal and lateral plane are derived from this nonlinear model. However, in the nonlinear case, the model is not separated into a longitudinal and lateral model (i.e., the full dynamic model is considered). Inputs that are not used for the particular clearance problem are fixed to the trim condition (steady-state, straight and level flight with no wind). This implies that for maximizing the normal load factor all controls for the lateral motion (i.e., the lateral load factor command, the roll angle command, and the lateral wind gust command) are fixed to the trim values (zero). Similarly, for the maximization of the aerodynamic angle of sideslip, the normal load factor command as well as the longitudinal and normal wind commands are fixed.

The results from the Cat III direct method are obtained using the same transcription methods and discretization step lengths as for the Cat II results (see Sec. 5.3). For the nonlinear analysis, the initial guesses for all quantities are set to the values corresponding to the trim condition. The model is the nominal model with fixed parameters. All numerical solutions presented in the following are obtained using the optimal control toolbox FALCON.m.*

Engineering Note 6.1.2: Convergence of Nonlinear Optimal Control Solvers

Regarding the numerical solution, it should be mentioned that for nonlinear problems, solvers are typically only guaranteed to converge to a global minimizer for convex problems. Moreover, numerical experience shows that for nonlinear problems there exists the possibility that

*www.fsd.ed.tum.de/software/falcon-m

the optimization "gets stuck" at a point of local infeasibility when solving the collocation defects within the prescribed bounds of the decision variables, a situation that can entail cumbersome root cause analyses in order to find the exact reason for impeding convergence. That being said, in many instances, practical experience suggests that these numerical difficulties can be traced back to strong nonlinearities or the violation of the differentiability requirements of these solvers.

Example 6.1.1: Worst-Case Input and Postoptimal Sensitivities (Cat III)

A comparison between the Cat II and III results regarding the maximization of the normal load factor $n_z(6\text{ s})$ in the longitudinal plane as well as the aerodynamic

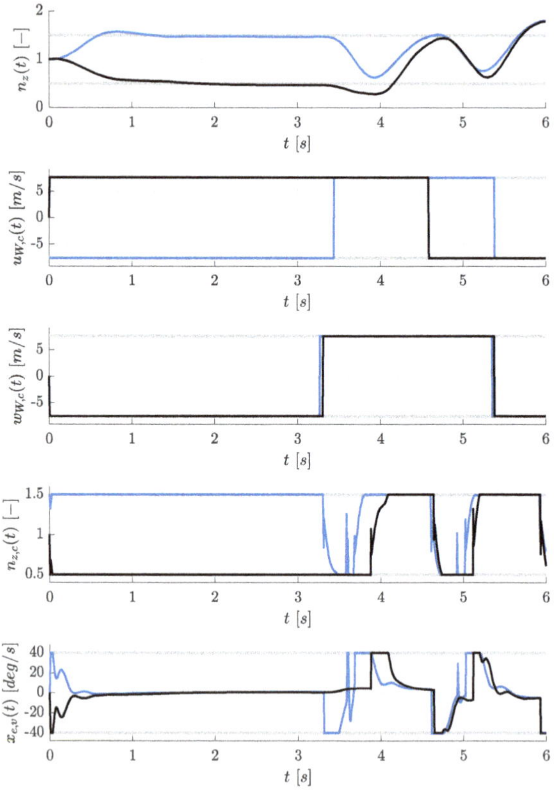

Fig. 6.1 Maximization of $n_z(6\text{ s})$.

angle of sideslip $\beta_A(6\,\mathrm{s})$ in the lateral plane is shown in Fig. 6.1 and Fig. 6.2. The figures compare the results using the nonlinear model (black lines) with those obtained from the linear models (blue lines).

In the longitudinal plane, the inputs for the normal load factor $n_{z,c}(t)$ as well as the normal wind command $w_{W,c}(t)$ and the longitudinal wind command $u_{W,c}(t)$ are used. In the lateral plane, the lateral load factor command $n_{y,c}(t)$, roll angle command $\Phi_c(t)$, and lateral wind input $v_{W,c}(t)$ are used. Additionally, state constraints for rate and position limits are imposed for all actuators related to the primary control surfaces (elevator, aileron, rudder).

Moreover, the postoptimal sensitivities regarding structural parameters and aerodynamic coefficients for the Cat II problems (see Sec. 5.3) and Cat III problems are compared in Fig. 6.3 and Fig. 6.4.

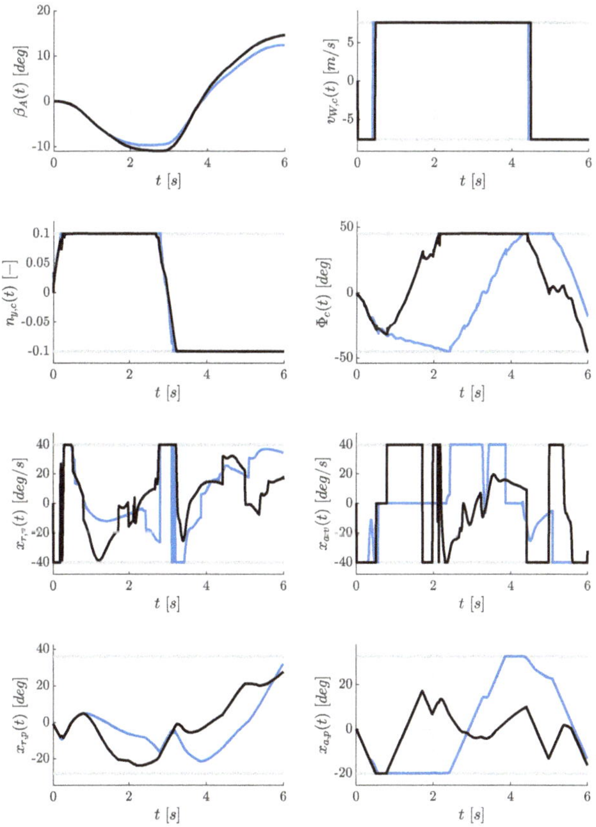

Fig. 6.2 Maximization of $\beta_A(6\,\mathrm{s})$.

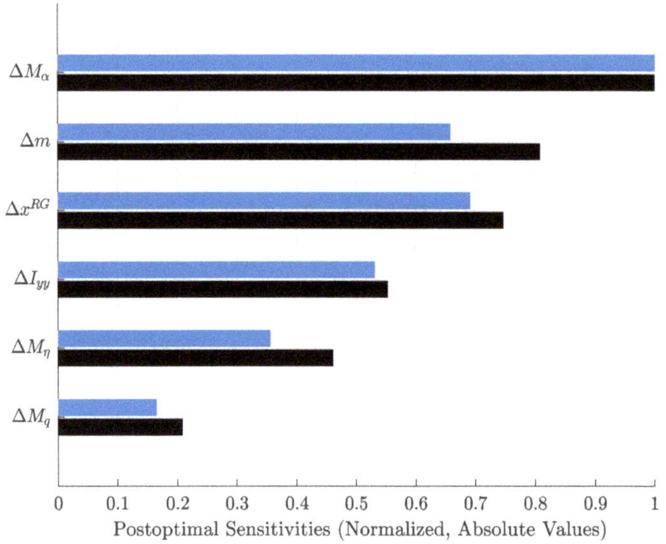

Fig. 6.3 Postoptimal sensitivities $\hat{n}_z(6\ \mathrm{s})$, Cat II/III (blue/black).

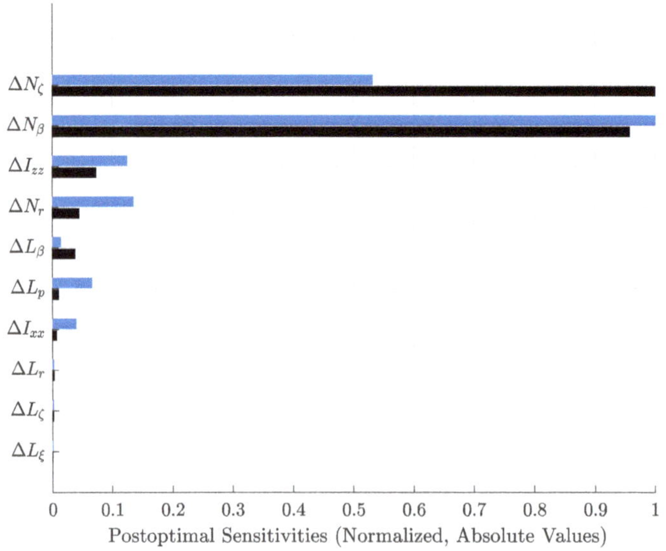

Fig. 6.4 Postoptimal sensitivities $\hat{\beta}_A(6\ \mathrm{s})$, Cat II/III (blue/black).

It is noteworthy that, similar to the linear case, the worst-case controls obtained from the solutions of the nonlinear clearance problems are of bang-bang type for the wind commands. This can be expected from the linear appearance of the wind commands in the state equations (see Sec. 3.3). Conversely, the other inputs exhibit also interior values (boundary controls) inside the respective control sets. The time intervals where interior values are observed correspond to the time intervals where the actuator states reach rate/position limits.

In both cases, that is, the maximization of the normal load factor $n_z(6\,\mathrm{s})$ (see Fig. 6.1) and the maximization of the aerodynamic angle of sideslip $\beta_A(6\,\mathrm{s})$ (see Fig. 6.2), the structure obtained for the nonlinear model clearly shares common features with the linear results. Moreover, the relative influence of the parameters on the worst-case cost appear to be remarkably similiar (see comparisons of the postoptimal sensitivities in Fig. 6.3 and Fig. 6.4). Recall that the deviations are for the absolute values. This represents an indicator that the search directions regarding the worst-case parameters obtained from a combined analysis in which parameters and controls are varied jointly can be expected to be similar.

Example 6.1.2: Combined Analysis (Cat III Box Constraints)

As for the parameter-dependent Cat II cases (see Sec. 5.5), the final time is increased to $t_f = 8\,\mathrm{s}$ for the combined optimization of parameters and controls. Table 6.1 and Table 6.2 show a comparison between the results obtained from the bilevel approach using a gradient-based method (SQP, see Table 5.1 and Table 5.4) and the solutions obtained from the nonlinear analysis.

TABLE 6.1 COMPARISON (CAT II/III) OF THE WORST-CASE PARAMETERS FOR MAXIMIZING $n_z(8\,\mathrm{s})$

Case	$\hat{n}_Z(8\,\mathrm{s})$	$\Delta\hat{M}_\alpha$	$\Delta\hat{M}_q$	$\Delta\hat{M}_\eta$	$\Delta\hat{m}$	$\Delta\hat{I}_{yy}$	$\Delta\hat{x}^{RG}$
Cat II	2.222	−0.100	−0.100	0.100	−0.100	0.100	−0.100
Cat III	2.349	−0.100	−0.100	0.100	−0.100	0.100	−0.100

TABLE 6.2 COMPARISON (CAT II/III) OF THE WORST-CASE PARAMETERS FOR MAXIMIZING $\beta_A(8\,\mathrm{s})$

Solver	$\hat{\beta}_A(8\,\mathrm{s})$	$\Delta\hat{L}_\beta$	$\Delta\hat{L}_p$	$\Delta\hat{N}_\beta$	$\Delta\hat{N}_\zeta$	$\Delta\hat{I}_{xx}$	$\Delta\hat{I}_{zz}$
Cat II	16.325 [*deg*]	−0.100	0.007	−0.100	0.100	0.100	0.100
Cat III	20.532 [*deg*]	−0.100	0.100	−0.100	0.100	−0.100	0.100

Note that even though the magnitude of the optimal cost function values are not equal for the Cat II and Cat III results, it is remarkable that the worst-case parameter combinations of the linear and nonlinear results are almost identical (merely, the values for $\Delta\hat{L}_p$ and $\Delta\hat{I}_{xx}$ in Table 6.2 show different values in comparison). This can be seen as an indicator that, regarding the parameters, the Cat II and III models behave qualitatively in a similar way. This observation together with the observation that the worst-case structure of the optimal control histories often share similarities is exploited in the Cat III hybrid approach presented in the following.

6.2 HYBRID APPROACH

Clearance Optimal Control Analysis (Cat III Hybrid)

The steps for the Cat III hybrid approach can be summarized as follows:

1. **Parameter Search**: Obtain the worst-case parameter combination from the bilevel analysis (see Sec. 5.5) for a parametric Cat II problem.
2. **Control Parameterization**: Solve a nominal Cat I problem (see Sec. 4.1) at the worst-case parameter combination and derive a control parameterization as well as associated initial guess based on the switching time points of the optimal control.
3. **Single Shooting**: Solve a black-box optimization problem (Cat III) with the nonlinear simulation model based on the derived control parameterization.

In practice, test inputs for flight control systems are often based on a combination of theoretical considerations and practical experience/engineering judgment. Such test inputs represent fixed input sequences expected to provoke worst-case responses for a particular criterion and can typically be parameterized using relatively few parameters. Inputs of this type are also used in the context of optimization-based flight control law clearance. For example, in Forssell [67], a fixed input is used, and the combination of worst-case parameters maximizing the normal load factor or angle of attack is sought by an optimization-based search. Similarly, in De Oliveira and Puyou [68], a low-dimensional input parameterization taken from random pilot input tests is used. This parameterization represents a sequence of pulses defined by parameters for amplitudes and pulse lengths (see Fig. 6.5).

The Cat III hybrid method presented in this chapter can be viewed as an automatized procedure based on the optimal control approach to determine a meaningful input parameterization together with an initial guess for a particular clearance criterion. This input can then be used as a starting point in a black-box (single shooting) optimization.

The motivation for this approach is twofold. On the one side, we seek to obtain a suitable control parameterization for a particular clearance task with a preferably low number of control parameters. On the other side, it is desirable

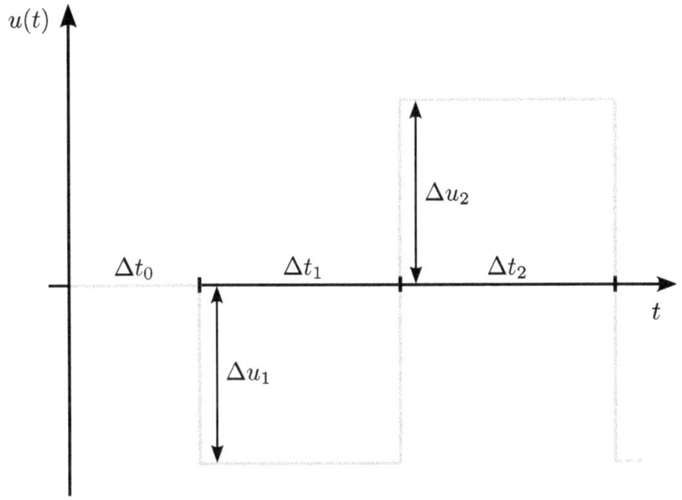

Fig. 6.5 Input parameterization.

to "warm-start" the single shooting method with a good initial guess for the parameterization obtained in this way. Both aspects are expected to facilitate the solution of the single shooting problem based on the full nonlinear-parametric model.

In the following, Cat I and Cat II models are used as reduced models for this purpose. The Cat II model is preferred in the first step to perform a parameter search under the bilevel approach presented in Sec. 5.5 in order to determine the worst-case parameter combination. The reasoning behind this strategy is that the degree of realism for the worst-case parameter combination is expected to be higher compared to a Cat I model if actuator limits are included in the analysis. Here, we assume that the reduced Cat II model shares the same tendency regarding the combination of worst case parameter values and optimal control actions with the full model. For determining a suitable control parameterization a Cat I problem is then solved at the worst-case parameter combination. This analysis is observed to be an excellent choice for our purpose as the bang-bang commands often seem to represent the most hazardous type of input signals also for the full model. Moreover, a parameterization based on the switching time points is easily determined in this case. Here, we assume that the switching structure (the sequence of bang-bang arcs in the case of a Cat I model) of the worst-case control functions from the reduced model is representative of the full model.

Note that the motivation behind this approach relies on the observation that linear models appear to represent good indicators for the parameter combination at which we can expect the worst case to occur in a nonlinear analysis and that the control structure obtained from a linear model is also effective for the nonlinear model. It should be noted that, under this approach, the number of parameters is greatly reduced compared to a search of the optimal control function "from scratch," and an initial guess is readily available.

Let the switching time points identified from the reduced model be collected in the vector $\mathbf{t} \in \mathbb{R}^{n_t}$. A simple method to determine these switching time points is to find the zero-crossings of the switching functions in the case of Cat I models. Computational experience suggests that this approach works robustly even if the control function exhibits numerical inaccuracies (such as chattering effects). The evaluation of the objective function $J : \mathbb{R}^{n_t} \to \mathbb{R}$ is then performed based on a simulation of the full model. As the values of $J(\mathbf{t})$ from the nonlinear model can be noisy due to the potentially complex structure of aircraft closed-loop systems, a gradient-free optimization method is used in the following. In particular, the Nelder–Mead simplex method (see Nelder and Mead [69] and Lagarias et al. [70]) appears to be effective for this task. This method represents a derivative-free optimization algorithm for the solution of unconstrained optimization problems, which merely requires the values of the cost function at particular points in the parameter space. The name "simplex" method originates from the fact that the method performs a set of operations (reflection, expansion, contraction) on a simplex body such that the "*simplex adapts itself to the local landscape, and contracts on to the final minimum*" [69]. Note that this method is not to be confused with Dantzig's simplex method (see Dantzig [54]) for solving LPs. An implementation is, for example, available in the MATLAB® function `fminsearch` designed for unconstrained minimization. Clearly, it is desirable for the control function parameterization using the switching time points to ensure that all entries in $\mathbf{t}$ lie between t_0 and t_f. To include these bounds in the unconstrained algorithm from `fminsearch`, the version from D'Errico [71] is used, which applies simple variable transformations to handle box-bounds. Moreover, it is required that the sequence of the switching structure, meaning the sequence of bang-bang arcs is maintained; after all, the amplitudes are fixed to the lower or upper bounds for such a signal. This can be ensured by sorting the vector $\mathbf{t}$ in ascending order before evaluating the cost function.

In all stages of the hybrid approach, we consider the criterion as a Mayer-type cost evaluated at the final time point of a fixed interval. This includes, in particular, the single shooting stage in which the switching time points are optimized. Let us illustrate the argument presented in the introduction of this part of the book (see Part II) regarding the modeling of the cost under the optimal control approach for the specific parameterization used in the hybrid method.

Consider, for example, a control function parameterization in the single shooting stage of the hybrid method with two switching time points for the maximization of an output y (see Fig. 6.6).

Suppose that the particular locations of the two switches are not tuned regarding the criterion evaluated for the nonlinear model and that the maximum occurs before the second switch. This can, for example, happen right in the first iteration of the single shooting stage in which the initial guess for the input parameterization of the worst-case input signal obtained from the linear Cat I model is

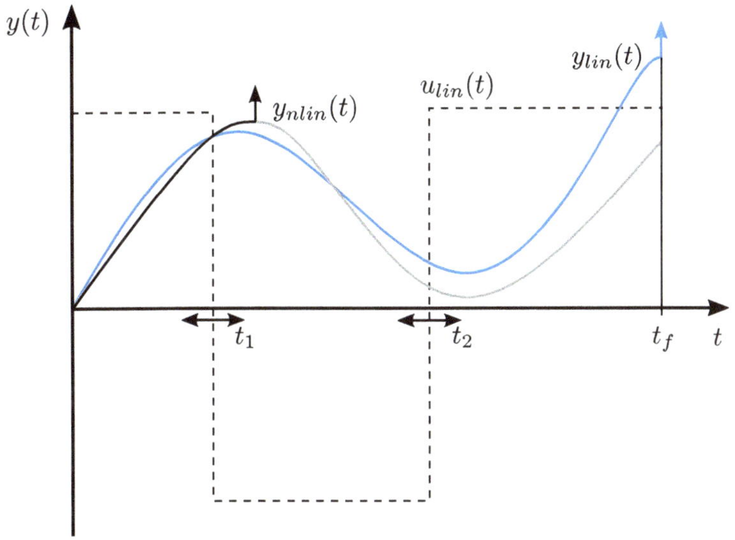

Fig. 6.6 Linear response [$y_{lin}(t)$, blue line] and nonlinear response [$y_{nlin}(t)$, black line/gray line] to the control input [$u_{lin}(t)$, black dashed line] parameterized by the time points t_1 and t_2. The grayed out part of the trajectory indicates the segment after the time point at which the maximum occurs.

directly applied to the nonlinear model. In such a situation, the switching time point occurring after the time point at which the maximum appears has (locally) no effect on the objective of the optimization problem. Conversely, if the output y is evaluated at the final time point, it is likely that the switching time points of the input show at least some influence on the final value of y, and the optimization method is "motivated" to exploit the full parameter space.

The Cat III hybrid approach is illustrated in the following using the linear models related to the longitudinal and lateral plane from Sec. 5.3. First, the nominal cases (i.e., without considering parameter uncertainties) are presented, and the Cat I solutions using the combined controls are parameterized based on the switching time points. In all examples, blue lines represent the initial control and response histories, whereas the black lines are the final results after the optimization of the switching time points. Additionally, the number of simulations n_s in the single shooting stage is provided for each result.

Example 6.2.1: Cat III Hybrid, Nominal

The optimal results for the Cat III hybrid approach are depicted in Fig. 6.7 for maximizing $n_z(6\,\text{s})$ and Fig. 6.8 for maximizing $\alpha_A(6\,\text{s})$ at nominal parameter

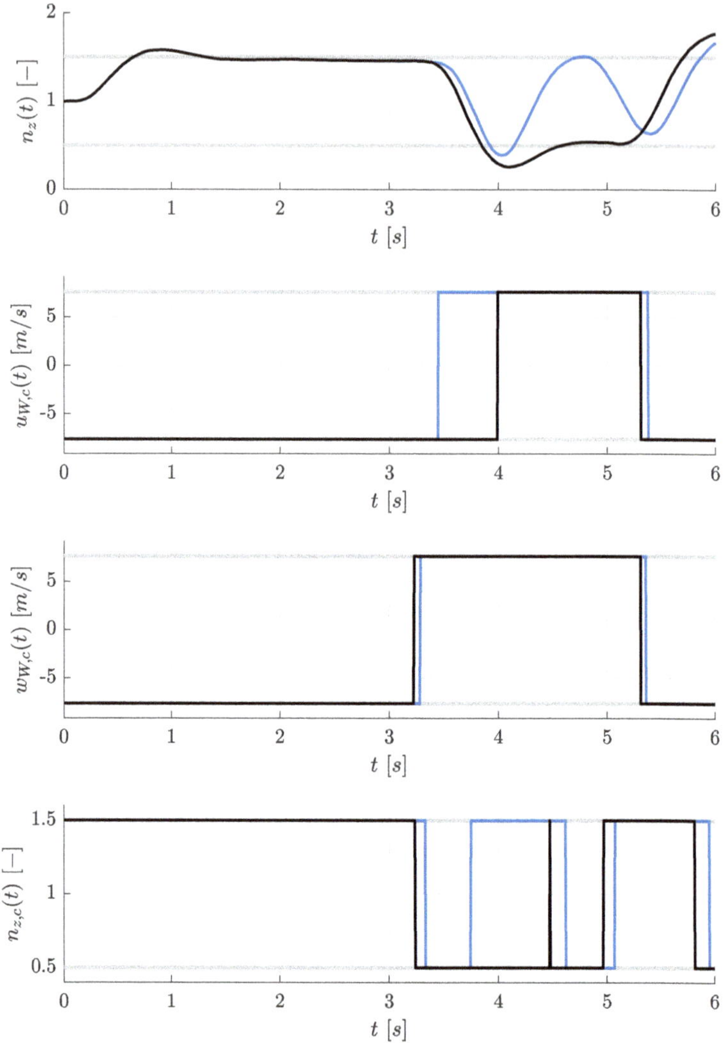

Fig. 6.7 Maximization of $n_z(6\,\mathrm{s})$, $n_s = 269$.

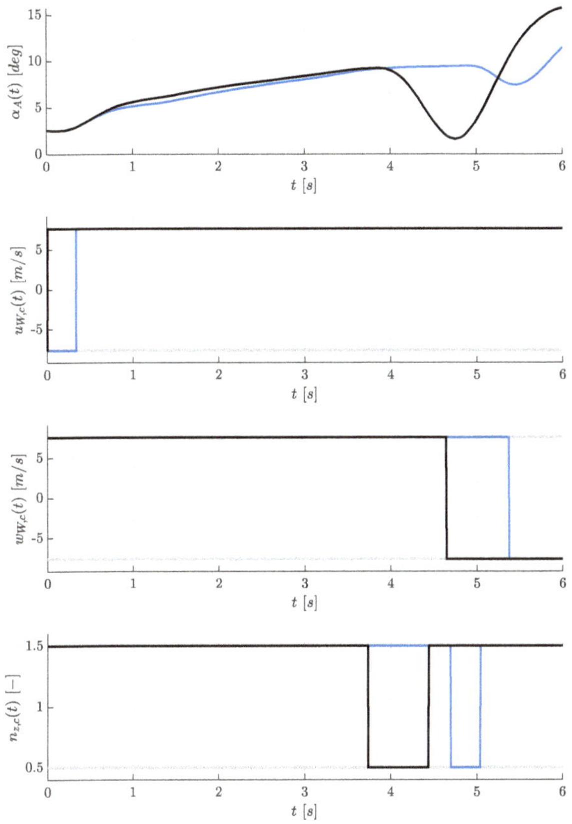

Fig. 6.8 Maximization of $\alpha_A(6\ \mathrm{s})$, $n_s = 223$.

values. In both cases, combined control inputs [normal load factor command $n_{z,c}(t)$, longitudinal wind velocity command $u_{W,c}(t)$, normal wind velocity command $w_{W,c}(t)$] are used.

Moreover, the optimal results for maximizing $n_y(6\ \mathrm{s})$ and maximizing $\beta_A(6\ \mathrm{s})$ at nominal parameter values are shown in Fig. 6.9 and Fig. 6.10. Here, the lateral load factor command $n_{y,c}(t)$, the roll angle command $\Phi_c(t)$, and the lateral wind velocity command $v_{W,c}(t)$ are used.

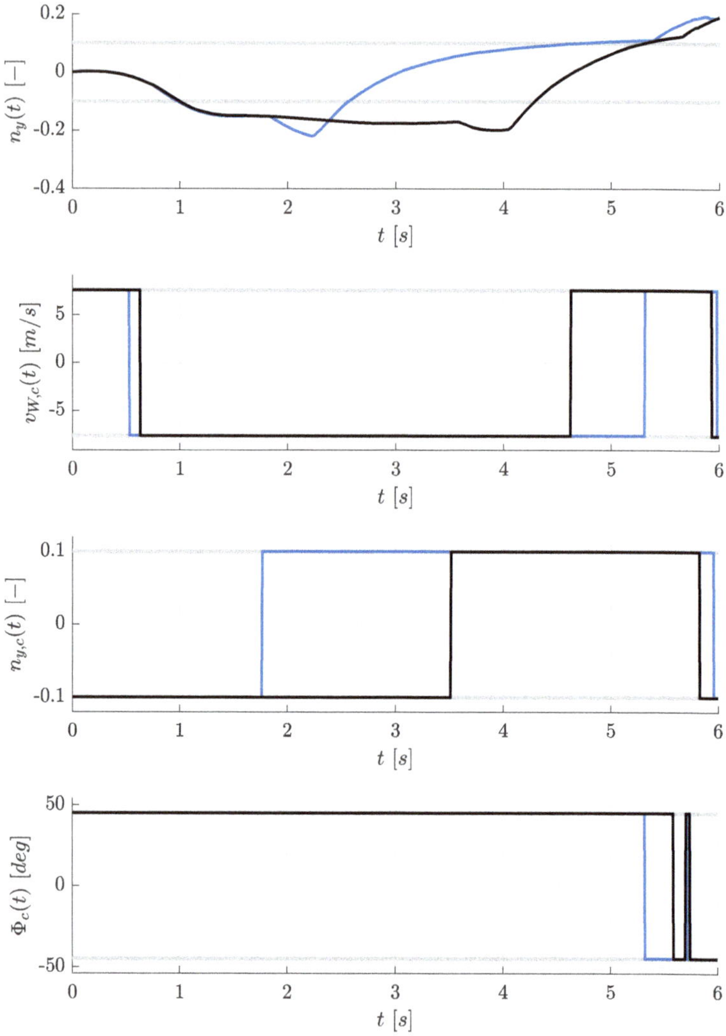

Fig. 6.9 Maximization of $n_y(6\,\mathrm{s})$, $n_s = 338$.

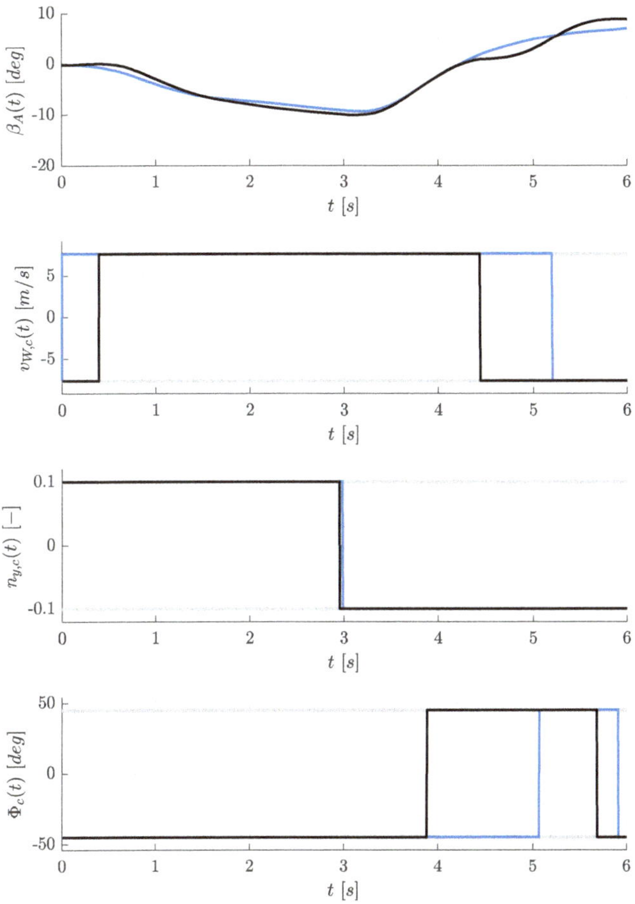

Fig. 6.10 Maximization of $\beta_A(6\,\mathrm{s})$, $n_s = 213$.

Note that in the initial command parameterizations derived from the Cat I solutions, the switching time points do not appear to be tuned regarding the final values of the criterion for the nonlinear model. Only for the maximization of $n_y(6\,\mathrm{s})$ do the initial and optimal values at the terminal time point appear to be similar (see Fig. 6.9). However, for all cases, the optimization algorithm seems to behave as expected and adjusts the switching time points maximizing the criteria at the end of the time interval.

Next, the parameter-dependent cases are investigated. Here, the results from Tables 5.2–5.4 (SQP) are used as starting points, and the optimal control functions at the worst-case parameter combinations are, as before, determined using a Cat I model and then parameterized using the zero-crossings of the switching functions. Again, the initial control and response are represented by blue lines, while the optimized final results are depicted by black lines in all plots.

Example 6.2.2: Cat III Hybrid, Parametric

The optimal results for the Cat III hybrid approach at worst-case parameter values obtained from the bilevel solution using Cat II–type problems in the lower level are depicted in Fig. 6.11 for maximizing $n_z(8\,\mathrm{s})$ and Fig. 6.12

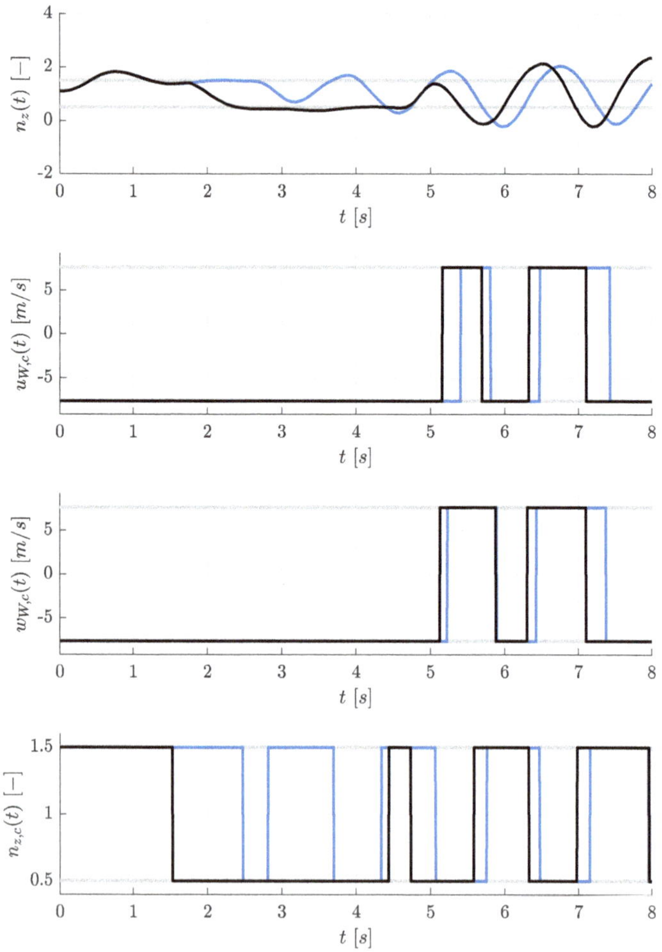

Fig. 6.11 Maximization of $n_z(8\,\mathrm{s})$, $n_s = 338$.

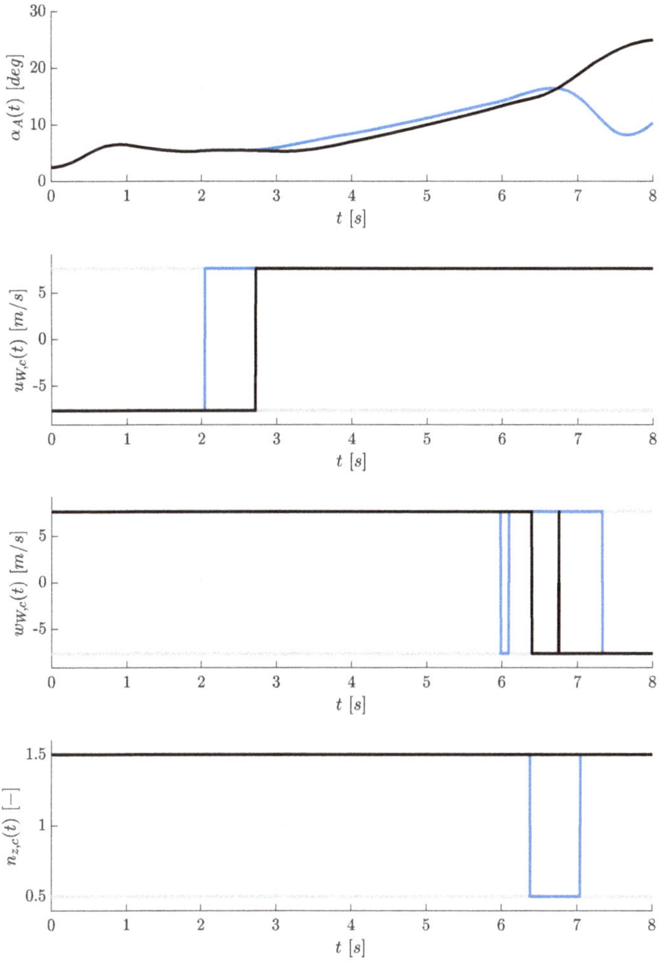

Fig. 6.12 Maximization of $\alpha_A(8\,\mathrm{s})$, $n_s = 213$.

for maximizing $\alpha_A(8\,\mathrm{s})$. In both cases, combined control inputs [normal load factor command $n_{z,c}(t)$, longitudinal wind velocity command $u_{W,c}(t)$, normal wind velocity command $w_{W,c}(t)$] are used.

Similarly, the optimal results for maximizing $n_y(8\,\mathrm{s})$ and maximizing $\beta_A(8\,\mathrm{s})$ at worst-case parameter values are shown in Fig. 6.13 and Fig. 6.14. Here, the lateral load factor command $n_{y,c}(t)$, the roll angle command $\Phi_c(t)$, and the lateral wind velocity command $v_{W,c}(t)$ are used.

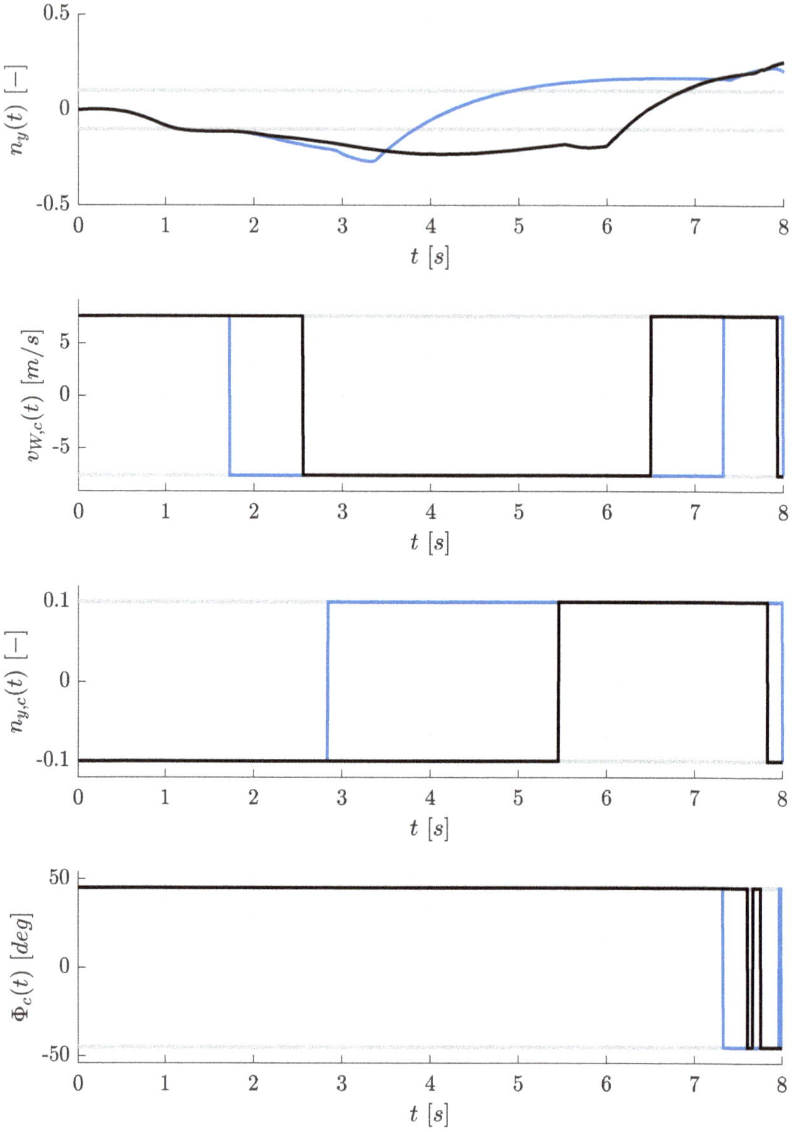

Fig. 6.13 Maximization of $n_y(8\,\mathrm{s})$, $n_s = 321$.

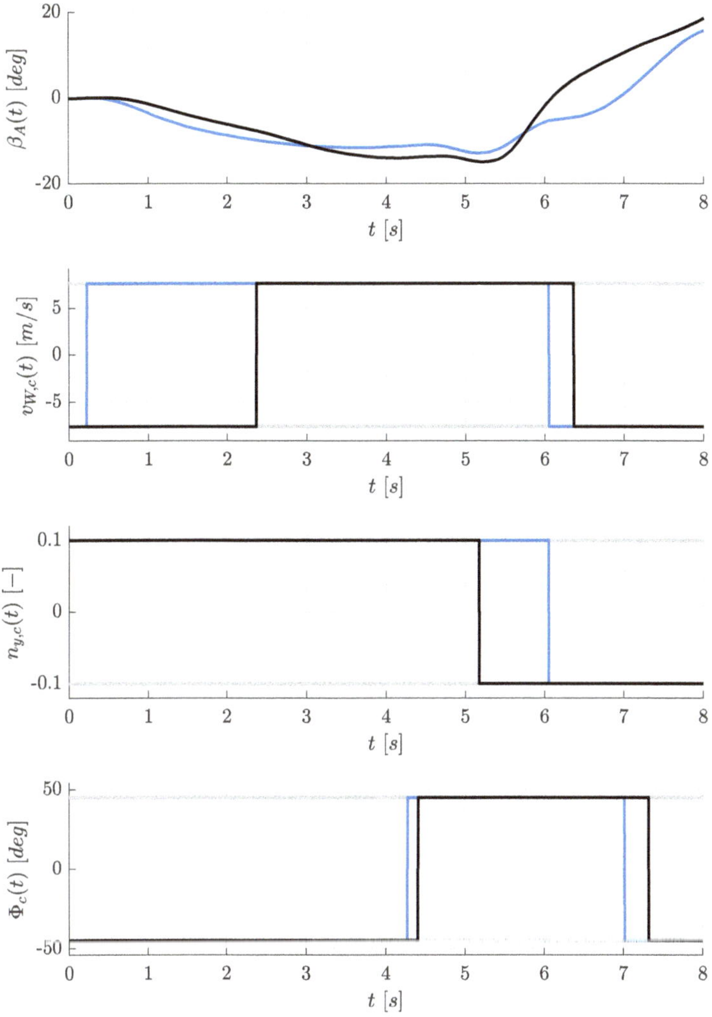

Fig. 6.14 Maximization of $\beta_A(8\,\text{s})$, $n_s = 187$.

Note that, as for the nominal examples, simply applying the worst-case control from the reduced model is less effective compared to the results obtained after "tuning" the switching time points in the single shooting stage. Observe further that for the maximization of the load factors $n_z(8\,\text{s})$ and $n_y(8\,\text{s})$ the identified worst-case solutions exceed the maximum commanded values by a noticeable amount.

Engineering Note 6.2.1: Sanity Checks: Hybrid Approach

The application of sanity checks is advisable in order to improve confidence in the worst-case results obtained from the Cat III hybrid approach. The following checks may be performed:

- Investigate the flatness of the slope for the criterion at the final time.
- Check if the most detrimental value of the criterion occurs at the final time and not in the interior of the time interval.
- Check how the optimization algorithm under the single shooting approach exploits the parameterization obtained from the reduced model (e.g., check if intervals are collapsed after the refinement).
- Apply different optimization algorithms for the refinement of the solution under the single shooting approach and check the consistency of the results.
- Perform the analysis for increasing values of the final time and check if the results obtained from the analyses yield consistent worst-case solutions.

If the preceding checks indicate that the solution is not satisfactory, the process may be repeated for a new final time, the analysis can be restarted from a randomized initial parameterization around the nominal values obtained from the single shooting approach, or a different optimization algorithm can be employed (e.g., the DIRECT method or the surrogate optimization method, see Sec. 5.5).

The examples presented in this section indicate that the Cat III hybrid approach is effective for the nonlinear simulation model and can be easily integrated into existing tool-chains. Merely, linearized models and a simulation model are required, which are essentially available at no cost as typically these models are required anyway for the development of flight control laws. Thus, the Cat III hybrid approach should be directly applicable to the solution of clearance problems in industry practice. In particular, it is observed that the Cat III hybrid approach allows for considerable automatization. On the one side, the solution using Cat I and Cat II models is reliable and efficient. On the other side, it is found that the automatic detection of the switching time points using the values of the switching functions works robustly. These aspects, in particular, allow for the application of this approach, for example, in the form of an automatic sanity check whenever the model is altered.

CHAPTER 7

Set-Based Clearance

Reachable set methods for dynamic systems provide the attainable set of points in a particular space of interest. This kind of information can be valuable for flight control law testing, as it permits a joint consideration of multiple criteria within the clearance task. Thus, this chapter is devoted to set-based approaches tailored to the optimal control methods for Cat I, II, and III types of models.

The computation of forward and backward reachable sets in the context of safety analysis constitutes an active research field. Because of its general applicability, considerable research interest has emerged in this field over the last few decades, also in areas beyond safety assessment and testing.

Several strategies exist for constructing reachable sets in the case of linear systems. For example, reachable sets can be obtained through set operations such as sums, differences, and overapproximations [72]. However, the associated computational effort can become intractable for larger problems, as the number of facets of the reachable sets tends to grow quickly. Conversely, basic geometrical bodies such as ellipsoids (see, e.g., Kurzhanski and Varaiya [73] and Kurzhanski and Varaiya [74]), zonotopes (see, e.g., Girard [75] and Althoff et al. [76]), and parallelotopes (see, e.g., Kostousova [77, 78]) can be used to obtain inner and outer approximations rather efficiently. Moreover, optimal control methods for reachable set approximations under input constraints are considered in Varaiya [79] and Baier et al. [80]. Conceptually, these optimal control approaches naturally fit the type of clearance methods considered in this book. Hence, following the ideas from these optimal control approaches, we present an adaptive method tailored to the linear clearance methods for nominal systems introduced in Chapter 4.

Under this approach, it is exploited that for linear systems originating from the origin in the state space and subject to convex control and state constraints, the reachable set is convex and compact. The main idea of this method is to compute both an inner and an outer approximation of the convex reachable set directly in the output space. These approximations are refined based on the successive solution of optimal control problems, maximizing the distance from the origin while constraining the endpoints in the output space to lie on fixed rays (see Fig. 7.1). Each endpoint obtained in this way is then added to the convex hull of the inner

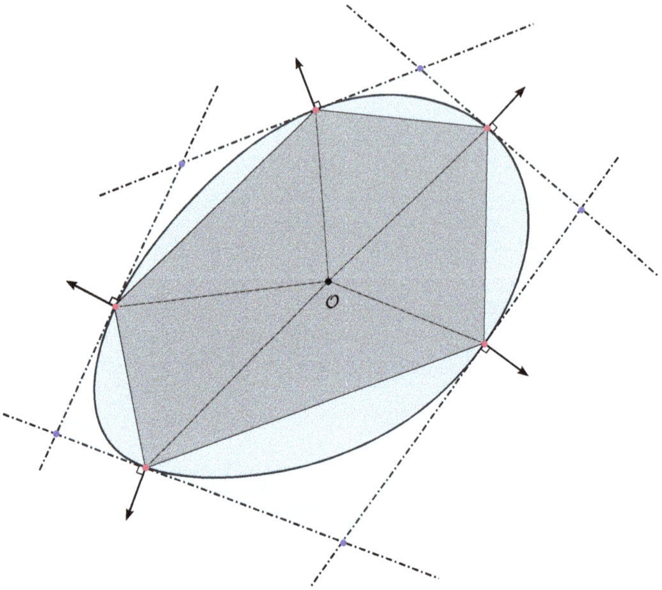

Fig. 7.1 Inner and outer approximations of a convex reachable set.

approximation. Moreover, at each endpoint, a tangent plane to the reachable set is constructed and added in the form of an inequality to the outer approximation. The construction of these tangent planes relies on the postoptimal sensitivity analysis from Chapter 5. Based on a conservative error estimate, these approximations are then refined adaptively until a given termination criterion is met.

Notably, as the reachable set under this method is directly constructed in the output space, the resulting algorithm is applicable to high state and control dimensions. Moreover, the error can be approximated conservatively based on the inner and outer approximations.

Reachability analysis for (nonlinear) Cat III–type models is considerably more involved compared to (linear) Cat I and II models. In particular, under Cat III, the reachable set is potentially nonconvex in general. A popular approach in this context is the application of level-set methods based on Hamilton–Jacobi equations [81, 82]. Unfortunately, the applicability of these approaches in higher-dimensional state spaces is computationally challenging. As in the case of dynamic programming, these approaches suffer from the curse of dimensionality due to the fact that the numerical solution of the respective partial differential equation requires not only a discretization in the temporal dimension but also in the state space. Nevertheless, these methods have been proven useful in certain aerospace applications [83, 84]. Notably, as for the linear case, there also exist approaches based on direct optimal control methods for reachability analysis.

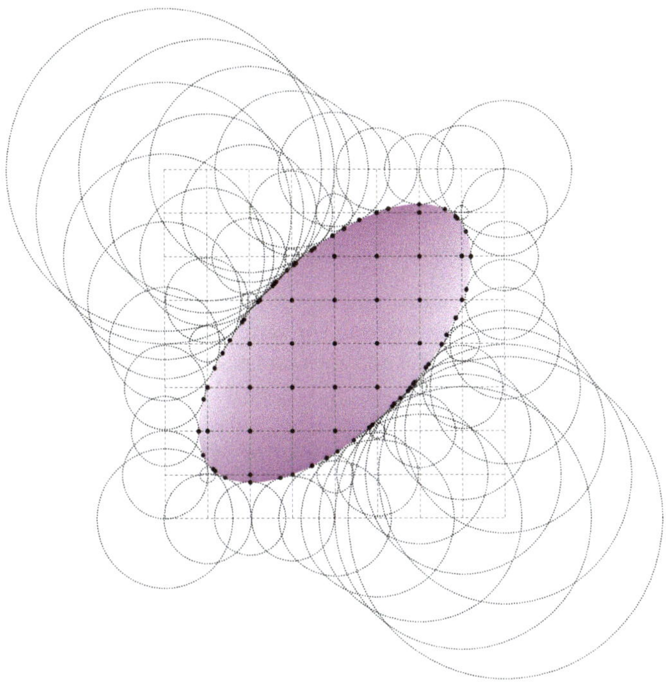

Fig. 7.2 Distance functions on a rectangular grid.

For example, the distance field on grids (DFOG) method [85] relies on the solution of optimal control problems on a grid in the space of interest. For each problem, the distance to one grid point is minimized to obtain points inside or on the boundary of the reachable set (see Fig. 7.2). This approach is conceptually close to the direct Cat III formulation introduced in Sec. 6.1. Moreover, the requirements on the type of dynamic system are fairly nonrestrictive, and input, as well as state constraints, can be imposed. Therefore, this method is considered for Cat III–type models.

This chapter is structured as follows: In the first section, a method for set-based analysis regarding (linear) Cat I and II clearance problems is presented. Herein, Sec. 7.1.1 outlines a strategy for obtaining an outer box approximation as well as an inner convex hull approximation to the reachable set. Following, Sec. 7.1.2 introduces a method for computing the distance from each vertex of an outer approximation to an inner approximation. This method allows a conservative error estimate for a given inner and outer approximation. Moreover, a procedure for the adaptive refinement of the reachable set's boundary is presented in Sec. 7.1.3. Finally, Sec. 7.2 describes the application of the DFOG method to Cat III–type clearance problems.

7.1 LINEAR

7.1.1 BOX AND CONVEX HULL APPROXIMATIONS

Outer and Inner Approximations from Extremal Points

Extremal points on the reachable set's boundary for each component of the output vector

$$\mathbf{y}(t) = \left[\mathbf{c}_1, \ldots, \mathbf{c}_{n_J}\right]^T \mathbf{x}(t)$$

are found from separately maximizing and minimizing each of the outputs at t_f starting from the origin in the state space for the linear system (4.2) subject to convex control and state constraints [Eqs. (4.3) and (4.4)]. Based on these extremal points, it is possible to construct an outer box approximation of the form $\mathbf{A}_O\mathbf{y} \leqslant \mathbf{b}_O$, $\mathbf{A}_O \in \mathbb{R}^{2n_J \times n_J}$, $\mathbf{b}_O \in \mathbb{R}^{2n_J}$ and, similarly, an inner convex hull approximation of the form $\mathbf{A}_I\mathbf{y} \leqslant \mathbf{b}_I$, $\mathbf{A}_I \in \mathbb{R}^{2n_J \times n_J}$, $\mathbf{b}_I \in \mathbb{R}^{2n_J}$.

Let us first consider an outer box approximation regarding the reachable points of $\mathbf{y}(t_f) = \left[\mathbf{c}_1, \ldots, \mathbf{c}_{n_J}\right]^T \mathbf{x}\left(t_f\right) = \mathbf{C}\mathbf{x}\left(t_f\right)$. Such an approximation to the reachable set can be obtained from the solution of $2n_J$ problems of the form (4.1) for $\mathbf{c} = \pm\mathbf{c}_1, \ldots, \pm\mathbf{c}_{n_J}$ (see Fig. 7.3).

Note that problems with a negated output vector in the cost effectively yield maximization problems for these outputs. Conversely, the other problems minimize the respective outputs by our convention to formulate the clearance optimal control problem as a minimization problem by default.

Let the optimal cost functions values of the maximization problems ($\mathbf{c} = -\mathbf{c}_1, \ldots, -\mathbf{c}_{n_J}$) be collected in

$$\bar{\mathbf{J}}^T = \left[\bar{J}_1, \ldots, \bar{J}_{n_J}\right]$$

Similarly, let us collect the optimal cost values of the minimization problems ($\mathbf{c} = \mathbf{c}_1, \ldots, \mathbf{c}_{n_J}$) in

$$\underline{\mathbf{J}}^T = \left[\underline{J}_1, \ldots, \underline{J}_{n_J}\right]$$

Note that each of the two inequalities

$$\mathbf{e}_i^T\mathbf{y} \leqslant -\bar{J}_i$$
$$\mathbf{e}_i^T\mathbf{y} \geqslant -\underline{J}_i$$

for $i = 1, \ldots, n_J$ separates a half-space containing no attainable points from a half-space containing the reachable set and closes a box in one output dimension. Thus,

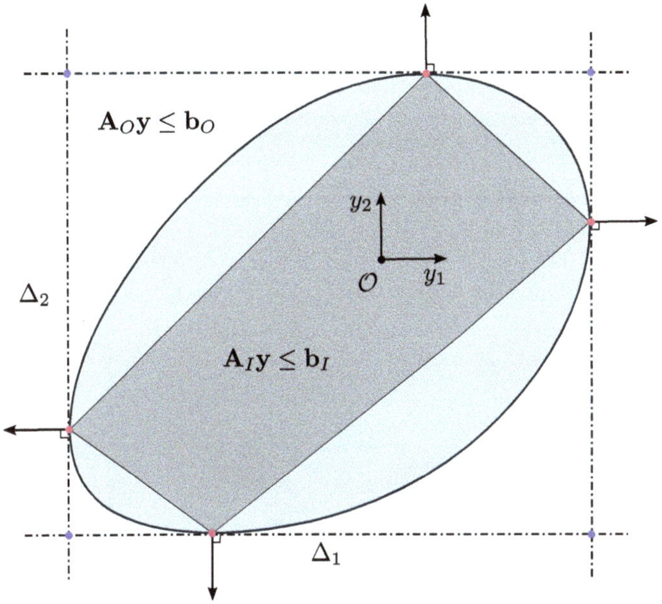

Fig. 7.3 Outer box and inner convex hull approximations.

the inequalities

$$\underbrace{\begin{bmatrix} \mathbf{I} \\ -\mathbf{I} \end{bmatrix}}_{\mathbf{A}_O} \mathbf{y} \leqslant \underbrace{\begin{bmatrix} -\bar{\mathbf{J}} \\ \underline{\mathbf{J}} \end{bmatrix}}_{\mathbf{b}_O}$$

define a bounding box to the reachable set in the output space. This bounding box has the following edge lengths in each dimension:

$$\boldsymbol{\Delta} = \left[\Delta_1, \ldots, \Delta_{n_J}\right]^T = -\bar{\mathbf{J}} - \underline{\mathbf{J}} \tag{7.1}$$

and constitutes a convex, closed, and bounded outer approximation to the reachable set.

Moreover, an inner approximation can be represented by the convex hull of the points obtained in the output space from the solution of the minimization and maximization problems (see Fig. 7.3):

$$\mathbf{A}_I\mathbf{y} \leqslant \mathbf{b}_I$$

Here, and in the sections that follow, the construction of both approximations (inner, outer) is illustrated using the nominal Cat II models from Sec. 5.3. The final time is set to $t_f = 8$ s, and the discretization step size for the trapezoidal collocation method is $h = 0.025$ s. Moreover, the values of all quantities under investigation are provided relative to the respective reference (trim) values.

Example 7.1.1: Outer Box and Inner Convex Hull Approximations (Cat II)

First, two-dimensional output spaces are considered for the reachable set approximation of the longitudinal and lateral Cat II model. In the case of the longitudinal model, the output combination under investigation is the normal load factor n_z together with the aerodynamic angle of attack α_A (see Fig. 7.4). For the

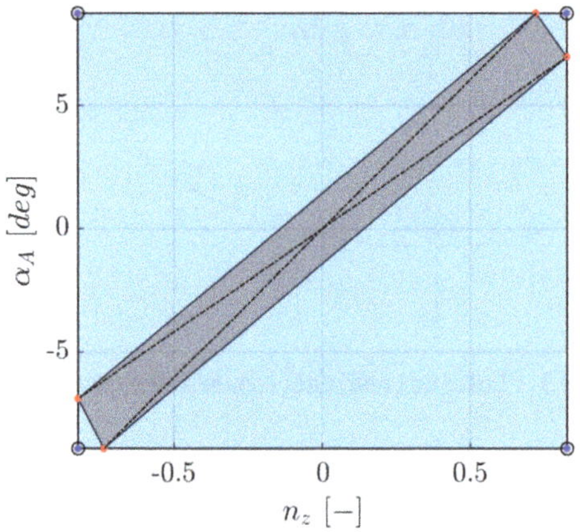

Fig. 7.4 Cat II $\alpha_A(8\text{ s}), n_z(8\text{ s})$.

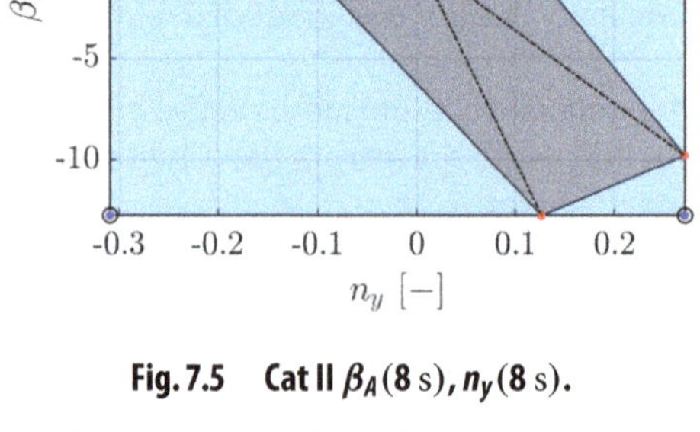

Fig. 7.5 Cat II $\beta_A(8\text{ s}), n_y(8\text{ s})$.

lateral model, the considered outputs are the lateral load factor n_y as well as the aerodynamic angle of sideslip β_A (see Fig. 7.5).

For both cases, four problems are solved to obtain the outer box approximations and the inner convex hull approximations of the reachable sets. If the dimension of the output spaces is increased by adding the pitch rate q_A for the longitudinal model and the yaw rate r_A for the lateral model, six problems need to be solved (see Fig. 7.6 and Fig. 7.7).

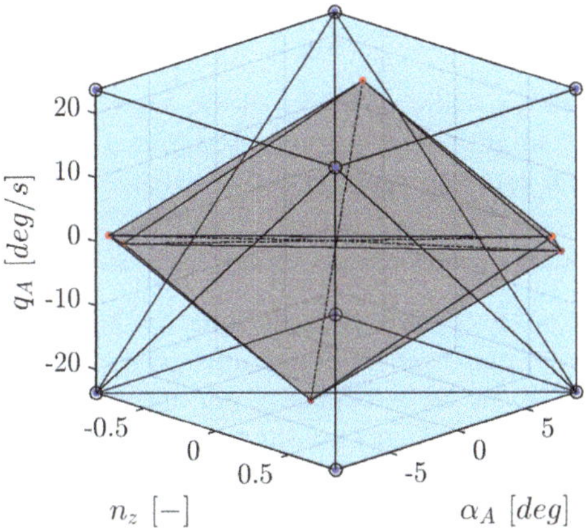

Fig. 7.6 Cat II α_A(8 s), n_z(8 s), q_A(8 s).

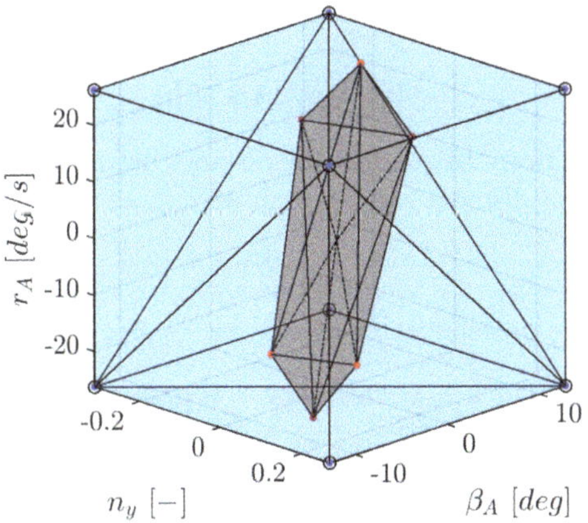

Fig. 7.7 Cat II β_A(8 s), n_y(8 s), r_A(8 s).

The numerical examples demonstrate that the approximations can be rather crude, showing considerable gaps between the outer and inner approximations. In most applications, refining these approximations and providing a suitable error estimate are both desirable. These topics are discussed in the subsequent sections.

7.1.2 ERROR ESTIMATION

Shortest Distance

A point $\mathbf{y} \in \mathbb{R}^{n_J}$ satisfying

$$\mathbf{A}_I \mathbf{y} \leqslant \mathbf{b}_I \tag{7.2}$$

located closest to a specific vertex $\mathbf{y}_O$ of the outer approximation can be found from the solution of a linear least-squares problem. The cost of this optimization problem represents the distance of $\mathbf{y}_O$ to the polytope [Eq. (7.2)].

Let $\mathbf{A}_O \mathbf{y} \leqslant \mathbf{b}_O$ be a convex, closed, and bounded outer approximation obtained from tangent planes at support points on the reachable set's boundary. Moreover, let $\mathbf{A}_I \mathbf{y} \leqslant \mathbf{b}_I$ be an associated inner approximation from the convex hull of these support points. Then, the shortest distance from a vertex $\mathbf{y}_O$ of the outer approximation to the inner approximation can be found by identifying the closest point to $\mathbf{y}_O$ from the set of points in the polytope [Eq. (7.2)] (see Fig. 7.8). For each vertex $\mathbf{y}_O$ from the set of vertices of the outer approximation, such a point can be found from the solution of the least-squares problem:

$$\begin{aligned} \underset{\mathbf{y} \in \mathbb{R}^{n_J}}{\text{minimize}} \quad & \left\| \mathbf{y} - \mathbf{y}_O \right\|_2 \\ \text{subject to} \quad & \mathbf{A}_I \mathbf{y} \leqslant \mathbf{b}_I \end{aligned}$$

using the Euclidean norm to quantify the distance between $\mathbf{y}$ and $\mathbf{y}_O$ in the cost.

Alternatively, this distance can be minimized in a scaled output space

$$\tilde{\mathbf{y}} = \mathbf{W}\mathbf{y}$$

and the least-squares problem can be stated as

$$\begin{aligned} \underset{\tilde{\mathbf{y}} \in \mathbb{R}^{n_J}}{\text{minimize}} \quad & \left\| \tilde{\mathbf{y}} - \mathbf{W}\mathbf{y}_O \right\|_2 \\ \text{subject to} \quad & \mathbf{A}_I \mathbf{W}^{-1} \tilde{\mathbf{y}} \leqslant \mathbf{b}_I \end{aligned}$$

with some suitable scaling matrix $\mathbf{W} \in \mathbb{R}^{n_J \times n_J}$. To provide an example, the outputs can be scaled using the edge lengths collected in $\boldsymbol{\Delta}$ from Eq. (7.1):

$$\tilde{y}_i = \frac{y_i}{\Delta_i}, \; i = 1, \ldots, n_J$$

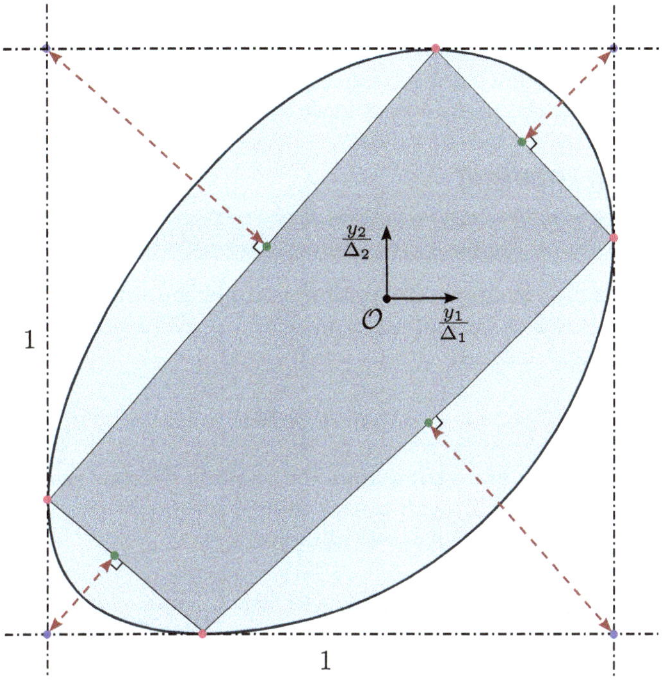

Fig. 7.8 Shortest distances (scaled output space).

Then, the scaling matrix is of the form

$$\mathbf{W} = \begin{bmatrix} \frac{1}{\Delta_1} & & \\ & \ddots & \\ & & \frac{1}{\Delta_{nJ}} \end{bmatrix}$$

Such scaling can be useful if the outputs span different orders of magnitude, for example, if one output represents the absolute velocity of the aircraft in meters per second and the other output is the angle of attack in radians.

The optimal cost values obtained from the solutions of the least-square problems for each vertex of the outer approximation then represent the smallest (scaled) distances for each such vertex to the convex boundary of the inner approximation. Moreover, the largest of these distances can be used as a conservative estimate for the approximation error of the reachable set's boundary.

Finally, if a (scaled) distance exceeds a predefined tolerance $\overline{\epsilon}$, refining the inner and outer approximations near the corresponding vertex may be beneficial. This adaptive refinement strategy can be implemented through successive optimization along rays in the output space, as discussed in the following section.

7.1.3 ADAPTIVE REFINEMENT

Maximization with Ray Constraint

Let a ray in the output space be parameterized by the distance $r \geqslant 0$ from the origin as well as the unit vector $\mathbf{w} : \mathbb{R}^{n_J - 1} \to \mathbb{R}^{n_J}$ depending on fixed angles collected in $\boldsymbol{\phi} \in \mathbb{R}^{n_J - 1}$:

$$\mathbf{y}(r, \boldsymbol{\phi}) = r\mathbf{w}(\boldsymbol{\phi}) \tag{7.3}$$

The maximization of r starting from the origin in the state space for the linear system [Eq. (4.2)] with convex control constraints [Eq. (4.3)] and state constraints [Eq. (4.4)] as well as subject to

$$\mathbf{Cx}\left(t_f\right) = r\mathbf{w}(\boldsymbol{\phi}) \tag{7.4}$$

yields the point $\hat{\mathbf{y}}(\boldsymbol{\phi})$ representing the intersection of the ray and the boundary of the reachable set. The columns of $\frac{d\hat{\mathbf{y}}}{d\boldsymbol{\phi}}$ describe a tangent plane to the reachable set at $\hat{\mathbf{y}}(\boldsymbol{\phi})$.

Consider a parameterization $\mathbf{y} : \mathbb{R} \times \mathbb{R}^{n_\phi} \to \mathbb{R}^{n_J}$, $n_\phi = n_J - 1$ of the form (7.3). For example, in two dimensions* a parameterization using polar coordinates is given as (see Fig. 7.9):

$$\mathbf{w}(\boldsymbol{\phi}) = \begin{bmatrix} \cos(\phi_1) \\ \sin(\phi_1) \end{bmatrix}$$

Using this parameterization, the refinement strategy of the inner and outer approximations is based on the solution of an appropriate optimal control problem. The trajectory of this problem starts at the origin in the state space, and the goal is to push the system away from the origin in the output space with the final output $\mathbf{Cx}\left(t_f\right)$ constrained to lie on a fixed ray [Eq. (7.3)]. This formulation is accomplished through the minimization of $-r$ for a fixed $\boldsymbol{\phi}$. Based on a trapezoidal transcription, the linear optimal control problem starting from $\mathbf{x}_0 = \mathbf{0}$ is

*Parameterizations of this type also exist in higher dimensions (see Appendix E).

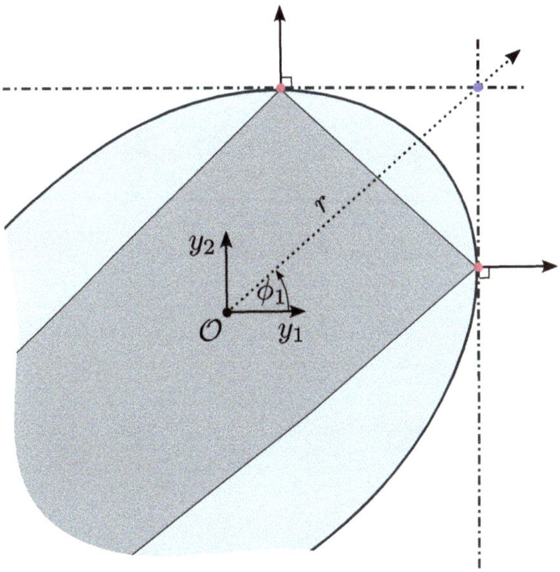

Fig. 7.9 Polar coordinates.

defined as:

$$\begin{aligned}
\underset{r,\mathbf{x}_i,\mathbf{u}_i,\mathbf{u}_0}{\text{minimize}} \quad & -r \\
\text{subject to} \quad & \mathbf{C}\mathbf{x}_N - r\mathbf{w}(\boldsymbol{\phi}) = \mathbf{0} \\
& \mathbf{x}_i = \mathbf{x}_{i-1} + h\left(\mathbf{A}\frac{\mathbf{x}_i + \mathbf{x}_{i-1}}{2} + \mathbf{B}\frac{\mathbf{u}_i + \mathbf{u}_{i-1}}{2}\right) \\
& \underline{\mathbf{x}} \leqslant \mathbf{x}_i \leqslant \overline{\mathbf{x}} \\
& \underline{\mathbf{u}} \leqslant \mathbf{u}_i \leqslant \overline{\mathbf{u}} \\
& \underline{\mathbf{u}} \leqslant \mathbf{u}_0 \leqslant \overline{\mathbf{u}} \\
& i = 1, \dots, N
\end{aligned}$$

In this problem, r is a free variable entering the cost as well as the equality constraint fixing the endpoint to the ray in the output space. The optimal solution of this problem for a particular $\boldsymbol{\phi}$ provides us with $\hat{r}(\boldsymbol{\phi})$, the largest attainable distance from the origin in the output space considering the ray constraint. Based on this distance, the corresponding point on the ray as well as the reachable set's boundary is

$$\hat{\mathbf{y}}(\boldsymbol{\phi}) = \hat{r}(\boldsymbol{\phi})\,\mathbf{w}(\boldsymbol{\phi})$$

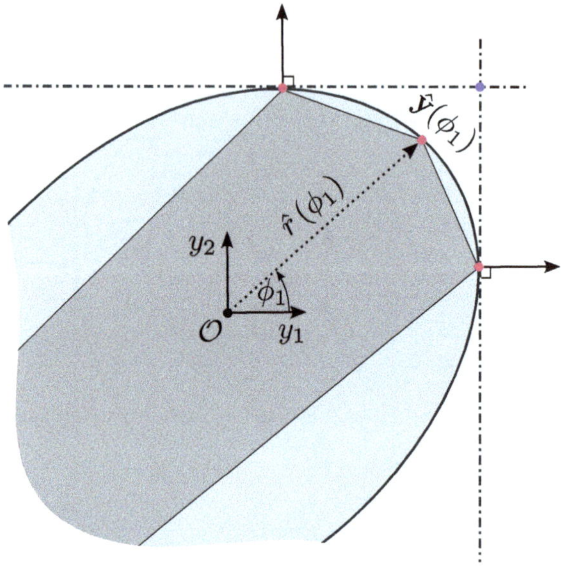

Fig. 7.10 Inner approximation.

On the one side, this point can be used to extend the convex hull forming the inner approximation (see Fig. 7.10), defined by all available points on the boundary of the reachable set. On the other side, the outer approximation can be extended through the construction of a tangent plane at this point.

This tangent plane is defined by the columns of the Jacobian:

$$\frac{d\hat{\mathbf{y}}}{d\boldsymbol{\phi}} = \hat{r}(\boldsymbol{\phi})\frac{\partial \mathbf{w}}{\partial \boldsymbol{\phi}} + \mathbf{w}(\boldsymbol{\phi})\frac{d\hat{r}}{d\boldsymbol{\phi}}$$

Regarding the computation of $\frac{d\hat{r}}{d\boldsymbol{\phi}}$, note that the only constraint in the LP depending on $\boldsymbol{\phi}$ is the equality constraint [Eq. (7.4)]. Let us denote the Lagrange multiplier associated with this equality constraint by $\boldsymbol{\kappa} \in \mathbb{R}^{n_J}$ and recall from Eq. (5.5) that the postoptimal sensitivity of the optimal cost $\hat{J}$ with respect to a parameter is equal to the partial derivative of the Lagrangian with respect to this parameter. As a direct consequence, the postoptimal sensitivity of $\hat{r}$ with respect to $\boldsymbol{\phi}$ is the negative partial derivative of the Lagrangian $\mathcal{L}$ with respect to $\boldsymbol{\phi}$:

$$\mathcal{L} = \ldots + \boldsymbol{\kappa}^T\left[\mathbf{C}\hat{\mathbf{x}}_N - \hat{r}(\boldsymbol{\phi})\,\mathbf{w}(\boldsymbol{\phi})\right] \quad \rightarrow \quad \frac{d\hat{r}}{d\boldsymbol{\phi}} = -\frac{d\hat{J}}{d\boldsymbol{\phi}} = -\frac{\partial \mathcal{L}}{\partial \boldsymbol{\phi}} = \hat{r}(\boldsymbol{\phi})\,\boldsymbol{\kappa}^T\frac{\partial \mathbf{w}}{\partial \boldsymbol{\phi}}$$

Thus, based on the first-order postoptimal sensitivities, a tangent plane for the reachable set supported at $\hat{\mathbf{y}}$ is obtained (see Fig. 7.11).

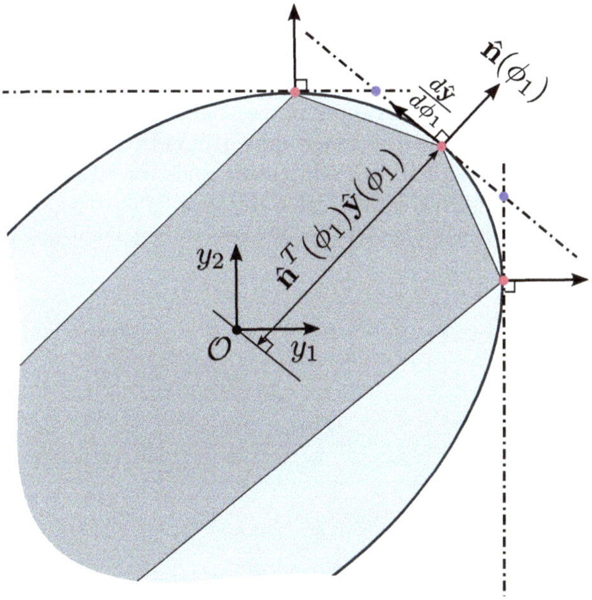

Fig. 7.11 Outer approximation.

Let $\hat{\mathbf{n}}(\boldsymbol{\phi})$ be a normal unit vector for this tangent plane pointing outside the reachable set. Then, the distance $d(\boldsymbol{\phi}) \geqslant 0$ from the origin to the tangent plane is $d(\boldsymbol{\phi}) = \hat{\mathbf{n}}^T(\boldsymbol{\phi})\hat{\mathbf{y}}(\boldsymbol{\phi})$. In normal form, the plane is given as

$$\hat{\mathbf{n}}^T(\boldsymbol{\phi})\,\mathbf{y} = d(\boldsymbol{\phi})$$

and the halfspace containing the reachable set is

$$\hat{\mathbf{n}}^T(\boldsymbol{\phi})\,\mathbf{y} \leqslant d(\boldsymbol{\phi})$$

This inequality can be added to the convex polytope defining the outer approximation, that is, the normal vector is appended to the rows in $\mathbf{A}_O$ and the corresponding distance is appended to $\mathbf{b}_O$:

$$\begin{bmatrix} \mathbf{A}_O \\ \hat{\mathbf{n}}^T(\boldsymbol{\phi}) \end{bmatrix} \mathbf{y} \leqslant \begin{bmatrix} \mathbf{b}_O \\ d(\boldsymbol{\phi}) \end{bmatrix}$$

Example 7.1.2: Adaptive Refinement (Cat II)

Let us illustrate the application of the refinement strategy from the example provided in Sec. 7.1.1 until the maximum of the scaled distances of all vertices from the outer approximation to the inner approximation is less than 0.01. These distances are computed after each refinement cycle for all vertices of the outer approximation. Each vertex with a scaled distance larger than 0.01 is refined in the next cycle. The refinement steps for the two-dimensional cases are shown in Figs. 7.12–7.16.

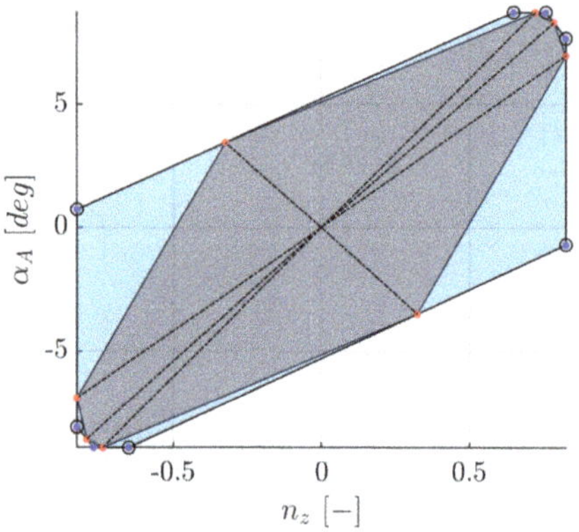

Fig. 7.12 Cat II (2D Lon), $n_{LP} = 8$.

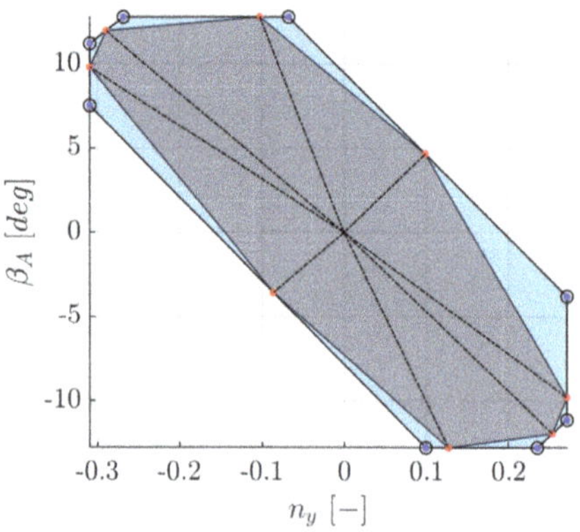

Fig. 7.13 Cat II (2D Lat), $n_{LP} = 8$.

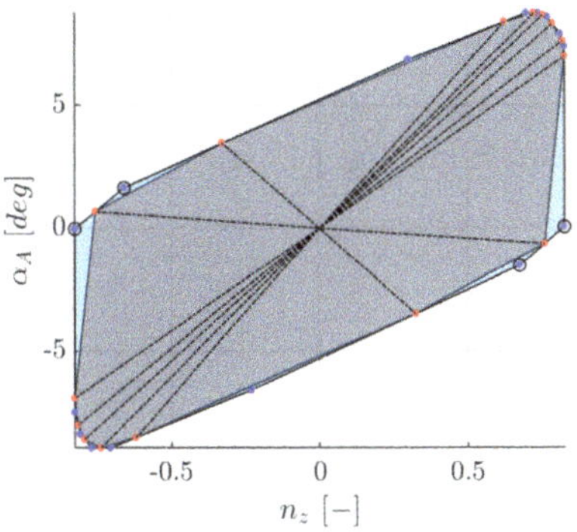

Fig. 7.14 Cat II (2D Lon), $n_{LP} = 15$.

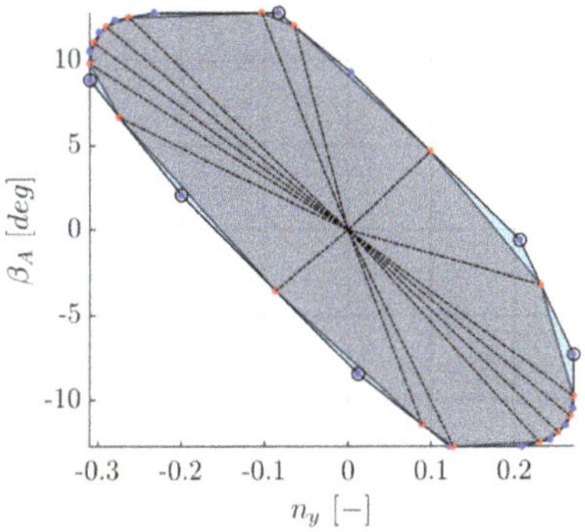

Fig. 7.15 Cat II (2D Lat), $n_{LP} = 16$.

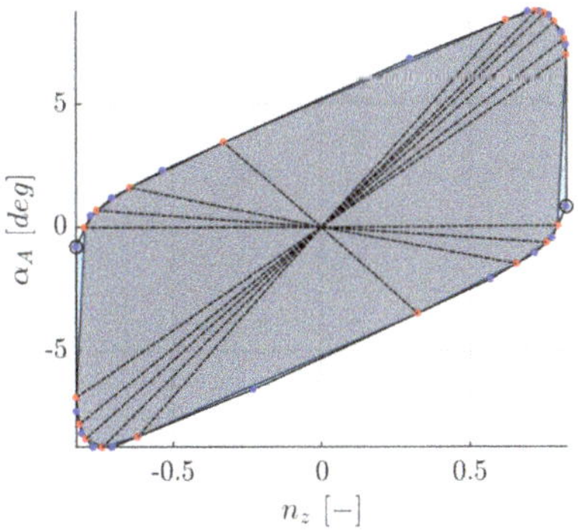

Fig. 7.16 Cat II (2D Lon), $n_{LP} = 19$.

In these figures, n_{LP} denotes the cumulative number of LP problems, and the vertices of the outer approximation (blue dots) that are selected to define the rays in the next refinement cycle are circled in black. The final outer set approximations are illustrated in Fig. 7.17 and Fig. 7.18.

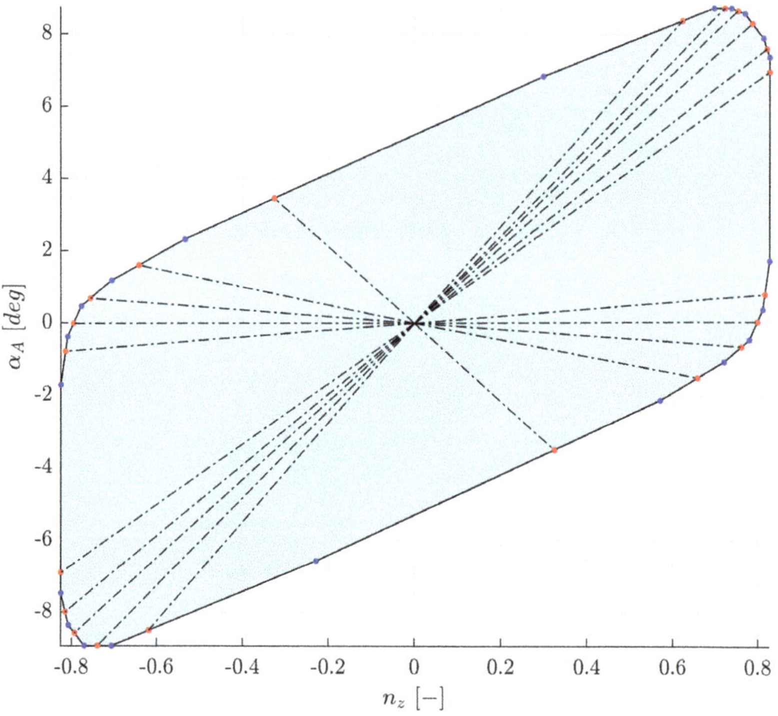

Fig. 7.17 Cat II (2D Lon), $n_{LP} = 21$.

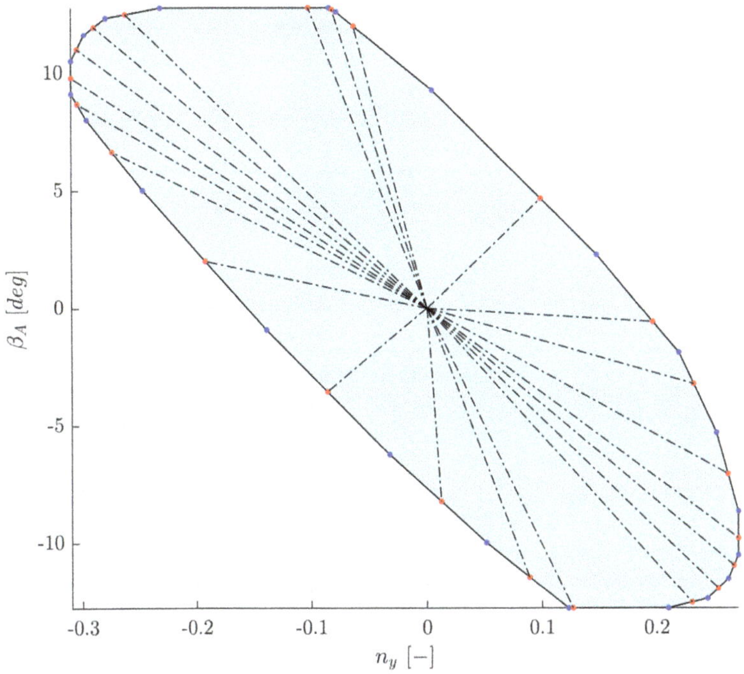

Fig. 7.18 Cat II (2D Lat), $n_{LP} = 22$.

Similarly, in three dimensions, the sequence of approximations is shown in Figs. 7.19–7.26.

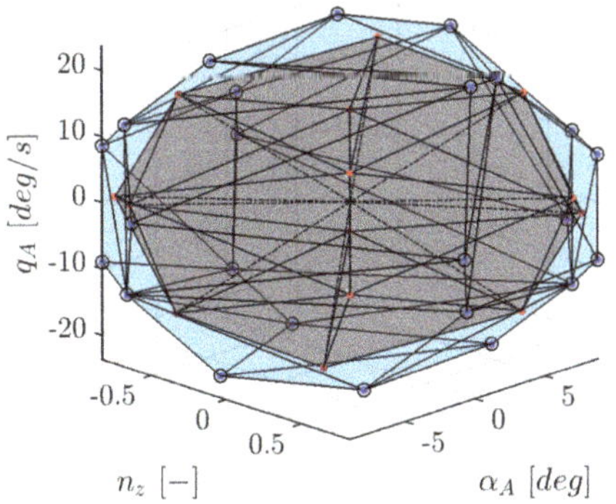

Fig. 7.19 Cat II (3D Lon) $n_{LP} = 14$.

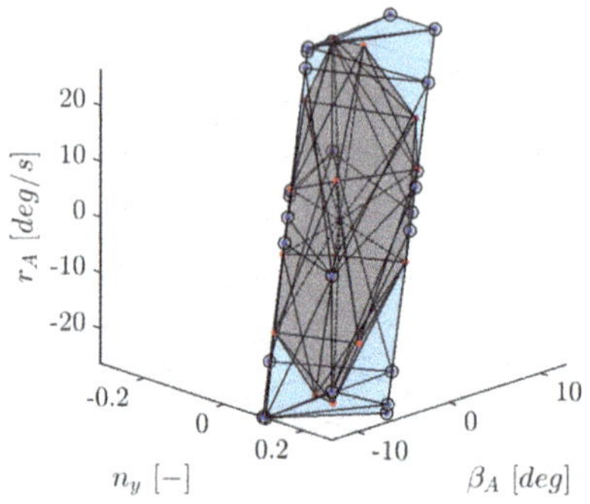

Fig. 7.20 Cat II (3D Lat) $n_{LP} = 14$.

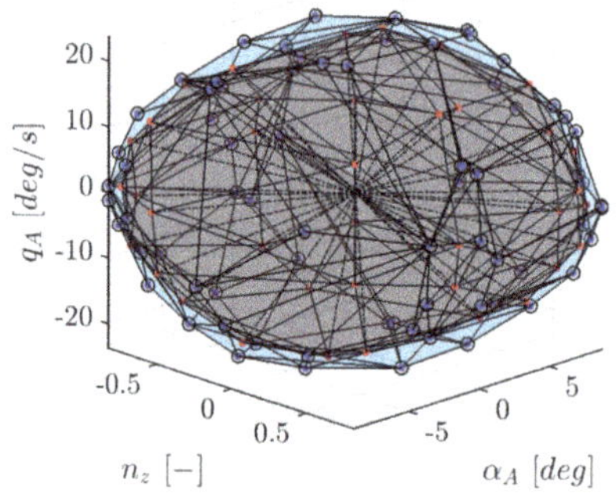

Fig. 7.21 Cat II (3D Lon) $n_{LP} = 38$.

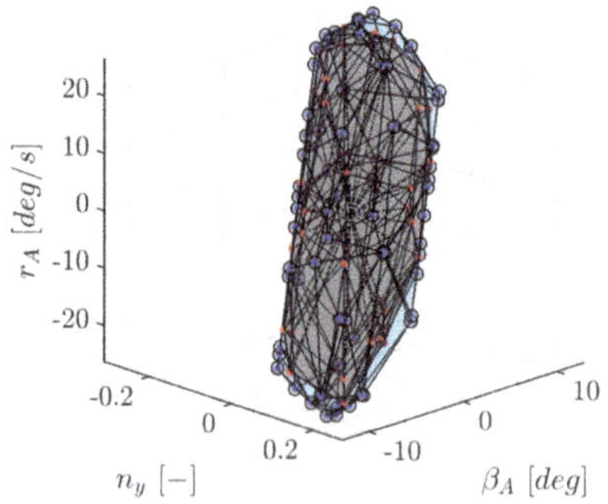

Fig. 7.22 Cat II (3D Lat) $n_{LP} = 38$.

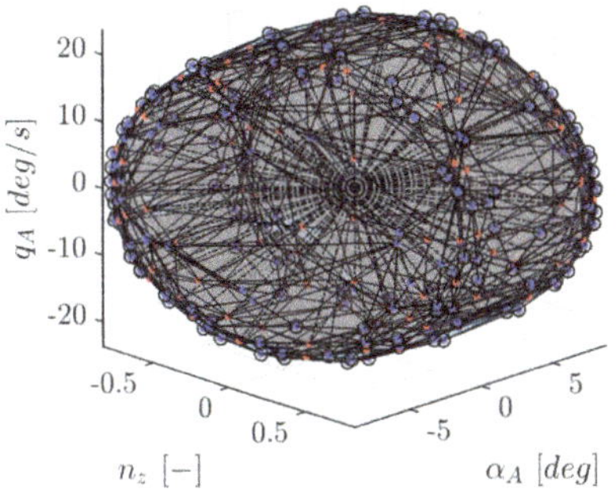

Fig. 7.23 Cat II (3D Lon) $n_{LP} = 110$.

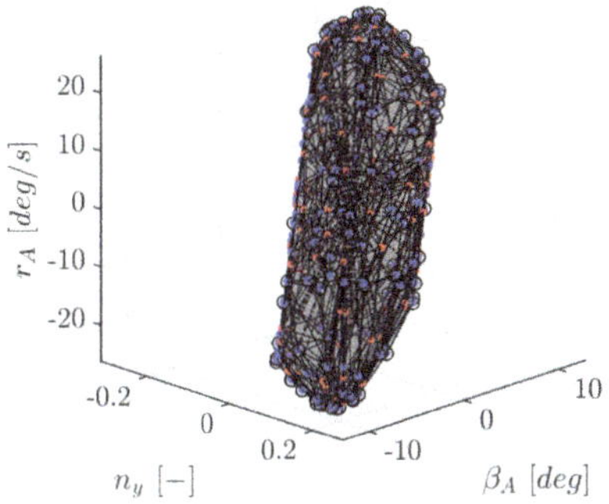

Fig. 7.24 Cat II (3D Lat) $n_{LP} = 110$.

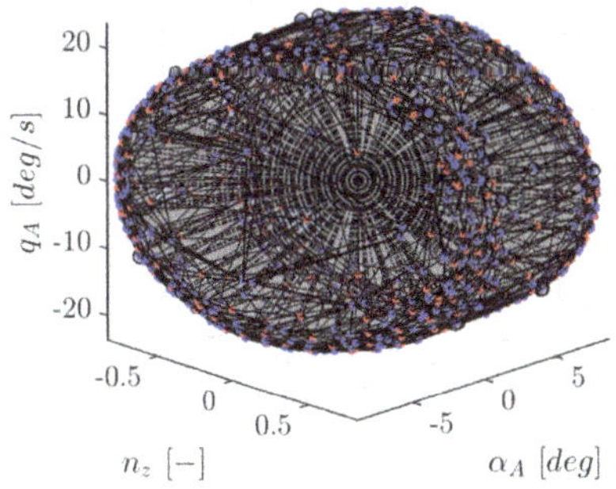

Fig. 7.25 Cat II (3D Lon) $n_{LP} = 288$.

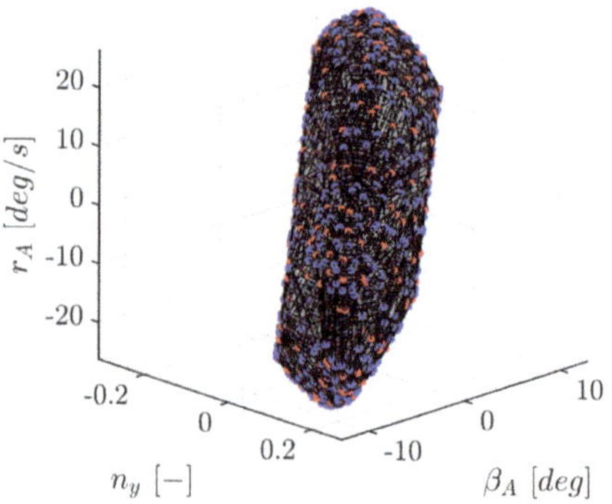

Fig. 7.26 Cat II (3D Lat) $n_{LP} = 263$.

The final outer set approximations are illustrated in Fig. 7.27 and Fig. 7.28.

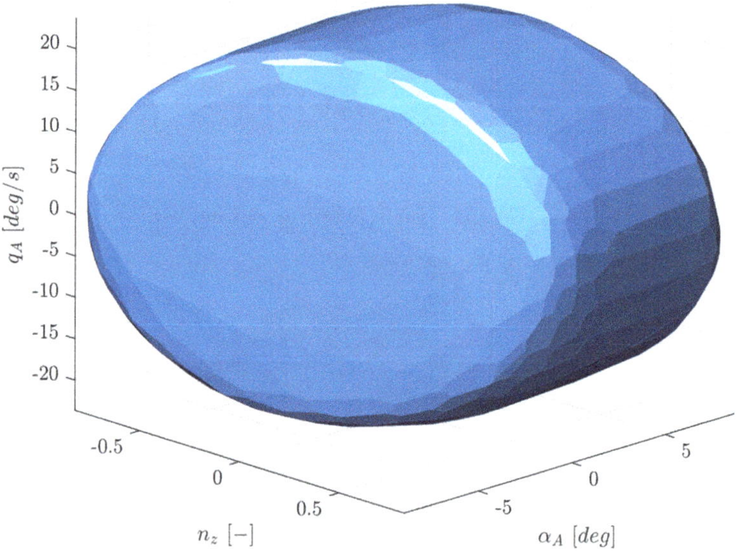

Fig. 7.27 Cat II (3D Lon) $n_{LP} = 349$.

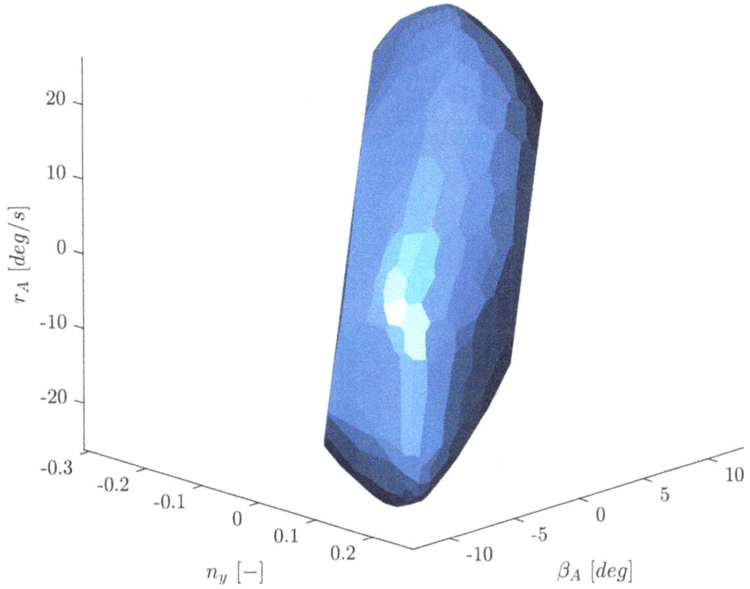

Fig. 7.28 Cat II (3D Lat) $n_{LP} = 300$.

7.2 NONLINEAR

Distance Field on Grids

The DFOG method uses a suitable optimal control formulation such that the "feasible set coincides with the reachable set and the optimal value of the problem yields the distance function" [85] to obtain a representation of the forward reachable set at the final time point t_f. Specifically, given a set of points $\mathcal{G}$ in the output space of $\mathbf{J} : \mathbb{R}^{n_x} \times \mathbb{R}^{n_p} \to \mathbb{R}^{n_J}$, a point in the reachable set associated with each $\mathbf{g} \in \mathcal{G}$ can be obtained by solving an appropriate optimal control problem minimizing

$$\left\| \mathbf{J}\left(\mathbf{x}\left(t_f\right), \mathbf{p}\right) - \mathbf{g} \right\|_2$$

For the DFOG method [85] the space of the outputs

$$\mathbf{y}(t) = \mathbf{J}\left(\mathbf{x}(t), \mathbf{p}\right)$$

is discretized by a suitable grid (e.g., equidistant Cartesian). Let $\mathcal{G}$ collect the grid points. Then, for each $\mathbf{g} \in \mathcal{G}$ a point $\hat{\mathbf{y}}\left(t_f\right) = \mathbf{J}\left(\hat{\mathbf{x}}\left(t_f\right), \hat{\mathbf{p}}\right)$ in the reachable set can be obtained from the solution of the following Cat III optimal control problem:

$$
\begin{aligned}
\underset{\mathbf{u}(t),\mathbf{p}}{\text{minimize}} \quad & \left\| \mathbf{J}\left[\mathbf{x}\left(t_f\right), \mathbf{p}\right] - \mathbf{g} \right\|_2 \\
\text{subject to} \quad & \dot{\mathbf{x}}(t) = \mathbf{f}\left[\mathbf{x}(t), \mathbf{u}(t), \mathbf{p}\right] \\
& \gamma\left[\mathbf{x}(t_0), \mathbf{p}\right] = \mathbf{0} \\
& \underline{\mathbf{p}} \leqslant \mathbf{p} \leqslant \overline{\mathbf{p}} \\
& \underline{\mathbf{x}} \leqslant \mathbf{x}(t) \leqslant \overline{\mathbf{x}} \\
& \underline{\mathbf{u}} \leqslant \mathbf{u}(t) \leqslant \overline{\mathbf{u}} \\
& t \in \left[t_0, t_f\right], \; t_0, t_f \text{ fixed}
\end{aligned}
$$

This problem can, for example, be transcribed using the trapezoidal collocation method to the following NLP:

$$
\begin{aligned}
\underset{\mathbf{x}_0,\mathbf{x}_i,\mathbf{u}_0,\mathbf{u}_i,\mathbf{p}}{\text{minimize}} \quad & \left\| \mathbf{J}\left(\mathbf{x}_N, \mathbf{p}\right) - \mathbf{g} \right\|_2 \\
\text{subject to} \quad & \mathbf{x}_i = \mathbf{x}_{i-1} + h\left(\frac{1}{2}\mathbf{f}\left(\mathbf{x}_{i-1}, \mathbf{u}_{i-1}, \mathbf{p}\right) + \frac{1}{2}\mathbf{f}\left(\mathbf{x}_i, \mathbf{u}_i, \mathbf{p}\right)\right) \\
& \gamma\left(\mathbf{x}_0, \mathbf{p}\right) = \mathbf{0} \\
& \underline{\mathbf{p}} \leqslant \mathbf{p} \leqslant \overline{\mathbf{p}} \\
& \underline{\mathbf{x}} \leqslant \mathbf{x}_i \leqslant \overline{\mathbf{x}} \\
& \underline{\mathbf{u}} \leqslant \mathbf{u}_i \leqslant \overline{\mathbf{u}} \\
& \underline{\mathbf{u}} \leqslant \mathbf{u}_0 \leqslant \overline{\mathbf{u}} \\
& i = 1, \ldots, N
\end{aligned}
$$

Engineering Note 7.2.1: Distance Field on Grids Computational Aspects

The reachable set approximation based on the DFOG method typically requires the solution of a large number of optimal control problems (depending on the cardinality of $\mathcal{G}$), which can be computationally demanding. Moreover, the solution to the NLP associated with each grid point is not necessarily global.

However, it is observed for the DFOG method "*that the lack of global optimality is often cured by considering many grid points*" [85]. Hence, the high number of optimal control problems required to be solved under this approach is expected to mitigate the local nature of the NLP solutions to a certain extent.

Also note that these problems are "embarrassingly parallel" as they are fully decoupled. This setting allows for a parallelization that can be expected to scale linearly in the number of grid points with a straightforward workload distribution. In addition, the initialization ("warm-start") of the optimal control problems (e.g., using an existing solution associated with the closest point) can significantly speed up the solution process.

Example 7.2.1: Tracking Error Estimation [86]

The system under investigation represents an F-16 closed-loop model featuring a Model Reference Adaptive Controller (MRAC). For this system, the plant is modeled by a short period approximation of the form

$$\dot{\mathbf{x}}_P(t) = \mathbf{A}_P \mathbf{x}_P(t) + \mathbf{B}_P \lambda_{M_\eta} u(t) \tag{7.5}$$

Here, the state vector of the plant $\mathbf{x}_P(t) = [\alpha_P(t), q_P(t)]^T$ collects the angle of attack $\alpha_P(t)$ as well as the pitch rate $q_P(t)$. Moreover, the input is denoted with u, and the system is subject to the parametric uncertainty $\lambda_{M_\eta} \in [0.75, 1.25]$ for the elevator effectiveness. In addition, the uncertain system matrix $\mathbf{A}_P$ is of the form

$$\mathbf{A}_P = \begin{bmatrix} Z_\alpha & 1 + Z_q \\ M_\alpha \lambda_{M_\alpha} & M_q \lambda_{M_q} \end{bmatrix}$$

and contains the entries $M_\alpha = -30.79$, $M_q = -3.75$, $Z_\alpha = -1.84$, and $Z_q = -0.09$, as well as the multiplicative uncertainties $\lambda_{M_\alpha} \in [0.75, 1.25]$ and $\lambda_{M_q} \in [0.75, 1.25]$ for the pitch stiffness and pitch damping. The uncertain input matrix

$$\mathbf{B}_P = \begin{bmatrix} 0 \\ M_\eta \end{bmatrix}$$

contains the coefficient $M_\eta = -12.75$. The reference model for the MRAC

$$\dot{\mathbf{x}}_M(t) = \mathbf{A}_M \mathbf{x}_M(t) + \mathbf{B}_M K_r r(t) \tag{7.6}$$

with the feedforward gain $K_r = -1.55$ has two states $\mathbf{x}_M(t) = [\alpha_M(t), q_M(t)]^T$, namely, the reference state for the angle of attack $\alpha_M(t)$ and the pitch rate $q_M(t)$.

These states should be tracked by the plant states $\alpha_P(t)$ and $q_P(t)$. The system matrix of the reference model $\mathbf{A}_M$ is of the form

$$\mathbf{A}_M = \begin{bmatrix} Z_\alpha & 1 + Z_q \\ M_{\alpha,M} & M_{q,M} \end{bmatrix}$$

with the desired values $M_{\alpha,M} = -11.38$ and $M_{q,M} = -4.16$. The reference command $r(t)$ for the angle of attack is subject to box-bounds $r(t) \in [-5\,\text{deg}, +5\,\text{deg}]$, and the input matrix $\mathbf{B}_M$ is defined as $\mathbf{B}_M = \mathbf{B}_P$. The reference model [Eq. (7.6)] exhibits desirable properties as it ensures stationary accuracy and has an ideal damping of $\frac{\sqrt{2}}{2}$. The task of the controller is to let the states of the plant states follow the states of the reference model, that is, $\mathbf{x}_P(t) \approx \mathbf{x}_M(t)$, as in this case these properties of the reference dynamics are transferred to the plant. The matching conditions

$$\mathbf{A}_M = \mathbf{A}_P + \mathbf{B}_P \lambda_{M_\eta} \mathbf{\Theta}_x^*(t) \tag{7.7}$$

$$\mathbf{B}_P K_r = \mathbf{B}_P \lambda_{M_\eta} \Theta_r^*(t) \tag{7.8}$$

relating the matrices of the plant [Eq. (7.5)] to those of the reference model [Eq. (7.6)] guarantee that this control objective is achievable. In these conditions, the so-called ideal parameters are denoted with $\mathbf{\Theta}_x^* \in \mathbb{R}^{1\times 2}$ and $\Theta_r^* \in \mathbb{R}$. Observe that these ideal parameters are uncertain as they depend on the multiplicative uncertainties λ_{M_α}, and λ_{M_q} through the plant matrix $\mathbf{A}_P$ and the uncertainty λ_{M_η}. The following adaptive control law (see Heise [87])

$$u(t) = \mathbf{\Theta}_x(t)\mathbf{x}_P(t) + \Theta_r(t)r(t) = \mathbf{\Theta}(t)\boldsymbol{\omega}(t) \tag{7.9}$$

$$\dot{\mathbf{\Theta}}^T(t) = -\mathbf{\Gamma}\boldsymbol{\omega}(t)\mathbf{e}_C^T(t)\mathbf{P}\mathbf{B}_P \tag{7.10}$$

with the regressor vector $\boldsymbol{\omega}^T(t) = [\mathbf{x}_P^T(t), r(t)]$, the parameter vector $\mathbf{\Theta}(t) = [\mathbf{\Theta}_x(t), \Theta_r(t)]$, as well as the error vector $\mathbf{e}_C(t) = \left[e_\alpha(t), e_q(t)\right]^T = \mathbf{x}_P(t) - \mathbf{x}_M(t)$, ensures that the plant asymptotically tracks the reference model; that is, $\lim_{t\to\infty} \mathbf{e}_C(t) = \mathbf{0}$.

In Eq. (7.10), the matrix $\mathbf{\Gamma} = diag([\Gamma, \Gamma, \Gamma])$ with the learning rate $\Gamma \in \mathbb{R}$ and the symmetric, positive definite matrix $\mathbf{P} \in \mathbb{R}^{2\times 2}$ obtained from the solution of the Lyapunov equation

$$\mathbf{A}_M^T\mathbf{P} + \mathbf{P}\mathbf{A}_M = -\mathbf{Q}$$

with $\mathbf{Q} = \mathbf{I}$ represent design parameters of the control law.

By adding and subtracting $\mathbf{B}_P K_r$ to the dynamic equations of the plant [Eq. (7.5)] as well as using the matching conditions [Eqs. (7.7) and (7.8)] yields:

$$\dot{\mathbf{x}}_P(t) = \mathbf{A}_M \mathbf{x}_P(t) + \mathbf{B}_P K_r r(t) + \mathbf{B}_P \lambda_{M_\eta}[u(t) - \boldsymbol{\Theta}^*(t)\boldsymbol{\omega}(t)],$$
$$\boldsymbol{\Theta}^*(t) = [\boldsymbol{\Theta}_x^*(t), \Theta_r^*(t)]$$

Moreover, inserting the adaptive control law [Eq. (7.9)], one obtains

$$\dot{\mathbf{x}}_P(t) = \mathbf{A}_M \mathbf{x}_P(t) + \mathbf{B}_P K_r r(t) + \mathbf{B}_P \lambda_{M_\eta}[\boldsymbol{\Theta}(t) - \boldsymbol{\Theta}^*(t)]\boldsymbol{\omega}(t)$$

Thus, the tracking error dynamics can be expressed as

$$\dot{\mathbf{e}}_C(t) = \dot{\mathbf{x}}_P(t) - \dot{\mathbf{x}}_M(t) = \mathbf{A}_M \mathbf{e}_C(t) + \mathbf{B}_P \lambda_{M_\eta}[\boldsymbol{\Theta}(t) - \boldsymbol{\Theta}^*(t)]\boldsymbol{\omega}(t)$$

Hence, the complete set of dynamic equations for the nonlinear closed-loop system can be written as

$$\dot{\mathbf{e}}_C(t) = \mathbf{A}_M \mathbf{e}_C(t) + \mathbf{B}_P \lambda_{M_\eta}(\boldsymbol{\Theta}(t) - \boldsymbol{\Theta}^*(t))\boldsymbol{\omega}(t)$$
$$\dot{\mathbf{x}}_M(t) = \mathbf{A}_M \mathbf{x}_M(t) + \mathbf{B}_P K_r r(t)$$
$$\dot{\boldsymbol{\Theta}}^T(t) = -\boldsymbol{\Gamma}\boldsymbol{\omega}(t)\mathbf{e}_C^T(t)\mathbf{P}\mathbf{B}_P$$

The quantification of the attainable set in the tracking error subspace for $e_\alpha(t)$ and $e_q(t)$ represents an important figure of merit to quantify the deviation between the states of the plant and the reference model. The results from the application of the DFOG method for $\Gamma = 10$ at $t_f = 1$ s estimating the reachable set for these tracking errors are depicted in Figs. 7.29–7.31.

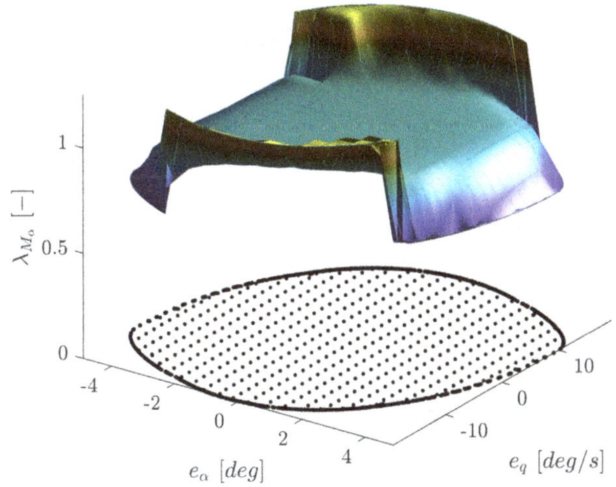

Fig. 7.29 $\lambda_{M_\alpha}, e_\alpha, e_q$.

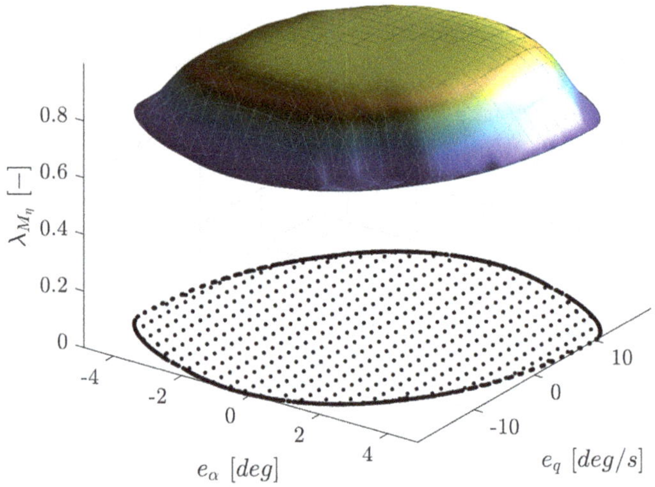

Fig. 7.30 $\lambda_{M_\eta}, e_\alpha, e_q$.

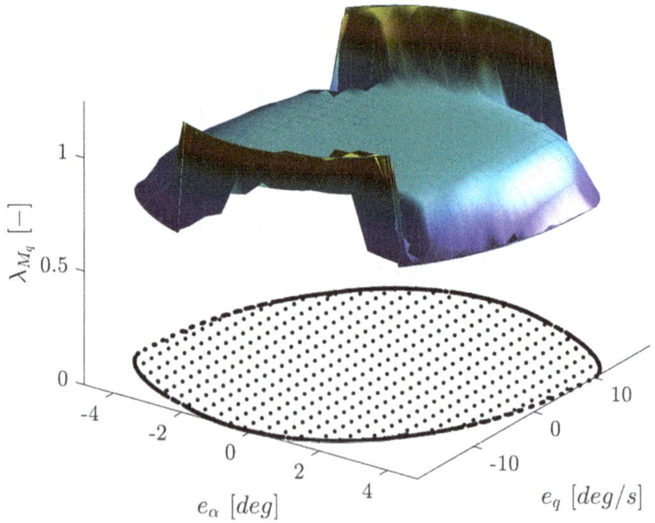

Fig. 7.31 $\lambda_{M_q}, e_\alpha, e_q$.

Summarizing, the application of the DFOG method for the estimation of reachable sets under the optimal control approach to flight control law clearance appears appealing as it

- allows a representation of the reachable set in a general setting (control constraints, state constraints, parameters, etc.),
- lends itself to the parallelization of all optimal control problems associated with each grid point, and
- mitigates the local nature of NLP solutions from the direct optimal control approach by considering a large number of problems.

Thus, it can be expected that the application of this optimal control approach in clearance tasks represents a valuable extension for Cat III–type problems.

PART III

Tools

The preceding part is primarily concerned with the description of computational methods for the optimal control approach to flight control law clearance. After all, "*a problem can never really be regarded as resolved until an efficient computational method of solution is derived*" ([88], p. 148). However, for computational methods to be successful in practice, efficient, accessible, and user-friendly software is required that can easily be integrated into existing tool-chains. Hence, the last part of this book highlights a selection of software tools to lower the entry barrier and equip practitioners with a starting point for the adoption of the methods from Part II in industry practice.

First, we discuss the COLIBRY* framework in Chapter 8. This tool constitutes an extensible library of optimization and optimal control-based testing methods. In particular, it features homogeneous user interfaces for preprocessing, solution, and postprocessing. For this framework, we focus on Cat I and II types of problems as well as the single shooting (Cat III) method. Note that the combination of these methods permits the implementation of the Cat III hybrid approach. It is worth mentioning that COLIBRY has also been employed for automated tests of large-scale aircraft closed-loop systems containing more than one hundred states in a continuous integration (CI) environment (see Braun et al. [89]). In this CI environment, optimal control-based tests are triggered automatically whenever changes in the closed-loop model are integrated on the remote server. This strategy has shown to be a valuable extension to the regular testing strategies, to perform sanity checks *early* and *often* during the development of flight control laws.

Following, Chapter 9 highlights relevant aspects of the direct optimal control framework FALCON.m.† This tool implements a segmented collocation method and constitutes a general solver for nonlinear-parametric optimal control problems. An integral part of this framework is a model builder module that can be used to create simulation models allowing for an efficient evaluation (including

*www.fsd.ed.tum.de/software/colibry

†www.fsd.ed.tum.de/software/falcon-m

derivatives). For this framework, we provide a generic description highlighting the most important concepts for implementing nonlinear-parametric clearance problems under the Cat III direct approach.

CHAPTER 8

The Counter-Optimization Library COLIBRY

COLIBRY [90, 91] (Counter-Optimization LIBRarY) is a MATLAB® software package featuring optimization- and optimal control-based testing methods. The term "counteroptimization" refers to the same branch of methods considered in this book for testing flight control laws (i.e., approaches employing optimization/optimal control strategies to search for violations of clearance criteria).

The main intention behind the user interface (UI) of COLIBRY is to provide an easy-to-use and structured modeling approach, enabling users to apply counteroptimization methods at a relatively high level of abstraction. In particular, it supports accessibility for users to a whole variety of methods without the need to learn a completely new interface for each particular method. Moreover, it allows for seamless integration with the MATLAB® and Simulink® ecosystem. From a software perspective, COLIBRY's architecture is designed to enable a modular extension of the toolbox with numerical optimization and optimal control solvers on the one side and application-centered counteroptimization methods on the other. This architecture ensures that the library concept behind COLIBRY supports extensibility through a high reuse of its core components (see Sec. 8.1). At the highest level, each "entry" in COLIBRY represents one counteroptimization method accessible by the user. The core classes in these entries are implemented against the same interfaces regarding the problem definition, validation, and solution, as well as the postprocessing capabilities, including visualization and report generation [91].

In this chapter, we cover the main modeling concepts to implement linear and linear-parametric Cat I and II types of problems as well as the single shooting method (Cat III) in COLIBRY (see Sec. 8.2). We would like to emphasize that this chapter is not intended as a substitute for the software manual (see Braun [90]) or demos provided with COLIBRY. Instead, we would like to establish the connection between the computational methods presented in this book and the modeling concepts in COLIBRY in a generic fashion to facilitate the transition from theory to practice. In particular, we strongly encourage readers to consult the demos provided in COLIBRY illustrating the library entries with several numerical examples.

8.1 LIBRARY CONCEPT

COLIBRY's software architecture is based on a multilayered library concept. The library entries at the top level represent counteroptimization methods available to the user. These top-level entries map the *engineering* counteroptimization problem through internal entries to bottom-level library entries, which solve a *mathematical* optimization problem. Entries at this bottom level do not depend on other entries within COLIBRY and essentially constitute wrappers for external mathematical libraries that are specialized in solving a particular type of mathematical problem (e.g., linear programs and NLP). Most importantly, all entries derive the core components (see Fig. 8.1) regarding problem setup and solution from the same abstract superclasses (see Braun et al. [91] for a detailed description of COLIBRY's library concept/software architecture):

- `colibry.Problem`: class of a problem to be solved
- `colibry.Solver`: class of a solver (takes a problem and produces a solution)
- `colibry.Solution`: class of a solution to a problem
- `colibry.Inspector`: class for checking the validity of a problem
- `colibry.Visualizer`: class for visualization tasks (problem and solution)

This library concept is intended to promote the reusability of entries. For example, the top-level entry `colibry.singleshooting` maps to the internal entry `colibry.blackboxnlp`, representing a general entry for solving a (black-box) parameter optimization problem. This entry then maps to bottom-level entries such as `colibry.fmincon` or `colibry.surrogateopt`. As the `colibry.blackboxnlp` entry provides a generic interface for solving black-box optimization problems, it can, for example, also be reused by the top-level `colibry.bilevellinearoptimalcontrol` entry for solving the parameter optimization problem in the upper level (see Sec. 5.5).

The central element regarding the problem definition is a problem object that needs to be instantiated for each COLIBRY entry. For example, in the case of

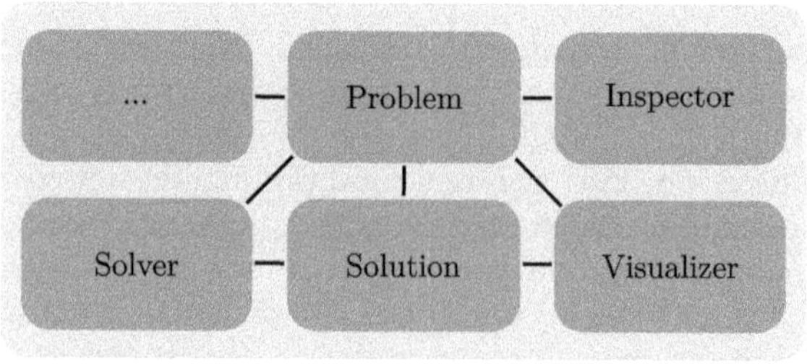

Fig. 8.1 Core library entry components.

the solution to a Cat I/II-type clearance problem, the user can instantiate an object of type `colibry.linearoptimalcontrol.Problem`. State, control, parameter, and output variables associated with a model can then be added to this object. Also, options regarding the method (length of the time interval, discretization type, discretization step size, etc.) or the particular numerical algorithm can be specified.

Regarding user implementations (e.g., related to model or objective functions), COLIBRY provides generic interface classes. For a particular analysis, the user can derive a new class from such an interface and implement the abstract methods of this class. It is worth mentioning that, if a library entry requires more than one type of implementation, the class implemented by the user can also inherit from more than one interface (multiple inheritance). An instance of this new user-defined class is then handed to the problem object. These interface classes ensure that the function prototypes are correctly implemented by the user. Additionally, during execution, checks ensure that for each call the values returned from the implementation are valid and consistent regarding the problem definition.

After a numerical solution is obtained, the user receives a solution object. This solution object can be used to access the solution data and perform postprocessing tasks. For example, in the case of the `colibry.linearoptimalcontrol` entry, the solution object contains optimal control, state, and output histories as well as the cost (including postoptimal sensitivities for parametric problems). Moreover, the associated visualizer and report-generation objects are available to generate plots or reports for the results of the analysis.

8.2 USER INTERFACES

The COLIBRY entry `colibry.linearoptimalcontrol` can be used to implement Cat I/II-type clearance problems for linear-parametric models. In addition to the system matrix $\mathbf{A} \in \mathbb{R}^{n_x \times n_x}$, input matrix $\mathbf{B} \in \mathbb{R}^{n_x \times n_u}$, and output matrix $\mathbf{C} \in \mathbb{R}^{n_J \times n_x}$, the user needs to supply a reference state vector $\mathbf{x}_r \in \mathbb{R}^{n_x}$, reference control vector $\mathbf{u}_r \in \mathbb{R}^{n_u}$, and reference output vector $\mathbf{y}_r \in \mathbb{R}^{n_J}$ for this type of model. Such a linear model, including the reference points, can be derived based on the linearization at a trim point from a nonlinear-parametric model with a right-hand side function

$$\dot{\mathbf{x}}(t;\mathbf{p}) = \mathbf{f}\left[\mathbf{x}(t), \mathbf{u}(t); \mathbf{p}\right]$$

and output function

$$\mathbf{y}(t;\mathbf{p}) = \mathbf{h}\left[\mathbf{x}(t); \mathbf{p}\right]$$

Both functions are allowed to depend on the parameter vector $\mathbf{p} \in \mathbb{R}^{n_p}$. Notably, also the reference values are allowed to depend on the parameter vector; that is,

$\mathbf{x}_r(\mathbf{p})$, $\mathbf{u}_r(\mathbf{p})$, as well as $\mathbf{y}_r(\mathbf{p}) = \mathbf{h}(\mathbf{x}_r(\mathbf{p}); \mathbf{p})$. These reference values are considered in COLIBRY automatically as offsets not only during the solution of the clearance problem but also in the postprocessor (e.g., for the visualization of the optimal control, state, and output histories). For a particular parameter vector $\mathbf{p}$, the linear system matrices for a nonlinear-parametric model can be obtained based on the Jacobians evaluated at the reference state vector $\mathbf{x}_r(\mathbf{p})$ and the corresponding reference control vector $\mathbf{u}_r(\mathbf{p})$:

$$\mathbf{A}(\mathbf{p}) = \frac{\partial \mathbf{f}}{\partial \mathbf{x}}[\mathbf{x}_r(\mathbf{p}), \mathbf{u}_r(\mathbf{p}); \mathbf{p}]$$

$$\mathbf{B}(\mathbf{p}) = \frac{\partial \mathbf{f}}{\partial \mathbf{u}}[\mathbf{x}_r(\mathbf{p}), \mathbf{u}_r(\mathbf{p}); \mathbf{p}]$$

$$\mathbf{C}(\mathbf{p}) = \frac{\partial \mathbf{h}}{\partial \mathbf{x}}[\mathbf{x}_r(\mathbf{p}); \mathbf{p}]$$

The MATLAB® function prototype for the model interface derived from `colibry.model.Linear` is defined in COLIBRY as follows:

```
provideLinearModel(obj, A, B, C, x_r, u_r, y_r, p)
%  > obj: Object                          (colibry.model.Linear)
% Outputs:
%  > A:   System Matrix                  — Matrix (colibry.Scalar)
%  > B:   Input Matrix                   — Matrix (colibry.Scalar)
%  > C:   Output Matrix                  — Matrix (colibry.Scalar)
%  > x_r: Reference State Vector         — Vector (colibry.Scalar)
%  > u_r: Reference Control Vector       — Vector (colibry.Scalar)
%  > y_r: Reference Output Vector        — Vector (colibry.Scalar)
% Inputs:
%  > p:   Parameters                     — Vector (colibry.Scalar)
```

To implement the interface, the user needs to derive a new class from the abstract superclass `colibry.model.Linear`:

```
classdef MyModel < colibry.model.Linear
    properties
        % custom properties...
    end
    methods
        function MyModel(obj, varargin)
            % custom constructor...
        end
        function provideLinearModel(obj, A, B, C, x_r, u_r,
            y_r, p)
```

```
                % implementation of the linear model...
            end
        end
    end
```

The information flow regarding the implementation in this class, considering the described influence of parameters on the linearized model and the reference vectors, is illustrated in Fig. 8.2.

The basic steps for setting up a linear parameter-dependent Cat I/II-type clearance problem in COLIBRY are as follows:

1. **Problem**: Create a `colibry.linearoptimalcontrol.Problem` instance.

   ```
   name    = % identifier of the problem
   problem = colibry.linearoptimalcontrol.Problem('Name',
       name);
   ```

   The instantiation of a problem object of this type allows the definition of Cat I/II-type clearance problems.

2. **Cost**: Create and add a `colibry.objective.MayerCost` instance to the problem.

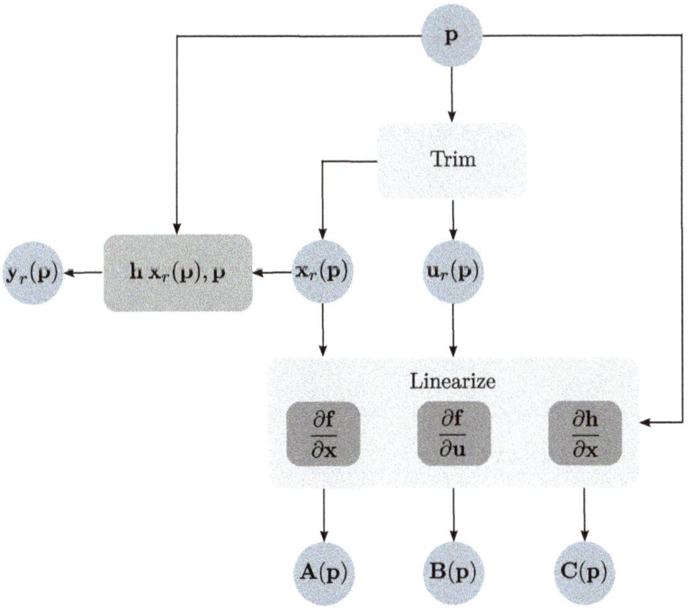

Fig. 8.2 Information flow for parametric trimming and linearization.

```
type = % "max" or "min" (string)
cost = colibry.objective.MayerCost(type);

out_id = % output identifier (string)
weight = % weight              (number)
cost.addComponentAndWeight(out_id, weight);

% other components and weights...

problem.setObjective(cost);
```

The standard cost type in COLIBRY for linear optimal control methods is a Mayer-type cost (i.e., a state-dependent cost evaluated at the final time point). The string passed to the constructor determines if the counteroptimization method maximizes or minimizes the objective. An arbitrary number of components and associated weights can be added to the Mayer-type cost. Each identifier of the added components needs to match one of the state-dependent outputs in the problem definition.

3. **Model**: Create and add a model instance to the problem.

```
model = % model object (child class of colibry.model.
    Linear)
problem.setModel(model);
```

The model object is required to be an instance of a user-defined class implemented against the `colibry.model.Linear` interface.

4. **States, Controls, and Outputs**: Add states, controls, and outputs to the problem.

```
name  = % identifier  (string)
lower = % lower bound (number)
upper = % upper bound (number)
trafo = % conversions for visualization (function_handle)
units = % plot  units for visualization (string)
label = % latex label for visualization (string)


problem.addState(name,...
    'Bounds',     [lower, upper],...
    'PlotUnit',   units,...
    'PlotConv',   trafo,...
    'PlotName',   label);

% other states as well as controls and outputs...
```

For all variables (states, controls, outputs), an identifier needs to be provided. In addition, a custom latex label (the identifier of the variable is set by default), units to be added to the label, as well as a transformation function handle can be specified. The transformation function is a convenience feature to convert between units (e.g., radians to degrees) in plots. Additionally, states can be bounded (Cat II), and controls must be bounded from below and above.

5. **Parameters**: Add parameters to the problem.

```
name  = % identifier  (string)
value = % fixed value (number)
problem.addParameter(name, 'Value', value);

% other parameters...
```

For problems of type `colibry.linearoptimalcontrol.Problem`, parameters are fixed to the provided values, and the first-order (postoptimal) cost sensitivities are computed after a successful solution.

6. **Final Time**: Set the final time of the problem.

```
final_time = % fixed final time (number)
problem.setFinalTime(final_time);
```

The time interval of the problem is defined as $[0, t_f]$, where t_f is the fixed final time.

7. **Options**: Set options.

```
step_size   = % discretization step size (number)
method      = % discretization method    (e.g. '
    BackwardEuler')
solver_type = % solver type              (e.g. 'clp')
problem.setOption('StepSize',              step_size);
problem.setOption('DiscretizationMethod', method);
problem.setOption('SolverType',            solver_type);

% other options...
```

Options represent general settings for the counteroptimization method.

8. **Solver Arguments**: Set solver arguments.

```
max_time = % maximum solution time         (number)
max_iter = % maximum number of iterations (number)
problem.setSolverArgument('MaxTime',        'max_time');
problem.setSolverArgument('MaxIterations', 'max_iter');

% other arguments...
```

Solver arguments represent specific settings for the numerical solver.

9. **Solution**: Solve the problem.

```
solution = solve(problem);
```

During the solution process, the solver first checks the validity of the problem object. Following, it maps the validated problem to bottom-level COLIBRY library entries and solves the problem. Finally, an instance of type `colibry.linearoptimalcontrol.Solution` is returned.

10. **Optimal Values**: Extract optimal values.

```
% optimal cost value
solution.OptimalCost.Value

% post-optimal sensitivities of the cost
solution.OptimalCost.Sensitivity

% optimal values of the state 'x_1'
solution.OptimalStateGridVector.x_1.Values

% similar for the values of controls and outputs...

% values of the switching function of the control 'u_1'
solution.SwitchingFunctionGridVector.u_1.Values
```

The raw values of the optimal solution can be extracted as MATLAB® arrays from the solution object.

11. **Visualization**: Visualize optimal state, control, and output histories.

```
% get visualizer instance
visualizer = solution.visualize();

% plot states
visualizer.plotStates();

% similar for controls, outputs...
```

The visualizer object can be used to create plots for all optimal state, control, and output histories.

12. **Report**: Generate a report for the counteroptimization problem.

```
solution.report();
```

The report generator creates a summary of the main results and exports it into an external file (Portable Document Format).

Recall that COLIBRY aims to provide a uniform UI across different counteroptimization methods. Hence, the library entries implementing the bilevel approach (see Sec. 5.5) for parametric problems as well as the single shooting (black-box optimization) approach essentially follow the same setup/steps as shown previously. In the case of the bilevel method, the only differences are that the problem must be of type `colibry.bilevellinearoptimalcontrol.Problem`,

```
problem = colibry.bilevellinearoptimalcontrol.Problem('Name',
    name);
```

lower and upper bounds on the parameter vector must be specified,

```
name  = % identifier    (string)
value = % nominal value (number)
lower = % lower bound   (number)
upper = % upper bound   (number)
problem.addParameter(name,...
    'Value',            value,...
    'Bounds',           [lower, upper]);

% other parameters...
```

and additional options as well as the solver arguments for the upper-level problem can be provided by the user:

```
sol_type = % upper-level solver type         (string)
alg_type = % upper-level algorithm type      (string)
max_time = % upper-level max solution time (number)
problem.setOption(         'UpLvlSolverType', sol_type);
problem.setSolverArgument('UpLvlAlgorithm',  alg_type);
problem.setSolverArgument('UpLvlMaxTime',    max_time);

% other options and solver arguments...
```

All other UI elements in the problem setup, preprocessing, and postprocessing remain the same. Similarly, the problem setup regarding the single shooting method in COLIBRY follows essentially the same steps.

Here, the differences are that the problem must be of type `colibry.singleshooting.Problem`,

```
problem = colibry.singleshooting.Problem('Name', name);
```

a control discretization step size needs to be specified,

```
step_size = % control step size (number)
problem.setOption('ControlStepSize', step_size);
```

and the class of the model instance has to be derived from the `colibry.model.Simulation` interface and implement the `simulate` function:

```
simulate(obj, x, y, u, p, t_f)
%  > obj: Object                (colibry.model.Simulation)
% Outputs:
%  > x:   States     — Vector (colibry.TimeSeries)
%  > y:   Outputs    — Vector (colibry.TimeSeries)
% Inputs:
%  > u:   Controls   — Vector (colibry.TimeSeries)
%  > p:   Parameters — Vector (colibry.Scalar)
%  > t_f: Final Time — Scalar (colibry.Scalar)
```

Note that a sequential solution of counteroptimization problems using Cat II (bilevel), Cat I, and the single shooting approach based on the switching time parameterization represent the Cat III hybrid method (see Sec. 6.2).

Engineering Note 8.2.1: Switching Time Parameterization (Cat III Hybrid)

Recall that the cost J in the Cat III hybrid method depends on the switching time points collected in $\mathbf{t} \in \mathbb{R}^{n_t}$:

$$J(\mathbf{t})$$

These switching time points define a control parameterization derived from the Cat I solution. A possible implementation of this type of parameterization in COLIBRY is to add the switching time points t_k, $k = 1, \ldots, n_t$ as parameters bounded by

$$0 \leqslant t_k \leqslant t_f$$

where t_f is the fixed final time. The control inputs for the simulation can then be constructed inside the `simulate` method based on these switching time points. Note that sorting the time points before the construction implicitly enforces monotonicity, thus avoiding the explicit introduction of the constraints $t_1 \leqslant t_2 \leqslant \ldots \leqslant t_{n_t}$.

CHAPTER 9

The FSD Optimal Control Toolbox FALCON.m

FALCON.m [29, 34, 35] (Fsd optimAL CONtrol) is a direct optimal control framework implemented in MATLAB®. Similar to COLIBRY, this framework is developed in an object-oriented programming paradigm and exposes classes in the user interface to describe, solve, and analyze general optimal control problems. In addition, it features consistency checks of the problem provided by the user as well as preprocessing and postprocessing capabilities such as scaling checks and the computation of costate estimates following a successful solution of the optimal control problem [35]. FALCON.m also contains builder modules, which can be used to create model, cost, and constraint functions [29]. In particular, FALCON.m provides the following features:

- **Transcription**: The two main simulation methods used for transcribing the optimal control problem from continuous time to a discretized version are the backward Euler and trapezoidal (default) discretization method (see Sec. 2.2.1).
- **Evaluation**: The FALCON.m model builder [29] relies on a subsystem-based approach leveraging symbolic computations, code generation, and compilation in order to enable the efficient evaluation of the model, cost, and constraint functions including derivatives (see Sec. 1.4).
- **Solvers**: FALCON.m provides the option to choose between several nonlinear programming solvers. The standard solver is IPOPT [24] (see Sec. 1.3).

In this chapter, we focus on a subset of the features available in FALCON.m that are deemed relevant to solve general nonlinear-parametric clearance problems using the Cat III direct approach (see Sec. 6.1). Just as in the preceeding chapter, we would like to emphasize that this chapter *highlights* specific capabilities of FALCON.m and provides implementation hints to facilitate an application of the optimal control approaches presented in this book. However, readers are encouraged to consult the software manual for a full feature description as well as the illustrative examples contained in the package.

9.1 MODEL BUILDER

The FALCON.m simulation model builder can be used for implementing the right-hand side and output function of an ordinary differential equation modeling the aircraft closed-loop system. It should be mentioned that there exist also other builder classes for implementing cost and constraint functions. However, for the clearance formulations considered in this book, it suffices to focus our attention solely on the simulation model builder.

FALCON.m defines the following MATLAB® evaluation prototype for the simulation model*:

$$[\dot{\mathbf{x}}, \mathbf{y}] = \mathbf{f}(\mathbf{x}, \mathbf{u}, \mathbf{p})$$

Note that in this definition the output function and the function computing the state time derivatives are combined into a single function. Additionally, FALCON.m exposes the following data types associated with the input and output arguments in its user interface:

- `falcon.State`: class of states $\mathbf{x}$
- `falcon.Control`: class of controls $\mathbf{u}$
- `falcon.Parameter`: class of parameters $\mathbf{p}$
- `falcon.Output`: class of outputs $\mathbf{y}$

No data type related to the state time derivatives $\dot{\mathbf{x}}$ is required.

As a mental model for the construction of models using FALCON.m builder classes described in the following, the reader may find it helpful to draw parallels to the subsystem-based modeling concepts in Simulink®.† A basic workflow for creating a simulation model in FALCON.m can be described as follows:

1. **Data Types**: Create MATLAB® arrays of FALCON.m data types for the state, control, parameter, and output vector.

```
x_ids = % cell array of identifiers (char)
x     = cellfun(@(x_id)falcon.State(x_id), x_ids);
% similar for parameters, controls, and outputs...
```

   For each data type, a name has to be provided. This name then becomes available as an identifier in the definition of subsystem arguments. Outputs can be used in other cost and constraint functions, or to impose path constraints. This can be useful if quantities are computed in the model that need to be constrained along the trajectory or contribute to the cost.

*As an extension to this standard prototype, an arbitrary number of constants can be added.

†Even though the model in FALCON.m is constructed programmatically with input/output connections being established purely based on provided identifiers, both Simulink®, as well as the FALCON.m builder classes, ultimately can be seen as instances of the abstract graph-based modeling approach discussed in Sec. 1.4.1.

2. **Builder Instance**: Create a `falcon.SimulationModelBuilder` instance.

```
mdl_id = % model identifier (char)
mdl    = falcon.SimulationModelBuilder(mdl_id, x, u, p);
```

The methods provided in this builder class represent the user interface exposed by FALCON.m for the construction of simulation models.

3. **Symbolic Subsystems**: Add functions that are callable with symbolic variables.

```
fhandle = % subsystem function
    (function_handle)
inputs  = % cell array of input  identifiers (char)
outputs = % cell array of output identifiers (char)
mdl.addSubsystem(fhandle,...
    'Inputs',  inputs,...
    'Outputs', outputs);
```

Symbolic subsystems are the default and recommended option. For symbolic subsystems, FALCON.m calls the provided function internally with symbolic variables to obtain symbolic expressions for each of the outputs. The derivatives together with the sparsity patterns are then computed symbolically. Moreover, the expressions obtained in this way are transformed into an executable form.

4. **Derivative Subsystems**: Add functions including derivative information.

```
fhandle = % subsystem function
    (function_handle)
inputs  = % cell array of input  identifiers (char)
outputs = % cell array of output identifiers (char)
pattern = % sparsity pattern (full matrix)
mdl.addDerivativeSubsystem(fhandle,...
    'Inputs',                    inputs,...
    'Outputs',                   outputs,...
    'OutputJacobianSparsity', pattern);
```

This type of subsystem represents a fallback option if the symbolic tool-chain cannot be applied. For derivative subsystems, FALCON.m expects the user to implement the derivatives. The reader may find the methods presented in Sec. 1.4.2 and Sec. 1.4.3 helpful for this purpose.

5. **Return Arguments**: Provide state time derivatives and outputs.

```
x_dot_ids = % cell array of derivative identifiers (char)
mdl.setStateDerivativeNames(x_dot_ids);
mdl.setModelOutputs(y);
```

Each state time derivative and output identifier is required to be provided as one of the output identifiers of exactly one subsystem.

6. **Build**: Build the model.

```
mdl_hdl = mdl.Build();
```

The build method performs consistency checks and ultimately produces a MATLAB® executable (MEX) function. In this function, the subsystems are joined and the derivatives are computed using the forward derivative procedure (see Sec. 1.4.1).

9.2 PROBLEM DESCRIPTION

In the following, we focus on the construction of a single-phase, Mayer-type problem in FALCON.m. This problem type can be used to model a general nonlinear-parametric clearance optimal control problem (Cat III direct, see Sec. 6.1). Here we assume that the user has defined an output vector $\mathbf{y} \in \mathbb{R}^{n_J}$, and it is desired to use a weighted combination of the outputs as the criterion to be minimized at a fixed final time t_f:

$$\sum_{i=1}^{n_J} w_i y_i \left(t_f\right), \; w_i \in \mathbb{R} \tag{9.1}$$

A basic workflow for setting up and solving a problem in FALCON.m is given as follows:

1. **Data Types**: Specify properties of states, controls, parameters, and outputs.

```
lower   = % lower bound              (number)
upper   = % upper bound              (number)
scale   = % scaling factor           (number)
index   = % index of vector element  (number)
x(index).setLowerBound(lower);
x(index).setUpperBound(upper);
x(index).setScaling(scale);
% similar for controls, parameters, and outputs...
```

Among other properties, lower and upper bounds as well as scaling factors can be set for the array elements of state, control, parameter, and output FALCON.m data types. This can be accomplished through setter methods.‡ A common rule of thumb in practice is to set the scaling factors such that the expected values of the quantities during the course of the optimization range between −0.5 and 0.5.

‡Here, the data types created for the simulation model builder can be reused. Alternatively, these values can be provided directly to the constructors.

2. **Problem Instance**: Create a `falcon.Problem` instance.

```
name    = % identifier (char)
problem = falcon.Problem(name);
```

The methods provided in the `falcon.Problem` class represent the user interface for a problem definition exposed by FALCON.m.

3. **Phase**: Add a new phase to the problem.

```
model = % model handle          (mdl_hdl from builder)
start = % start time            (number)
final = % final time            (number)
ngrid = % normalized time grid (number array)
phase = problem.addNewPhase(model, x, ngrid, start, final)
    ;
```

When a new phase is added to the problem, a discretized state grid is created for this phase. Control grids need to be added in a separate step.

4. **Control**: Add a new control grid.

```
ngrid = % normalized time grid (number array)
phase.addNewControlGrid(u, ngrid);
```

If no normalized time grid is specified, FALCON.m uses the normalized grid of the state grid by default. If the control grid discretization is coarser than the state grid discretization, intermediate values are calculated by interpolation.

5. **Initial Condition**: Specify the initial condition of the trajectory.

```
lower = % lower bounds on initial states (number array)
upper = % upper bounds on initial states (number array)
phase.setInitialBoundaries(lower, upper);
```

An equality boundary condition can be imposed by setting the lower and upper bounds to the same value or by providing only a single vector containing the initial values to the setter.

6. **Cost**: Specify the Mayer-type cost.

```
w = % weights for all outputs (number array)
problem.addNewLinearPointCost(y, phase, 1, 'Weight', w);
```

The weighted combination of outputs [see Eq. (9.1)] is evaluated at the final point of the phase.

7. **Solution**: Solve the problem.

```
max_iters = % maximum number of iterations      (number)
tolerance = % feasibility/optimality tolerance (number)
```

```
problem.setMaxIterations(max_iters);
problem.setMajorOptTol(tolerance);
problem.setMajorFeasTol(tolerance);
problem.Solve();
```

The standard solver used in FALCON.m is IPOPT (see Sec. 1.3 regarding a description of the solver output).

8. **Optimal Values**: Extract optimal values.

```
% optimal state grid values
phase.StateGrid.Values

% optimal control grid values
phase.ControlGrids.Values

% optimal output grid values
phase.Model.ModelOutputGrid.Values
```

Optimal results are accessible through the public property "`Values`" in the respective grids.

Blending Functions (Two-Dimensional Example)

The following example illustrates the specific form of combined steps:

$$\tilde{\boldsymbol{\beta}}\left(x_1, x_2; \mathcal{K}_1, \mathcal{K}_2, s\right) = \begin{bmatrix} \tilde{\beta}_1\left(x_1, x_2; \mathcal{K}_1, \mathcal{K}_2, s\right) \\ \vdots \\ \tilde{\beta}_9\left(x_1, x_2; \mathcal{K}_1, \mathcal{K}_2, s\right) \end{bmatrix}$$

in two dimensions ($d = 2$) for the coordinates x_1 and x_2. The constant $s > 0$ controlling the smoothness of the step functions as well as the knot grids $\mathcal{K}_1$ and $\mathcal{K}_2$ (each with two elements, i.e., $k_1^1 < k_2^1$ for $\mathcal{K}_1$ and $k_1^2 < k_2^2$ for $\mathcal{K}_2$) are assumed given. Based on these grids, the combined steps are constructed based on Eqs. (1.17) as follows:

$$\begin{aligned}
\tilde{\boldsymbol{\beta}}\left(x_1, x_2; \mathcal{K}_1, \mathcal{K}_2, s\right) &= \boldsymbol{\beta}\left(x_1; \mathcal{K}_1, s\right) \otimes \boldsymbol{\beta}\left(x_2; \mathcal{K}_2, s\right) \\
&= \begin{bmatrix} \rceil\!\lfloor\left(x_1; k_1^1, s\right) \\ \rfloor\!\lceil\left(x_1; k_1^1, s\right) \rceil\!\lfloor\left(x_1; k_2^1, s\right) \\ \rfloor\!\lceil\left(x_1; k_2^1, s\right) \end{bmatrix} \otimes \begin{bmatrix} \rceil\!\lfloor\left(x_2; k_1^2, s\right) \\ \rfloor\!\lceil\left(x_2; k_1^2, s\right) \rceil\!\lfloor\left(x_2; k_2^2, s\right) \\ \rfloor\!\lceil\left(x_2; k_2^2, s\right) \end{bmatrix} \\
&= \begin{bmatrix}
\rceil\!\lfloor\left(x_1; k_1^1, s\right) \rceil\!\lfloor\left(x_2; k_1^2, s\right) \\
\rceil\!\lfloor\left(x_1; k_1^1, s\right) \rfloor\!\lceil\left(x_2; k_1^2, s\right) \rceil\!\lfloor\left(x_2; k_2^2, s\right) \\
\rceil\!\lfloor\left(x_1; k_1^1, s\right) \rfloor\!\lceil\left(x_2; k_2^2, s\right) \\
\rfloor\!\lceil\left(x_1; k_1^1, s\right) \rceil\!\lfloor\left(x_1; k_2^1, s\right) \rceil\!\lfloor\left(x_2; k_1^2, s\right) \\
\rfloor\!\lceil\left(x_1; k_1^1, s\right) \rceil\!\lfloor\left(x_1; k_2^1, s\right) \rfloor\!\lceil\left(x_2; k_1^2, s\right) \rceil\!\lfloor\left(x_2; k_2^2, s\right) \\
\rfloor\!\lceil\left(x_1; k_1^1, s\right) \rceil\!\lfloor\left(x_1; k_2^1, s\right) \rfloor\!\lceil\left(x_2; k_2^2, s\right) \\
\rfloor\!\lceil\left(x_1; k_2^1, s\right) \rceil\!\lfloor\left(x_2; k_1^2, s\right) \\
\rfloor\!\lceil\left(x_1; k_2^1, s\right) \rfloor\!\lceil\left(x_2, k_1^2, s\right) \rceil\!\lfloor\left(x_2; k_2^2, s\right) \\
\rfloor\!\lceil\left(x_1; k_2^1, s\right) \rfloor\!\lceil\left(x_2; k_2^2, s\right)
\end{bmatrix}
\end{aligned}$$

Explicitly, inserting the definition of the step functions (1.17), the vector $\tilde{\boldsymbol{\beta}}$ is equal to

$$\begin{bmatrix} \{1 - \tanh[(x_1 - k_1^1)/s]\}\{1 - \tanh[(x_2 - k_1^2)]/s\}/4 \\ \{1 - \tanh[(x_1 - k_1^1)/s]\}\{1 + \tanh[(x_2 - k_1^2)]/s\}\{1 - \tanh[(x_2 - k_2^2)]/s\}/8 \\ \{1 - \tanh[(x_1 - k_1^1)/s]\}\{1 + \tanh[(x_2 - k_2^2)]/s\}/4 \\ \{1 + \tanh[(x_1 - k_1^1)]/s\}\{1 - \tanh[(x_1 - k_2^1)/s]\}\{1 - \tanh[(x_2 - k_1^2)]/s\}/8 \\ \{1 + \tanh[(x_1 - k_1^1)]/s\}\{1 - \tanh[(x_1 - k_2^1)/s]\}\{1 + \tanh[(x_2 - k_1^2)]/s\} \\ \quad \times\{1 - \tanh[(x_2 - k_2^2)]/s\}/16 \\ \{1 + \tanh[(x_1 - k_1^1)]/s\}\{1 - \tanh[(x_1 - k_2^1)/s]\}\{1 + \tanh[(x_2 - k_2^2)]/s\}/8 \\ \{1 + \tanh[(x_1 - k_2^1)]/s\}\{1 - \tanh[(x_2 - k_1^2)]/s\}/4 \\ \{1 + \tanh[(x_1 - k_2^1)]/s\}\{1 + \tanh[(x_2 - k_1^2)]/s\}\{1 - \tanh[(x_2 - k_2^2)]/s\}/8 \\ \{1 + \tanh[(x_1 - k_2^1)]/s\}\{1 + \tanh[(x_2 - k_2^2)]/s\}/4 \end{bmatrix}$$

The combined steps $\tilde{\beta}_{1,\ldots,9}$ are illustrated in Fig. A.1.

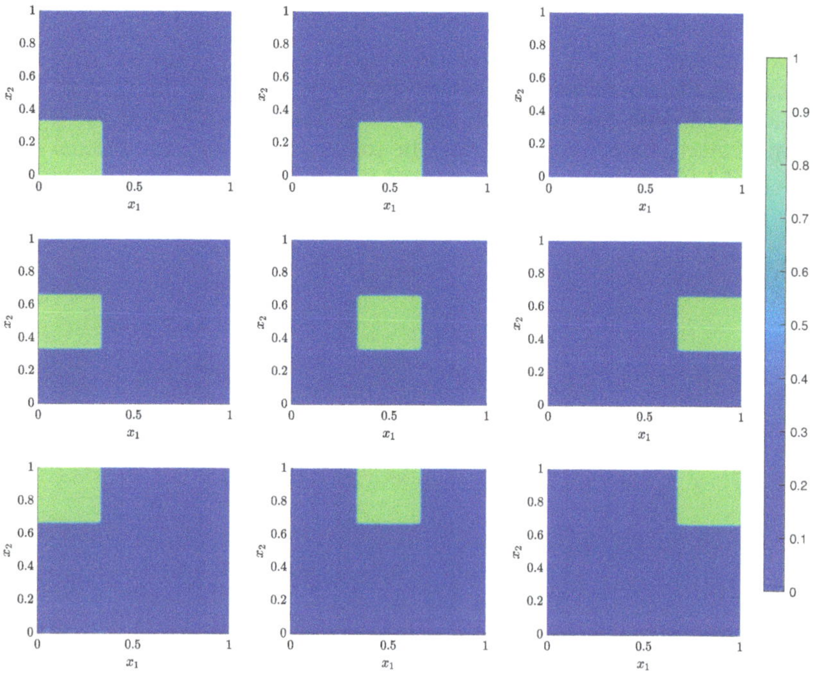

Fig. A.1 Combined steps $\tilde{\beta}_{1,\ldots,9}$ (components of the vector $\tilde{\beta}$ arranged left to right, top to bottom) for $\mathcal{K}_{1/2} = \{1/3, 2/3\}$, $s = 0.01$.

Collocation Methods

In the following, the Butcher tableau for collocation methods with three points in integral form is derived. The interpolating polynomial for the state derivatives with three basis functions is

$$l_1(t) = \frac{\left(t_i - t + c_2 h\right)\left(t_i - t + c_3 h\right)}{h^2 (c_1 - c_2)(c_1 - c_3)}$$

$$l_2(t) = \frac{\left(t_i - t + c_1 h\right)\left(t_i - t + c_3 h\right)}{h^2 (c_2 - c_1)(c_2 - c_3)}$$

$$l_3(t) = \frac{\left(t_i - t + c_1 h\right)\left(t_i - t + c_2 h\right)}{h^2 (c_1 - c_3)(c_2 - c_3)}$$

The indefinite integrals are

$$\int_{t_i}^{t} l_1(\theta)\, d\theta = \frac{\begin{array}{l}(t - t_i)\left(2t^2 - 4tt_i + 2t_i^2 - 3c_2 ht - 3c_3 ht + 3c_2 ht_i\right.\\ \left. + 3c_3 ht_i + 6c_2 c_3 h^2\right)\end{array}}{6h^2 (c_1 - c_2)(c_1 - c_3)}$$

$$\int_{t_i}^{t} l_2(\theta)\, d\theta = \frac{\begin{array}{l}(t - t_i)\left(2t^2 - 4tt_i + 2t_i^2 - 3c_1 ht - 3c_3 ht + 3c_1 ht_i\right.\\ \left. + 3c_3 ht_i + 6c_1 c_3 h^2\right)\end{array}}{6h^2 (c_2 - c_1)(c_2 - c_3)}$$

$$\int_{t_i}^{t} l_3(\theta)\, d\theta = \frac{\begin{array}{l}(t - t_i)\left(2t^2 - 4tt_i + 2t_i^2 - 3c_1 ht - 3c_2 ht + 3c_1 ht_i\right.\\ \left. + 3c_2 ht_i + 6c_1 c_2 h^2\right)\end{array}}{6h^2 (c_1 - c_3)(c_2 - c_3)}$$

The dense output is

$$\begin{aligned}
\boldsymbol{\sigma}_i(t) &= \tilde{\mathbf{x}}(t_i) \\
&+ \mathbf{f}\left[\boldsymbol{\sigma}_i\left(t_{i,1}\right), t_{i,1}\right] \frac{(t-t_i)\left(2t^2 - 4tt_i + 2t_i^2 - 3c_2ht - 3c_3ht + 3c_2ht_i + 3c_3ht_i + 6c_2c_3h^2\right)}{6h^2(c_1-c_2)(c_1-c_3)} \\
&+ \mathbf{f}\left[\boldsymbol{\sigma}_i\left(t_{i,2}\right), t_{i,2}\right] \frac{(t-t_i)\left(2t^2 - 4tt_i + 2t_i^2 - 3c_1ht - 3c_3ht + 3c_1ht_i + 3c_3ht_i + 6c_1c_3h^2\right)}{6h^2(c_2-c_1)(c_2-c_3)} \\
&+ \mathbf{f}\left[\boldsymbol{\sigma}_i\left(t_{i,3}\right), t_{i,3}\right] \frac{(t-t_i)\left(2t^2 - 4tt_i + 2t_i^2 - 3c_1ht - 3c_2ht + 3c_1ht_i + 3c_2ht_i + 6c_1c_2h^2\right)}{6h^2(c_1-c_3)(c_2-c_3)}
\end{aligned}$$

Evaluation at the collocation point $t_{i,1} = t_i + hc_1$ results in

$$\begin{aligned}
\boldsymbol{\sigma}_i\left(t_{i,1}\right) &= \tilde{\mathbf{x}}(t_i) \\
&+ \mathbf{f}\left[\boldsymbol{\sigma}_i\left(t_{i,1}\right), t_{i,1}\right] \overbrace{\frac{\left(t_{i,1}-t_i\right)\left(2t_{i,1}^2 - 4t_{i,1}t_i + 2t_i^2 - 3c_2ht_{i,1} - 3c_3ht_{i,1} + 3c_2ht_i + 3c_3ht_i + 6c_2c_3h^2\right)}{6h^2(c_1-c_2)(c_1-c_3)}}^{h\frac{c_1\left(3c_1c_2+3c_1c_3-6c_2c_3-2c_1{}^2\right)}{6(c_2-c_1)(c_1-c_3)}} \\
&+ \mathbf{f}\left[\boldsymbol{\sigma}_i\left(t_{i,2}\right), t_{i,2}\right] \underbrace{\frac{\left(t_{i,1}-t_i\right)\left(2t_{i,1}^2 - 4t_{i,1}t_i + 2t_i^2 - 3c_1ht_{i,1} - 3c_3ht_{i,1} + 3c_1ht_i + 3c_3ht_i + 6c_1c_3h^2\right)}{6h^2(c_2-c_1)(c_2-c_3)}}_{h\frac{c_1{}^2(c_1-3c_3)}{6(c_1-c_2)(c_2-c_3)}} \\
&+ \mathbf{f}\left[\boldsymbol{\sigma}_i\left(t_{i,3}\right), t_{i,3}\right] \underbrace{\frac{\left(t_{i,1}-t_i\right)\left(2t_{i,1}^2 - 4t_{i,1}t_i + 2t_i^2 - 3c_1ht_{i,1} - 3c_2ht_{i,1} + 3c_1ht_i + 3c_2ht_i + 6c_1c_2h^2\right)}{6h^2(c_1-c_3)(c_2-c_3)}}_{h\frac{c_1{}^2(c_1-3c_2)}{6(c_3-c_1)(c_2-c_3)}}
\end{aligned}$$

Evaluation at the collocation point $t_{i,2} = t_i + hc_2$ results in

$$\boldsymbol{\sigma}_i\left(t_{i,2}\right) = \tilde{\mathbf{x}}\left(t_i\right) + \mathbf{f}\left[\boldsymbol{\sigma}_i\left(t_{i,1}\right), t_{i,1}\right] \overbrace{\frac{\left(t_{i,2} - t_i\right)\left(2t_{i,2}^2 - 4t_{i,2}t_i + 2t_i^2 - 3c_2ht_{i,2} - 3c_3ht_{i,2} + 3c_2ht_i + 3c_3ht_i + 6c_2c_3h^2\right)}{6h^2\left(c_1 - c_2\right)\left(c_1 - c_3\right)}}^{h\frac{c_2{}^2\left(c_2 - 3c_3\right)}{6\left(c_1 - c_2\right)\left(c_3 - c_1\right)}}$$

$$+ \mathbf{f}\left[\boldsymbol{\sigma}_i\left(t_{i,2}\right), t_{i,2}\right] \underbrace{\frac{\left(t_{i,2} - t_i\right)\left(2t_{i,2}^2 - 4t_{i,2}t_i + 2t_i^2 - 3c_1ht_{i,2} - 3c_3ht_{i,2} + 3c_1ht_i + 3c_3ht_i + 6c_1c_3h^2\right)}{6h^2\left(c_2 - c_1\right)\left(c_2 - c_3\right)}}_{h\frac{c_2\left(3c_1c_2 - 6c_1c_3 + 3c_2c_3 - 2c_2{}^2\right)}{6\left(c_1 - c_2\right)\left(c_2 - c_3\right)}}$$

$$+ \mathbf{f}\left[\boldsymbol{\sigma}_i\left(t_{i,3}\right), t_{i,3}\right] \underbrace{\frac{\left(t_{i,2} - t_i\right)\left(2t_{i,2}^2 - 4t_{i,2}t_i + 2t_i^2 - 3c_1ht_{i,2} - 3c_2ht_{i,2} + 3c_1ht_i + 3c_2ht_i + 6c_1c_2h^2\right)}{6h^2\left(c_1 - c_3\right)\left(c_2 - c_3\right)}}_{h\frac{c_2{}^2\left(3c_1 - c_2\right)}{6\left(c_1 - c_3\right)\left(c_2 - c_3\right)}}$$

Evaluation at the collocation point $t_{i,3} = t_i + hc_3$ results in

$$\boldsymbol{\sigma}_i\left(t_{i,3}\right) = \tilde{\mathbf{x}}\left(t_i\right) + \mathbf{f}\left[\boldsymbol{\sigma}_i\left(t_{i,1}\right), t_{i,1}\right] \overbrace{\frac{\left(t_{i,3} - t_i\right)\left(2t_{i,3}^2 - 4t_{i,3}t_i + 2t_i^2 - 3c_2ht_{i,3} - 3c_3ht_{i,3} + 3c_2ht_i + 3c_3ht_i + 6c_2c_3h^2\right)}{6h^2\left(c_1 - c_2\right)\left(c_1 - c_3\right)}}^{h\frac{c_3{}^2\left(3c_2 - c_3\right)}{6\left(c_1 - c_2\right)\left(c_1 - c_3\right)}}$$

$$+ \mathbf{f}\left[\boldsymbol{\sigma}_i\left(t_{i,2}\right), t_{i,2}\right] \underbrace{\frac{\left(t_{i,3} - t_i\right)\left(2t_{i,3}^2 - 4t_{i,3}t_i + 2t_i^2 - 3c_1ht_{i,3} - 3c_3ht_{i,3} + 3c_1ht_i + 3c_3ht_i + 6c_1c_3h^2\right)}{6h^2\left(c_2 - c_1\right)\left(c_2 - c_3\right)}}_{h\frac{c_3{}^2\left(3c_1 - c_3\right)}{6\left(c_1 - c_2\right)\left(c_3 - c_2\right)}}$$

$$+\ \mathbf{f}\left[\boldsymbol{\sigma}_i\left(t_{i,3}\right), t_{i,3}\right]\underbrace{\frac{\left(t_{i,3}-t_i\right)\left(2t_{i,3}^2-4t_{i,3}t_i+2t_i^2-3c_1ht_{i,3}-3c_2ht_{i,3}+3c_1ht_i+3c_2ht_i+6c_1c_2h^2\right)}{6h^2\left(c_1-c_3\right)\left(c_2-c_3\right)}}_{h\frac{c_3\left(6c_1c_2-3c_1c_3-3c_2c_3+2c_3^2\right)}{6\left(c_1-c_3\right)\left(c_2-c_3\right)}}$$

Evaluation at the next point $t_{i+1}=t_i+h$ results in

$$\boldsymbol{\sigma}_i\left(t_{i+1}\right)=\tilde{\mathbf{x}}\left(t_i\right)$$

$$+\ \mathbf{f}\left[\boldsymbol{\sigma}_i\left(t_{i,1}\right), t_{i,1}\right]\overbrace{\frac{\left(t_{i+1}-t_i\right)\left(2t_{i+1}^2-4t_{i+1}t_i+2t_i^2-3c_2ht_{i+1}-3c_3ht_{i+1}+3c_2ht_i+3c_3ht_i+6c_2c_3h^2\right)}{6h^2\left(c_1-c_2\right)\left(c_1-c_3\right)}}^{h\frac{\left(3c_2+3c_3-6c_2c_3-2\right)}{6\left(c_1-c_2\right)\left(c_3-c_1\right)}}$$

$$+\ \mathbf{f}\left[\boldsymbol{\sigma}_i\left(t_{i,2}\right), t_{i,2}\right]\underbrace{\frac{\left(t_{i+1}-t_i\right)\left(2t_{i+1}^2-4t_{i+1}t_i+2t_i^2-3c_1ht_{i+1}-3c_3ht_{i+1}+3c_1ht_i+3c_3ht_i+6c_1c_3h^2\right)}{6h^2\left(c_2-c_1\right)\left(c_2-c_3\right)}}_{h\frac{\left(3c_1+3c_3-6c_1c_3-2\right)}{6\left(c_1-c_2\right)\left(c_2-c_3\right)}}$$

$$+\ \mathbf{f}\left[\boldsymbol{\sigma}_i\left(t_{i,3}\right), t_{i,3}\right]\underbrace{\frac{\left(t_{i+1}-t_i\right)\left(2t_{i+1}^2-4t_{i+1}t_i+2t_i^2-3c_1ht_{i+1}-3c_2ht_{i+1}+3c_1ht_i+3c_2ht_i+6c_1c_2h^2\right)}{6h^2\left(c_1-c_3\right)\left(c_2-c_3\right)}}_{h\frac{\left(3c_1+3c_2-6c_1c_2-2\right)}{6\left(c_1-c_3\right)\left(c_3-c_2\right)}}$$

The parameterized Butcher tableau is

$$\begin{array}{c|ccc}
c_1 & \frac{c_1\left(3c_1c_2+3c_1c_3-6c_2c_3-2c_1^2\right)}{6\left(c_2-c_1\right)\left(c_1-c_3\right)} & \frac{c_1^2\left(c_1-3c_3\right)}{6\left(c_1-c_2\right)\left(c_2-c_3\right)} & \frac{c_1^2\left(c_1-3c_2\right)}{6\left(c_3-c_1\right)\left(c_2-c_3\right)} \\
c_2 & \frac{c_2^2\left(c_2-3c_3\right)}{6\left(c_1-c_2\right)\left(c_3-c_1\right)} & \frac{c_2\left(3c_1c_2-6c_1c_3+3c_2c_3-2c_2^2\right)}{6\left(c_1-c_2\right)\left(c_2-c_3\right)} & \frac{c_2^2\left(3c_1-c_2\right)}{6\left(c_1-c_3\right)\left(c_2-c_3\right)} \\
c_3 & \frac{c_3^2\left(3c_2-c_3\right)}{6\left(c_1-c_2\right)\left(c_1-c_3\right)} & \frac{c_3^2\left(3c_1-c_3\right)}{6\left(c_1-c_2\right)\left(c_3-c_2\right)} & \frac{c_3\left(6c_1c_2-3c_1c_3-3c_2c_3+2c_3^2\right)}{6\left(c_1-c_3\right)\left(c_2-c_3\right)} \\
\hline
 & \frac{\left(3c_2+3c_3-6c_2c_3-2\right)}{6\left(c_1-c_2\right)\left(c_3-c_1\right)} & \frac{\left(3c_1+3c_3-6c_1c_3-2\right)}{6\left(c_1-c_2\right)\left(c_2-c_3\right)} & \frac{\left(3c_1+3c_2-6c_1c_2-2\right)}{6\left(c_1-c_3\right)\left(c_3-c_2\right)}
\end{array}$$

The Lobatto points $c_1 = 0$, $c_2 = 1/2$, $c_3 = 1$ yield

$$
\begin{array}{c|ccc}
0 & 0 & 0 & 0 \\
\frac{1}{2} & \frac{5}{24} & \frac{1}{3} & -\frac{1}{24} \\
1 & \frac{1}{6} & \frac{2}{3} & \frac{1}{6} \\
\hline
 & \frac{1}{6} & \frac{2}{3} & \frac{1}{6}
\end{array}
$$

The Radau points (flipped) $c_1 = 2/5 - \sqrt{6}/10$, $c_2 = 2/5 + \sqrt{6}/10$, $c_3 = 1$ yield

$$
\begin{array}{c|ccc}
\frac{2}{5} - \frac{\sqrt{6}}{10} & \frac{11}{45} - \frac{7\sqrt{6}}{360} & \frac{37}{225} - \frac{169\sqrt{6}}{1800} & -\frac{2}{225} + \frac{\sqrt{6}}{75} \\
\frac{2}{5} + \frac{\sqrt{6}}{10} & \frac{37}{225} + \frac{169\sqrt{6}}{1800} & \frac{11}{45} + \frac{7\sqrt{6}}{360} & -\frac{2}{225} - \frac{\sqrt{6}}{75} \\
1 & \frac{4}{9} - \frac{\sqrt{6}}{36} & \frac{4}{9} + \frac{\sqrt{6}}{36} & \frac{1}{9} \\
\hline
 & \frac{4}{9} - \frac{\sqrt{6}}{36} & \frac{4}{9} + \frac{\sqrt{6}}{36} & \frac{1}{9}
\end{array}
$$

The Gauss points $c_1 = 1/2 - \sqrt{15}/10$, $c_2 = 1/2$, $c_3 = 1/2 + \sqrt{15}/10$ yield

$$
\begin{array}{c|ccc}
\frac{1}{2} - \frac{\sqrt{15}}{10} & \frac{5}{36} & \frac{2}{9} - \frac{\sqrt{15}}{15} & \frac{5}{36} - \frac{\sqrt{15}}{30} \\
\frac{1}{2} & \frac{5}{36} + \frac{\sqrt{15}}{24} & \frac{2}{9} & \frac{5}{36} - \frac{\sqrt{15}}{24} \\
\frac{1}{2} + \frac{\sqrt{15}}{10} & \frac{5}{36} + \frac{\sqrt{15}}{30} & \frac{2}{9} + \frac{\sqrt{15}}{15} & \frac{5}{36} \\
\hline
 & \frac{5}{18} & \frac{4}{9} & \frac{5}{18}
\end{array}
$$

APPENDIX C

Common Subexpression Elimination

Consider the following source code for the evaluation of the expression $\sin(a)\cos(a^2)a + \sin(a) + \cos(a^2)$ without common subexpression elimination

Source File: `without_cse.c`

```
#include "math.h"
double without_cse(double a){
     return sin(a)  * cos(a * a)  * a + sin(a)  + cos(a * a);
}
```

and with common subexpression elimination

Source File: `with_cse.c`

```
#include "math.h"
double with_cse(double a){
          double b = a * a;
          double sa = sin(a);
          double cb = cos(b);
          double sa_times_cb_times_a = sa * cb * a;
          double sa_plus_cb = sa + cb;
          return sa_times_cb_times_a + sa_plus_cb;
     }
```

The assembly instructions generated by a C-compiler (GCC 11.2.0) with optimization option -O2 (performing nearly all available optimizations) yields:

Compiler Output: `gcc -S with_cse.c -O2`

```
      .file   "with_cse.c"
      .text
      .p2align 4
```

```
    .globl  with_cse
    .type   with_cse, @function
with_cse:
.LFB0:
    .cfi_startproc
    endbr64
    subq    $24, %rsp
    .cfi_def_cfa_offset 32
    movsd   %xmm0, (%rsp)
    call    sin@PLT
    movsd   (%rsp), %xmm2
    movsd   %xmm0, 8(%rsp)
    movapd  %xmm2, %xmm0
    mulsd   %xmm2, %xmm0
    call    cos@PLT
    movsd   8(%rsp), %xmm1
    movsd   (%rsp), %xmm2
    addq    $24, %rsp
    .cfi_def_cfa_offset 8
    movapd  %xmm0, %xmm3
    movapd  %xmm1, %xmm0
    addsd   %xmm3, %xmm1
    mulsd   %xmm3, %xmm0
    mulsd   %xmm2, %xmm0
    addsd   %xmm1, %xmm0
    ret
    .cfi_endproc
.LFE0:
    .size   with_cse, .-with_cse
    .ident  "GCC: (Ubuntu 11.2.0-7ubuntu2) 11.2.0"
    .section    .note.GNU-stack,"",@progbits
    .section    .note.gnu.property,"a"
    .align 8
    .long   1f - 0f
    .long   4f - 1f
    .long   5
0:
    .string "GNU"
1:
    .align 8
    .long   0xc0000002
    .long   3f - 2f
2:
    .long   0x3
```

```
3:
    .align 8
4:
```

Compiler Output: `gcc -S without_cse.c -O2`

```
    .file   "without_cse.c"
    .text
    .p2align 4
    .globl  without_cse
    .type   without_cse, @function
without_cse:
.LFB0:
    .cfi_startproc
    endbr64
    subq    $24, %rsp
    .cfi_def_cfa_offset 32
    movsd   %xmm0, (%rsp)
    call    sin@PLT
    movsd   (%rsp), %xmm1
    movsd   %xmm0, 8(%rsp)
    movapd  %xmm1, %xmm0
    mulsd   %xmm1, %xmm0
    call    cos@PLT
    movsd   8(%rsp), %xmm3
    movsd   (%rsp), %xmm1
    addq    $24, %rsp
    .cfi_def_cfa_offset 8
    movapd  %xmm0, %xmm2
    movapd  %xmm3, %xmm0
    mulsd   %xmm2, %xmm0
    mulsd   %xmm1, %xmm0
    addsd   %xmm3, %xmm0
    addsd   %xmm2, %xmm0
    ret
    .cfi_endproc
.LFE0:
    .size   without_cse, .-without_cse
    .ident  "GCC: (Ubuntu 11.2.0-7ubuntu2) 11.2.0"
    .section    .note.GNU-stack,"",@progbits
    .section    .note.gnu.property,"a"
    .align 8
    .long   1f - 0f
    .long   4f - 1f
```

```
        .long   5
0:
        .string "GNU"
1:
        .align 8
        .long   0xc0000002
        .long   3f - 2f
2:
        .long   0x3
3:
        .align 8
4:
```

The reader is not expected to follow the exact sequence and effect of the instructions for the interpretation of this example. Merely observe that for both functions the same instructions are generated (preliminary optimization with common subexpression elimination in this example could be considered a non-value-added activity). Recall that the previous result is obtained for the optimization option (-O2). If optimization is turned off (-O0) the produced assembly instructions are not directly comparable for both functions:

Compiler Output: `gcc -S with_cse.c -O0`

```
        .file   "with_cse.c"
        .text
        .globl  with_cse
        .type   with_cse, @function
with_cse:
.LFB0:
        .cfi_startproc
        endbr64
        pushq   %rbp
        .cfi_def_cfa_offset 16
        .cfi_offset 6, -16
        movq    %rsp, %rbp
        .cfi_def_cfa_register 6
        subq    $64, %rsp
        movsd   %xmm0, -56(%rbp)
        movsd   -56(%rbp), %xmm0
        mulsd   %xmm0, %xmm0
        movsd   %xmm0, -40(%rbp)
        movq    -56(%rbp), %rax
        movq    %rax, %xmm0
        call    sin@PLT
```

```
    movq    %xmm0, %rax
    movq    %rax, -32(%rbp)
    movq    -40(%rbp), %rax
    movq    %rax, %xmm0
    call    cos@PLT
    movq    %xmm0, %rax
    movq    %rax, -24(%rbp)
    movsd   -32(%rbp), %xmm0
    mulsd   -24(%rbp), %xmm0
    movsd   -56(%rbp), %xmm1
    mulsd   %xmm1, %xmm0
    movsd   %xmm0, -16(%rbp)
    movsd   -32(%rbp), %xmm0
    addsd   -24(%rbp), %xmm0
    movsd   %xmm0, -8(%rbp)
    movsd   -16(%rbp), %xmm0
    addsd   -8(%rbp), %xmm0
    movq    %xmm0, %rax
    movq    %rax, %xmm0
    leave
    .cfi_def_cfa 7, 8
    ret
    .cfi_endproc
.LFE0:
    .size   with_cse, .-with_cse
    .ident  "GCC: (Ubuntu 11.2.0-7ubuntu2) 11.2.0"
    .section    .note.GNU-stack,"",@progbits
    .section    .note.gnu.property,"a"
    .align 8
    .long   1f - 0f
    .long   4f - 1f
    .long   5
0:
    .string "GNU"
1:
    .align 8
    .long   0xc0000002
    .long   3f - 2f
2:
    .long   0x3
3:
    .align 8
4:
```

Compiler Output: `gcc -S without_cse.c -O0`

```
	.file	"without_cse.c"
	.text
	.globl	without_cse
	.type	without_cse, @function
without_cse:
.LFB0:
	.cfi_startproc
	endbr64
	pushq	%rbp
	.cfi_def_cfa_offset 16
	.cfi_offset 6, -16
	movq	%rsp, %rbp
	.cfi_def_cfa_register 6
	subq	$16, %rsp
	movsd	%xmm0, -8(%rbp)
	movq	-8(%rbp), %rax
	movq	%rax, %xmm0
	call	sin@PLT
	movsd	%xmm0, -16(%rbp)
	movsd	-8(%rbp), %xmm0
	movapd	%xmm0, %xmm1
	mulsd	%xmm0, %xmm1
	movq	%xmm1, %rax
	movq	%rax, %xmm0
	call	cos@PLT
	mulsd	-16(%rbp), %xmm0
	mulsd	-8(%rbp), %xmm0
	movsd	%xmm0, -16(%rbp)
	movq	-8(%rbp), %rax
	movq	%rax, %xmm0
	call	sin@PLT
	movapd	%xmm0, %xmm3
	addsd	-16(%rbp), %xmm3
	movsd	%xmm3, -16(%rbp)
	movsd	-8(%rbp), %xmm0
	mulsd	%xmm0, %xmm0
	movq	%xmm0, %rax
	movq	%rax, %xmm0
	call	cos@PLT
	addsd	-16(%rbp), %xmm0
	movq	%xmm0, %rax
	movq	%rax, %xmm0
```

```
	leave
	.cfi_def_cfa 7, 8
	ret
	.cfi_endproc
.LFE0:
	.size	without_cse, .-without_cse
	.ident	"GCC: (Ubuntu 11.2.0-7ubuntu2) 11.2.0"
	.section	.note.GNU-stack,"",@progbits
	.section	.note.gnu.property,"a"
	.align 8
	.long	1f - 0f
	.long	4f - 1f
	.long	5
0:
	.string "GNU"
1:
	.align 8
	.long	0xc0000002
	.long	3f - 2f
2:
	.long	0x3
3:
	.align 8
4:
```

Optimal Control Cost Transformations

In certain clearance applications, it can be useful to minimize the minimum or maximum of the criterion $J[\mathbf{x}(t)]$ over a fixed time interval $t \in [t_0, t_f]$.

The minimization of the maximum of the criterion on the time interval $[t_0, t_f]$, that is,

$$\min \max_{t\in[t_0,t_f]} J[\mathbf{x}(t)]$$

represents a so-called *Tschebyscheff* problem and can be modeled through the introduction of a parameter p with the path constraint

$$J[\mathbf{x}(t)] \leqslant p$$

and the minimization of the parameter p.

The minimization of the minimum of the criterion on the time interval $[t_0, t_f]$, that is,

$$\min \min_{t\in[t_0,t_f]} J[\mathbf{x}(t)]$$

can be modeled through the introduction of a parameter p bounded by $t_0 \leqslant p \leqslant t_f$ and the minimization of

$$J[\mathbf{x}(p)]$$

Ray Parameterization

A parameterization of the form

$$\mathbf{y}(r, \boldsymbol{\phi}) = r\mathbf{w}(\boldsymbol{\phi})$$

with $r \geqslant 0$ and $n_\phi = n_J - 1$ angles $\phi_1, \ldots, \phi_{n_\phi}$ in dimension $n_J > 1$ can be based on the spherical unit vector [92]

$$\mathbf{w}(\boldsymbol{\phi}) = \begin{bmatrix} \cos(\phi_1) \\ \sin(\phi_1)\cos(\phi_2) \\ \sin(\phi_1)\sin(\phi_2)\cos(\phi_3) \\ \vdots \\ \sin(\phi_1)\sin(\phi_2)\sin(\phi_3)\ldots\sin(\phi_{n_\phi-1})\cos(\phi_{n_\phi}) \\ \sin(\phi_1)\sin(\phi_2)\sin(\phi_3)\ldots\sin(\phi_{n_\phi-1})\sin(\phi_{n_\phi}) \end{bmatrix}$$

or, more compactly,

$$w_i(\boldsymbol{\phi}) = \begin{cases} \cos(\phi_i)\prod_{j=1}^{i-1}\sin(\phi_j), & 1 \leqslant i < n_J \\ \prod_{j=1}^{n_\phi}\sin(\phi_j), & i = n_J \end{cases}$$

Aircraft Model

The responses of the aircraft dynamics to step inputs for the inner-loop controller at a typical envelope point are illustrated in Fig. F.1. Notice the nonminimum phase behavior of the normal load factor n_z and n_y as well as the high damping of the roll angle response Φ. Additionally, the response of the second-order wind model for the vertical wind state w_W is depicted in this figure.

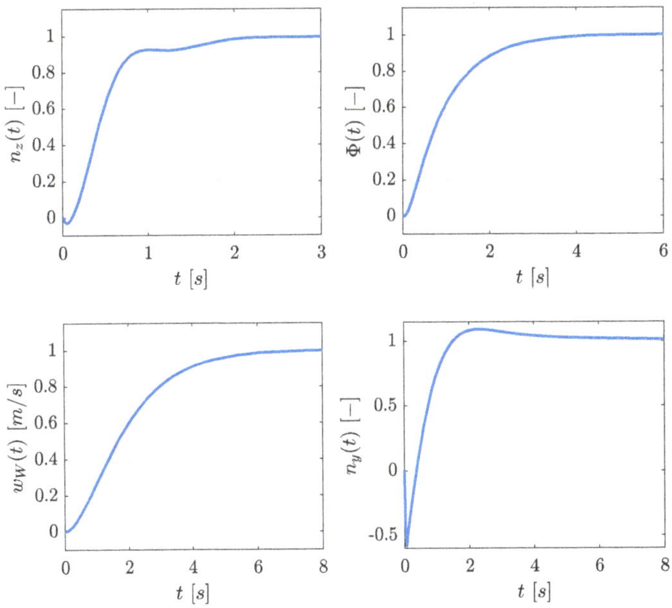

Fig. F.1 Step responses to a unit step.

Moreover, pole-zero plots for this aircraft model regarding the normal load factor, lateral load factor, and the roll angle are shown in Figs. F.2–F.4.

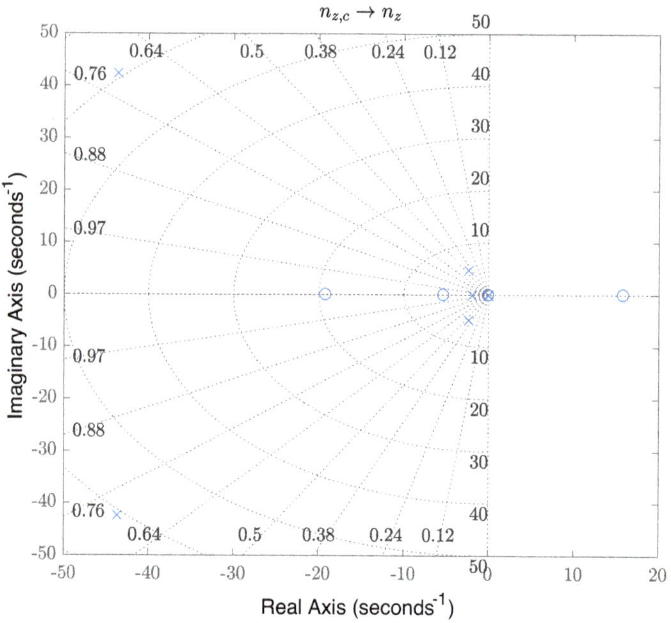

Fig. F.2 Pole-zero plot $n_{z,c} \to n_z$.

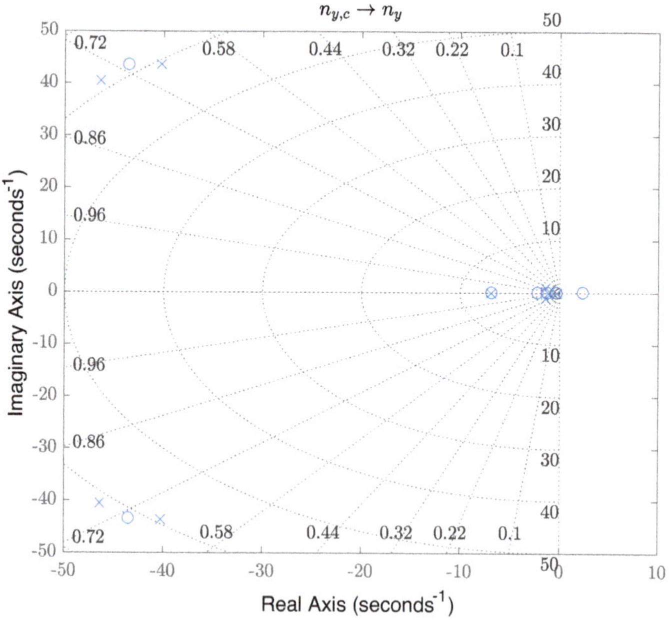

Fig. F.3 Pole-zero plot $n_{y,c} \to n_y$.

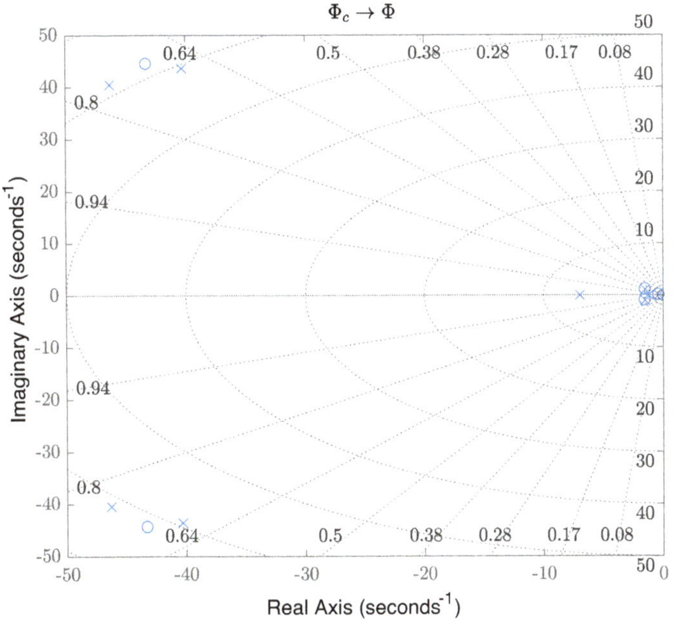

Fig. F.4 Pole-zero plot $\Phi_c \rightarrow \Phi$.

Hermite–Simpson Linear Nominal

Compared to the trapezoidal collocation method, additional state and control variables $\tilde{\mathbf{x}}_i$ and $\tilde{\mathbf{u}}_i$ for $i = 1, \ldots, N$ associated with the segment midpoints are introduced for the Hermite–Simpson discretization method. Together with $\mathbf{x}_0 = \mathbf{0}$, the Hermite constraints for the midpoints are

$$\tilde{\mathbf{x}}_i = \frac{1}{2}(\mathbf{x}_{i-1} + \mathbf{x}_i) + h\left(\mathbf{A}\frac{\mathbf{x}_{i-1} - \mathbf{x}_i}{8} + \mathbf{B}\frac{\mathbf{u}_{i-1} - \mathbf{u}_i}{8}\right) \tag{G.1}$$

and the Simpson constraints for each interval are

$$\mathbf{x}_i = \mathbf{x}_{i-1} + h\left(\mathbf{A}\frac{\mathbf{x}_{i-1} + 4\tilde{\mathbf{x}}_i + \mathbf{x}_i}{6} + \mathbf{B}\frac{\mathbf{u}_{i-1} + 4\tilde{\mathbf{u}}_i + \mathbf{u}_i}{6}\right) \tag{G.2}$$

for $i = 1, \ldots, N$. The discretized nominal Cat II problem is

$$\begin{aligned}
\underset{\mathbf{x}_i, \tilde{\mathbf{x}}_i, \mathbf{u}_0, \mathbf{u}_i, \tilde{\mathbf{u}}_i}{\text{minimize}} \quad & j(\mathbf{x}_N) \\
\text{subject to} \quad & \tilde{\mathbf{x}}_i = \frac{1}{2}(\mathbf{x}_{i-1} + \mathbf{x}_i) + h\left(\mathbf{A}\frac{\mathbf{x}_{i-1} - \mathbf{x}_i}{8} + \mathbf{B}\frac{\mathbf{u}_{i-1} - \mathbf{u}_i}{8}\right) \\
& \mathbf{x}_i = \mathbf{x}_{i-1} + h\left(\mathbf{A}\frac{\mathbf{x}_{i-1} + 4\tilde{\mathbf{x}}_i + \mathbf{x}_i}{6} + \mathbf{B}\frac{\mathbf{u}_{i-1} + 4\tilde{\mathbf{u}}_i + \mathbf{u}_i}{6}\right) \\
& \underline{\mathbf{x}} \leqslant \mathbf{x}_i, \tilde{\mathbf{x}}_i \leqslant \overline{\mathbf{x}} \\
& \underline{\mathbf{u}} \leqslant \mathbf{u}_i, \tilde{\mathbf{u}}_i \leqslant \overline{\mathbf{u}} \\
& \underline{\mathbf{u}} \leqslant \mathbf{u}_0 \leqslant \overline{\mathbf{u}} \\
& i = 1, \ldots, N
\end{aligned}$$

The cost vector $\mathbf{v} \in \mathbb{R}^n$ and optimization variable vector $\mathbf{z} \in \mathbb{R}^n$ with corresponding lower bound $\underline{\mathbf{z}} \in \mathbb{R}^n$ and upper bound $\overline{\mathbf{z}} \in \mathbb{R}^n$ are

$$\mathbf{v} = \begin{bmatrix} \mathbf{0} \\ \mathbf{0} \\ \mathbf{0} \\ \mathbf{0} \\ \mathbf{0} \\ \vdots \\ \mathbf{0} \\ \mathbf{0} \\ \mathbf{c} \\ \mathbf{0} \end{bmatrix}, \quad \underline{\mathbf{z}} = \begin{bmatrix} \underline{\mathbf{u}} \\ \underline{\mathbf{x}} \\ \underline{\mathbf{u}} \\ \underline{\mathbf{x}} \\ \underline{\mathbf{u}} \\ \vdots \\ \underline{\mathbf{x}} \\ \underline{\mathbf{u}} \\ \underline{\mathbf{x}} \\ \underline{\mathbf{u}} \end{bmatrix}, \quad \mathbf{z} = \begin{bmatrix} \mathbf{u}_0 \\ \tilde{\mathbf{x}}_1 \\ \tilde{\mathbf{u}}_1 \\ \mathbf{x}_1 \\ \mathbf{u}_1 \\ \vdots \\ \tilde{\mathbf{x}}_N \\ \tilde{\mathbf{u}}_N \\ \mathbf{x}_N \\ \mathbf{u}_N \end{bmatrix}, \quad \overline{\mathbf{z}} = \begin{bmatrix} \overline{\mathbf{u}} \\ \overline{\mathbf{x}} \\ \overline{\mathbf{u}} \\ \overline{\mathbf{x}} \\ \overline{\mathbf{u}} \\ \vdots \\ \overline{\mathbf{x}} \\ \overline{\mathbf{u}} \\ \overline{\mathbf{x}} \\ \overline{\mathbf{u}} \end{bmatrix}$$

The equality constraint matrix is defined as

$$\mathbf{M} = \begin{bmatrix} \mathbf{H} \\ \mathbf{S} \end{bmatrix}$$

with the matrix $\mathbf{H} \in \mathbb{R}^{n_h \times n}$, $n_h = Nn_x$:

$$\mathbf{H} = \begin{bmatrix} -\frac{h\mathbf{B}}{8} & \mathbf{I}\ \mathbf{0} & \frac{h\mathbf{A}}{8} - \frac{\mathbf{I}}{2} & \frac{h\mathbf{B}}{8} & \mathbf{0}\ \mathbf{0} & \mathbf{0} & \mathbf{0} & \mathbf{0}\ \mathbf{0} & \mathbf{0} & \mathbf{0} & \cdots \\ \mathbf{0} & \mathbf{0}\ \mathbf{0} & -\frac{h\mathbf{A}}{8} - \frac{\mathbf{I}}{2} & -\frac{h\mathbf{B}}{8} & \mathbf{I}\ \mathbf{0} & \frac{h\mathbf{A}}{8} - \frac{\mathbf{I}}{2} & \frac{h\mathbf{B}}{8} & \mathbf{0}\ \mathbf{0} & \mathbf{0} & \mathbf{0} & \cdots \\ \mathbf{0} & \mathbf{0}\ \mathbf{0} & \mathbf{0} & \mathbf{0} & \mathbf{0}\ \mathbf{0} & -\frac{h\mathbf{A}}{8} - \frac{\mathbf{I}}{2} & -\frac{h\mathbf{B}}{8} & \mathbf{I}\ \mathbf{0} & \frac{h\mathbf{A}}{8} - \frac{\mathbf{I}}{2} & \frac{h\mathbf{B}}{8} & \\ \vdots & \vdots\ \vdots & \vdots & \vdots & \vdots\ \vdots & & \ddots & & & & \end{bmatrix}$$

for the linear Hermite constraints [Eq. (G.1)] and the matrix $\mathbf{S} \in \mathbb{R}^{n_h \times n}$:

$$\mathbf{S} = -\begin{bmatrix} \frac{h\mathbf{B}}{6} & \frac{2h\mathbf{A}}{3} & \frac{2h\mathbf{B}}{3} & \frac{h\mathbf{A}}{6} - \mathbf{I} & \frac{h\mathbf{B}}{6} & \mathbf{0} & \mathbf{0} & \mathbf{0} & \mathbf{0} & \mathbf{0} & \mathbf{0} & \mathbf{0} & \mathbf{0} & \cdots \\ \mathbf{0} & \mathbf{0} & \mathbf{0} & \frac{h\mathbf{A}}{6} + \mathbf{I} & \frac{h\mathbf{B}}{6} & \frac{2h\mathbf{A}}{3} & \frac{2h\mathbf{B}}{3} & \frac{h\mathbf{A}}{6} - \mathbf{I} & \frac{h\mathbf{B}}{6} & \mathbf{0} & \mathbf{0} & \mathbf{0} & \mathbf{0} & \cdots \\ \mathbf{0} & \mathbf{0} & \mathbf{0} & \mathbf{0} & \mathbf{0} & \mathbf{0} & \mathbf{0} & \frac{h\mathbf{A}}{6} + \mathbf{I} & \frac{h\mathbf{B}}{6} & \frac{2h\mathbf{A}}{3} & \frac{2h\mathbf{B}}{3} & \frac{h\mathbf{A}}{6} - \mathbf{I} & \frac{h\mathbf{B}}{6} & \\ \vdots & \vdots & \vdots & \vdots & \vdots & \vdots & \vdots & \ddots & & & & & & \end{bmatrix}$$

for the linear Simpson constraints [Eq. (G.2)].

Hermite–Simpson Linear Parametric

The parametric form of the Hermite constraints [Eq. (G.1)] for the midpoints is

$$\tilde{\mathbf{x}}_i = \frac{1}{2}\left(\mathbf{x}_{i-1} + \mathbf{x}_i\right) + h\left[\mathbf{A}\left(\mathbf{p}\right)\frac{\mathbf{x}_{i-1} - \mathbf{x}_i}{8} + \mathbf{B}\left(\mathbf{p}\right)\frac{\mathbf{u}_{i-1} - \mathbf{u}_i}{8}\right] \tag{H.1}$$

and the Simpson constraints [Eq. (G.2)] for each interval are

$$\begin{aligned}\mathbf{x}_i = \mathbf{x}_{i-1} + h\Bigg[&\mathbf{A}\left(\mathbf{p}\right)\frac{\mathbf{x}_{i-1} + 4\tilde{\mathbf{x}}_i + \mathbf{x}_i - 6\mathbf{x}_r\left(\mathbf{p}\right)}{6} \\ &+ \mathbf{B}\left(\mathbf{p}\right)\frac{\mathbf{u}_{i-1} + 4\tilde{\mathbf{u}}_i + \mathbf{u}_i - 6\mathbf{u}_r\left(\mathbf{p}\right)}{6}\Bigg]\end{aligned} \tag{H.2}$$

for $i = 1, \ldots, N$. The discretized Cat II problem depending on the parameter $\mathbf{p}$ is

$$\begin{aligned}\underset{\mathbf{x}_0,\mathbf{x}_i,\tilde{\mathbf{x}}_i,\mathbf{u}_0,\mathbf{u}_i,\tilde{\mathbf{u}}_i}{\text{minimize}} \quad & \mathbf{c}^T\left(\mathbf{p}\right)\mathbf{x}_N \\ \text{subject to} \quad & \tilde{\mathbf{x}}_i = \frac{1}{2}\left(\mathbf{x}_{i-1} + \mathbf{x}_i\right) + h\left[\mathbf{A}\left(\mathbf{p}\right)\frac{\mathbf{x}_{i-1} - \mathbf{x}_i}{8} + \mathbf{B}\left(\mathbf{p}\right)\frac{\mathbf{u}_{i-1} - \mathbf{u}_i}{8}\right] \\ & \mathbf{x}_i = \mathbf{x}_{i-1} + h\Bigg[\mathbf{A}\left(\mathbf{p}\right)\frac{\mathbf{x}_{i-1} + 4\tilde{\mathbf{x}}_i + \mathbf{x}_i - 6\mathbf{x}_r\left(\mathbf{p}\right)}{6} \\ & \qquad + \mathbf{B}\left(\mathbf{p}\right)\frac{\mathbf{u}_{i-1} + 4\tilde{\mathbf{u}}_i + \mathbf{u}_i - 6\mathbf{u}_r\left(\mathbf{p}\right)}{6}\Bigg] \\ & \mathbf{x}_r\left(\mathbf{p}\right) - \mathbf{x}_0 = \mathbf{0} \\ & \underline{\mathbf{x}} \leqslant \mathbf{x}_i, \tilde{\mathbf{x}}_i \leqslant \overline{\mathbf{x}} \\ & \underline{\mathbf{u}} \leqslant \mathbf{u}_i, \tilde{\mathbf{u}}_i \leqslant \overline{\mathbf{u}} \\ & \underline{\mathbf{u}} \leqslant \mathbf{u}_0 \leqslant \overline{\mathbf{u}} \\ & i = 1, \ldots, N\end{aligned}$$

The cost vector $\mathbf{v}(\mathbf{p}) : \mathbb{R}^{n_p} \to \mathbb{R}^n$ and optimization variable vector $\mathbf{z} \in \mathbb{R}^n$ with corresponding lower bound $\underline{\mathbf{z}} \in \mathbb{R}^n$ and upper bound $\overline{\mathbf{z}} \in \mathbb{R}^n$ are

$$\mathbf{v} = \begin{bmatrix} \mathbf{0} \\ \mathbf{0} \\ \mathbf{0} \\ \mathbf{0} \\ \mathbf{0} \\ \mathbf{0} \\ \vdots \\ \mathbf{0} \\ \mathbf{0} \\ \mathbf{c}(\mathbf{p}) \\ \mathbf{0} \end{bmatrix}, \quad \underline{\mathbf{z}} = \begin{bmatrix} -\infty \\ \underline{\mathbf{u}} \\ \underline{\mathbf{x}} \\ \underline{\mathbf{u}} \\ \underline{\mathbf{x}} \\ \underline{\mathbf{u}} \\ \vdots \\ \underline{\mathbf{x}} \\ \underline{\mathbf{u}} \\ \underline{\mathbf{x}} \\ \underline{\mathbf{u}} \end{bmatrix}, \quad \mathbf{z} = \begin{bmatrix} \mathbf{x}_0 \\ \mathbf{u}_0 \\ \tilde{\mathbf{x}}_1 \\ \tilde{\mathbf{u}}_1 \\ \mathbf{x}_1 \\ \mathbf{u}_1 \\ \vdots \\ \tilde{\mathbf{x}}_N \\ \tilde{\mathbf{u}}_N \\ \mathbf{x}_N \\ \mathbf{u}_N \end{bmatrix}, \quad \overline{\mathbf{z}} = \begin{bmatrix} \infty \\ \overline{\mathbf{u}} \\ \overline{\mathbf{x}} \\ \overline{\mathbf{u}} \\ \overline{\mathbf{x}} \\ \overline{\mathbf{u}} \\ \vdots \\ \overline{\mathbf{x}} \\ \overline{\mathbf{u}} \\ \overline{\mathbf{x}} \\ \overline{\mathbf{u}} \end{bmatrix}$$

The equality constraint matrix is defined as

$$\mathbf{M}(\mathbf{p}) = \begin{bmatrix} \mathbf{H}(\mathbf{p}) \\ \mathbf{S}(\mathbf{p}) \end{bmatrix}$$

with the matrix $\mathbf{H} : \mathbb{R}^{n_p} \to \mathbb{R}^{n_h \times n}$, $n_h = N n_x$:

$$\mathbf{H}(\mathbf{p}) = \begin{bmatrix} -\frac{h\mathbf{A}(\mathbf{p})}{8} - \frac{1}{2}\mathbf{I} & -\frac{h\mathbf{B}(\mathbf{p})}{8} & \mathbf{I} & \mathbf{0} & \frac{h\mathbf{A}(\mathbf{p})}{8} - \frac{1}{2}\mathbf{I} & \frac{h\mathbf{B}(\mathbf{p})}{8} & \mathbf{0} & \mathbf{0} & \mathbf{0} & \mathbf{0} & \cdots \\ \mathbf{0} & \mathbf{0} & \mathbf{0} & \mathbf{0} & -\frac{h\mathbf{A}(\mathbf{p})}{8} - \frac{1}{2}\mathbf{I} & -\frac{h\mathbf{B}(\mathbf{p})}{8} & \mathbf{I} & \mathbf{0} & \frac{h\mathbf{A}(\mathbf{p})}{8} - \frac{1}{2}\mathbf{I} & \frac{h\mathbf{B}(\mathbf{p})}{8} & \\ \vdots & \vdots & \vdots & \vdots & & & & & \ddots & & \end{bmatrix}$$

for the linear Hermite constraints [Eq. (H.1)] and the matrix $\mathbf{S} : \mathbb{R}^{n_p} \to \mathbb{R}^{(n_h+n_x)\times n}$:

$$\mathbf{S}(\mathbf{p}) = -\begin{bmatrix} -\mathbf{I} & \mathbf{0} & \mathbf{0} & \mathbf{0} & \mathbf{0} & \mathbf{0} & \mathbf{0} & \mathbf{0} & \mathbf{0} & \mathbf{0} & \cdots \\ \frac{h\mathbf{A}(\mathbf{p})}{6} + \mathbf{I} & \frac{h\mathbf{B}(\mathbf{p})}{6} & \frac{2h\mathbf{A}(\mathbf{p})}{3} & \frac{2h\mathbf{B}(\mathbf{p})}{3} & \frac{h\mathbf{A}(\mathbf{p})}{6} - \mathbf{I} & \frac{h\mathbf{B}(\mathbf{p})}{6} & \mathbf{0} & \mathbf{0} & \mathbf{0} & \mathbf{0} & \cdots \\ \mathbf{0} & \mathbf{0} & \mathbf{0} & \mathbf{0} & \frac{h\mathbf{A}(\mathbf{p})}{6} + \mathbf{I} & \frac{h\mathbf{B}(\mathbf{p})}{6} & \frac{2h\mathbf{A}(\mathbf{p})}{3} & \frac{2h\mathbf{B}(\mathbf{p})}{3} & \frac{h\mathbf{A}(\mathbf{p})}{6} - \mathbf{I} & \frac{h\mathbf{B}(\mathbf{p})}{6} & \\ \vdots & \vdots & \vdots & \vdots & \ddots & & & & & & \end{bmatrix}$$

for the linear Simpson constraints [Eq. (H.2)]. The corresponding right-side vector $\mathbf{b}(\mathbf{p}) : \mathbb{R}^{n_p} \rightarrow \mathbb{R}^{2n_h+n_x}$ is of the form

$$\mathbf{b}(\mathbf{p}) = \begin{bmatrix} \mathbf{0} \\ \vdots \\ \mathbf{0} \\ \mathbf{x}_r(\mathbf{p}) \\ h\mathbf{A}(\mathbf{p})\,\mathbf{x}_r(\mathbf{p}) + h\mathbf{B}(\mathbf{p})\,\mathbf{u}_r(\mathbf{p}) \\ \vdots \\ h\mathbf{A}(\mathbf{p})\,\mathbf{x}_r(\mathbf{p}) + h\mathbf{B}(\mathbf{p})\,\mathbf{u}_r(\mathbf{p}) \end{bmatrix}$$

Hermite–Simpson Nonlinear

The Hermite constraints for the midpoints are

$$\tilde{\mathbf{x}}_i = \frac{1}{2}\left(\mathbf{x}_{i-1} + \mathbf{x}_i\right) + h\left[\frac{1}{8}\mathbf{f}\left(\mathbf{x}_{i-1}, \mathbf{u}_{i-1}, \mathbf{p}\right) - \frac{1}{8}\mathbf{f}\left(\mathbf{x}_i, \mathbf{u}_i, \mathbf{p}\right)\right]$$

and the Simpson constraints for each interval are

$$\mathbf{x}_i = \mathbf{x}_{i-1} + h\left[\frac{1}{6}\mathbf{f}\left(\mathbf{x}_{i-1}, \mathbf{u}_{i-1}, \mathbf{p}\right) + \frac{4}{6}\mathbf{f}\left(\tilde{\mathbf{x}}_i, \tilde{\mathbf{u}}_i, \mathbf{p}\right) + \frac{1}{6}\mathbf{f}\left(\mathbf{x}_i, \mathbf{u}_i, \mathbf{p}\right)\right]$$

for $i = 1, \ldots, N$. Here, $\tilde{\mathbf{x}}_i$ and $\tilde{\mathbf{u}}_i$ represent additional state and control variables associated with the segment midpoints. The discretized Cat III problem is

$$\begin{aligned}
\underset{\mathbf{x}_0, \mathbf{x}_i, \tilde{\mathbf{x}}_i, \mathbf{u}_0, \mathbf{u}_i, \tilde{\mathbf{u}}_i, \mathbf{p}}{\text{minimize}} \quad & J\left(\mathbf{x}_N, \mathbf{p}\right) \\
\text{subject to} \quad & \tilde{\mathbf{x}}_i = \frac{1}{2}\left(\mathbf{x}_{i-1} + \mathbf{x}_i\right) + h\left[\frac{1}{8}\mathbf{f}\left(\mathbf{x}_{i-1}, \mathbf{u}_{i-1}, \mathbf{p}\right) - \frac{1}{8}\mathbf{f}\left(\mathbf{x}_i, \mathbf{u}_i, \mathbf{p}\right)\right] \\
& \mathbf{x}_i = \mathbf{x}_{i-1} + h\left[\frac{1}{6}\mathbf{f}\left(\mathbf{x}_i, \mathbf{u}_i, \mathbf{p}\right) + \frac{4}{6}\mathbf{f}\left(\tilde{\mathbf{x}}_i, \tilde{\mathbf{u}}_i, \mathbf{p}\right) + \frac{1}{6}\mathbf{f}\left(\mathbf{x}_{i-1}, \mathbf{u}_{i-1}, \mathbf{p}\right)\right] \\
& \gamma\left(\mathbf{x}_0, \mathbf{p}\right) = \mathbf{0} \\
& \underline{\mathbf{p}} \leqslant \mathbf{p} \leqslant \overline{\mathbf{p}} \\
& \underline{\mathbf{x}} \leqslant \mathbf{x}_i, \tilde{\mathbf{x}}_i \leqslant \overline{\mathbf{x}} \\
& \underline{\mathbf{u}} \leqslant \mathbf{u}_i, \tilde{\mathbf{u}}_i \leqslant \overline{\mathbf{u}} \\
& \underline{\mathbf{u}} \leqslant \mathbf{u}_0 \leqslant \overline{\mathbf{u}} \\
& i = 1, \ldots, N
\end{aligned}$$

This nonlinear programming problem features $n = (2N + 1)(n_x + n_u) + n_p$ decision variables and $n_a = 2Nn_x + n_\gamma$ equality constraints. The decision variable vector $\mathbf{z} \in \mathbb{R}^n$ and the corresponding lower bounds $\underline{\mathbf{z}} \in \mathbb{R}^n$ as well as upper bounds $\overline{\mathbf{z}} \in \mathbb{R}^n$ are

$$\underline{\mathbf{z}} = \begin{bmatrix} \underline{\mathbf{p}} \\ -\infty \\ \underline{\mathbf{u}} \\ \underline{\mathbf{x}} \\ \underline{\mathbf{u}} \\ \underline{\mathbf{x}} \\ \underline{\mathbf{u}} \\ \vdots \\ \underline{\mathbf{x}} \\ \underline{\mathbf{u}} \\ \underline{\mathbf{x}} \\ \underline{\mathbf{u}} \end{bmatrix}, \qquad \mathbf{z} = \begin{bmatrix} \mathbf{p} \\ \mathbf{x}_0 \\ \mathbf{u}_0 \\ \tilde{\mathbf{x}}_1 \\ \tilde{\mathbf{u}}_1 \\ \mathbf{x}_1 \\ \mathbf{u}_1 \\ \vdots \\ \tilde{\mathbf{x}}_N \\ \tilde{\mathbf{u}}_N \\ \mathbf{x}_N \\ \mathbf{u}_N \end{bmatrix}, \qquad \overline{\mathbf{z}} = \begin{bmatrix} \overline{\mathbf{p}} \\ \infty \\ \overline{\mathbf{u}} \\ \overline{\mathbf{x}} \\ \overline{\mathbf{u}} \\ \overline{\mathbf{x}} \\ \overline{\mathbf{u}} \\ \vdots \\ \overline{\mathbf{x}} \\ \overline{\mathbf{u}} \\ \overline{\mathbf{x}} \\ \overline{\mathbf{u}} \end{bmatrix}$$

and the equality constraints can be collected in

$$\mathbf{a}(\mathbf{z}) = \begin{bmatrix} \mathbf{h}(\mathbf{z}) \\ \mathbf{s}(\mathbf{z}) \end{bmatrix}$$

with

$$\mathbf{h}(\mathbf{z}) = \begin{bmatrix} \tilde{\mathbf{x}}_1 - \frac{1}{2}(\mathbf{x}_0 - \mathbf{x}_1) - h\left[\frac{1}{8}\mathbf{f}(\mathbf{x}_0, \mathbf{u}_0, \mathbf{p}) - \frac{1}{8}\mathbf{f}(\mathbf{x}_1, \mathbf{u}_1, \mathbf{p})\right] \\ \vdots \\ \tilde{\mathbf{x}}_N - \frac{1}{2}(\mathbf{x}_{N-1} - \mathbf{x}_N) - h\left[\frac{1}{8}\mathbf{f}(\mathbf{x}_{N-1}, \mathbf{u}_{N-1}, \mathbf{p}) - \frac{1}{8}\mathbf{f}(\mathbf{x}_N, \mathbf{u}_N, \mathbf{p})\right] \end{bmatrix}$$

and

$$\mathbf{s}(\mathbf{z}) = \begin{bmatrix} \boldsymbol{\gamma}(\mathbf{x}_0, \mathbf{p}) \\ \mathbf{x}_1 - \mathbf{x}_0 - h\left[\frac{1}{6}\mathbf{f}(\mathbf{x}_0, \mathbf{u}_0, \mathbf{p}) + \frac{4}{6}\mathbf{f}(\tilde{\mathbf{x}}_1, \tilde{\mathbf{u}}_1, \mathbf{p}) + \frac{1}{6}\mathbf{f}(\mathbf{x}_1, \mathbf{u}_1, \mathbf{p})\right] \\ \vdots \\ \mathbf{x}_N - \mathbf{x}_{N-1} - h\left[\frac{1}{6}\mathbf{f}(\mathbf{x}_{N-1}, \mathbf{u}_{N-1}, \mathbf{p}) + \frac{4}{6}\mathbf{f}(\tilde{\mathbf{x}}_N, \tilde{\mathbf{u}}_N, \mathbf{p}) + \frac{1}{6}\mathbf{f}(\mathbf{x}_N, \mathbf{u}_N, \mathbf{p})\right] \end{bmatrix}$$

The Jacobians of the equality constraints are

$$\frac{\partial \mathbf{h}}{\partial \mathbf{z}} = \begin{bmatrix} -\frac{h\frac{\partial \mathbf{f}_0}{\partial \mathbf{p}} + h\frac{\partial \mathbf{f}_1}{\partial \mathbf{p}}}{8} & -\frac{h\frac{\partial \mathbf{f}_0}{\partial \mathbf{x}_0}}{8} + \frac{1}{2}\mathbf{I} & -\frac{h\frac{\partial \mathbf{f}_0}{\partial \mathbf{u}_0}}{8} & \mathbf{I} & \mathbf{0} & \frac{h\frac{\partial \mathbf{f}_1}{\partial \mathbf{x}_1}}{8} + \frac{1}{2}\mathbf{I} & \frac{h\frac{\partial \mathbf{f}_1}{\partial \mathbf{u}_1}}{8} & \mathbf{0} & \mathbf{0} & \mathbf{0} & \mathbf{0} & \dots \\ -\frac{h\frac{\partial \mathbf{f}_1}{\partial \mathbf{p}} + h\frac{\partial \mathbf{f}_2}{\partial \mathbf{p}}}{8} & \mathbf{0} & \mathbf{0} & \mathbf{0} & \mathbf{0} & -\frac{h\frac{\partial \mathbf{f}_1}{\partial \mathbf{x}_1}}{8} + \frac{1}{2}\mathbf{I} & -\frac{h\frac{\partial \mathbf{f}_1}{\partial \mathbf{u}_1}}{8} & \mathbf{I} & \mathbf{0} & \frac{h\frac{\partial \mathbf{f}_2}{\partial \mathbf{x}_2}}{8} + \frac{1}{2}\mathbf{I} & \frac{h\frac{\partial \mathbf{f}_2}{\partial \mathbf{u}_2}}{8} & \\ \vdots & \vdots & \vdots & \vdots & \vdots & & \ddots & & & & & \end{bmatrix}$$

and

$$\frac{\partial \mathbf{s}}{\partial \mathbf{z}} = -\begin{bmatrix} \frac{\partial \boldsymbol{\gamma}}{\partial \mathbf{p}} & \frac{\partial \boldsymbol{\gamma}}{\partial \mathbf{x}_0} & \mathbf{0} & \mathbf{0} & \mathbf{0} & \mathbf{0} & \mathbf{0} & \mathbf{0} & \mathbf{0} & \mathbf{0} & \mathbf{0} & \dots \\ \frac{h\frac{\partial \mathbf{f}_0}{\partial \mathbf{p}} + 4h\frac{\partial \tilde{\mathbf{f}}_1}{\partial \mathbf{p}} + h\frac{\partial \mathbf{f}_1}{\partial \mathbf{p}}}{6} & \frac{h\frac{\partial \mathbf{f}_0}{\partial \mathbf{x}_0}}{6} + \mathbf{I} & \frac{h\frac{\partial \mathbf{f}_0}{\partial \mathbf{u}_0}}{6} & \frac{4h\frac{\partial \tilde{\mathbf{f}}_1}{\partial \tilde{\mathbf{x}}_1}}{6} & \frac{4h\frac{\partial \tilde{\mathbf{f}}_1}{\partial \tilde{\mathbf{u}}_1}}{6} & \frac{h\frac{\partial \mathbf{f}_1}{\partial \mathbf{x}_1}}{6} - \mathbf{I} & \frac{h\frac{\partial \mathbf{f}_1}{\partial \mathbf{u}_1}}{6} & \mathbf{0} & \mathbf{0} & \mathbf{0} & \mathbf{0} & \dots \\ \frac{h\frac{\partial \mathbf{f}_1}{\partial \mathbf{p}} + 4h\frac{\partial \tilde{\mathbf{f}}_2}{\partial \mathbf{p}} + h\frac{\partial \mathbf{f}_2}{\partial \mathbf{p}}}{6} & \mathbf{0} & \mathbf{0} & \mathbf{0} & \mathbf{0} & \frac{h\frac{\partial \mathbf{f}_1}{\partial \mathbf{x}_1}}{6} + \mathbf{I} & \frac{h\frac{\partial \mathbf{f}_1}{\partial \mathbf{u}_1}}{6} & \frac{4h\frac{\partial \tilde{\mathbf{f}}_2}{\partial \tilde{\mathbf{x}}_2}}{6} & \frac{4h\frac{\partial \tilde{\mathbf{f}}_2}{\partial \tilde{\mathbf{u}}_2}}{6} & \frac{h\frac{\partial \mathbf{f}_2}{\partial \mathbf{x}_2}}{6} - \mathbf{I} & \frac{h\frac{\partial \mathbf{f}_2}{\partial \mathbf{u}_2}}{6} & \\ \vdots & \vdots & \vdots & \vdots & \vdots & & \ddots & & & & & \end{bmatrix}$$

using the short notation $\mathbf{f}_i = \mathbf{f}\left(\mathbf{x}_i, \mathbf{u}_i, \mathbf{p}\right), \tilde{\mathbf{f}}_i = \tilde{\mathbf{f}}\left(\tilde{\mathbf{x}}_i, \tilde{\mathbf{u}}_i, \mathbf{p}\right)$.

References

[1] Diepolder, J., "Optimal Control Based Clearance of Flight Control Laws," Ph.D. Dissertation, Technical Univ. of Munich, Munich, 2021.

[2] Khorikov, V., *Unit Testing: Principles, Practices, and Patterns*, Manning Publications, Shelter Island, NY, 2020, p. 4.

[3] Dijkstra, E. W., "Notes on Structured Programming," *Structured Programming*, edited by O. -J. Dahl, E. W. Dijkstra, and C. A. R. Hoare, Academic Press, London, 1972, pp. 1–82.

[4] Fielding, C., and Varga, A. (eds.), *Advanced Techniques for Clearance of Flight Control Laws*, Lecture Notes in Control and Information Sciences, Vol. 283, Springer Berlin, Heidelberg, 2002.

[5] Varga, A., Hansson, A., and Puyou, G. (eds.), *Optimization Based Clearance of Flight Control Laws*, Lecture Notes in Control and Information Sciences, Vol. 416, Springer Berlin, Heidelberg, 2012.

[6] Menon, P. P., Kim, J., Bates, D. G., and Postlethwaite, I., "Improved Clearance of Flight Control Laws Using Hybrid Optimisation," *IEEE Conference on Cybernetics and Intelligent Systems*, Singapore, 2004, pp. 677–682.

[7] Menon, P. P., Bates, D. G., and Postlethwaite, I., "A Deterministic Hybrid Optimisation Algorithm for Nonlinear Flight Control Systems Analysis," *American Control Conference*, Institute of Electrical and Electronics Engineers, Minneapolis, MN, 2006, pp. 333–338.

[8] Menon, P. P., Kim, J., Bates, D. G., and Postlethwaite, I., "Clearance of Nonlinear Flight Control Laws Using Hybrid Evolutionary Optimization," *IEEE Transactions on Evolutionary Computation*, Vol. 10, No. 6, 2006, pp. 689–699.

[9] Skoogh, D., Eliasson, P., Berefelt, F., Amiree, R., Tourde, D., and Forssell, L., "Clearance of Flight Control Laws for Time Varying Pilot Input Signals," *IFAC Proceedings Volumes*, Vol. 42, No. 6, 2009, pp. 343–348.

[10] Joos, H.-D., "Application of Parallel Nonlinear Programming Methods for Worst-Case Parameter Search," *Optimization Based Clearance of Flight Control Laws*, edited by A. Varga, A. Hansson, and G. Puyou, Lecture Notes in Control and Information Sciences, Vol. 416, Springer Berlin, Heidelberg, 2012, pp. 233–252.

[11] Joos, H.-D., "Worst-Case Parameter Search Based Clearance Using Parallel Nonlinear Programming Methods," *Optimization Based Clearance of Flight Control Laws*, edited by A. Varga, A. Hansson, and G. Puyou, Lecture Notes in

Control and Information Sciences, Vol. 416, Springer Berlin, Heidelberg, 2012, pp. 149–159.

[12] Skoogh, D., and Berefelt, F., "Application of Nonlinear Programming Methods for Determination of Worst-Case Pilot Inputs," *Optimization Based Clearance of Flight Control Laws*, edited by A. Varga, A. Hansson, and G. Puyou, Lecture Notes in Control and Information Sciences, Vol. 416, Springer Berlin, Heidelberg, 2012, pp. 299–315.

[13] Skoogh, D., and Berefelt, F., "Nonlinear Programming Methods for Worst-Case Pilot Input Determination," *Optimization Based Clearance of Flight Control Laws*, edited by A. Varga, A. Hansson, and G. Puyou, Lecture Notes in Control and Information Sciences, Vol. 416, Springer Berlin, Heidelberg, 2012, pp. 203–217.

[14] Puyou, G., de Oliveira, R. F., and Berard, A., "Evaluation of Clearance Techniques in an Industrial Context," *Optimization Based Clearance of Flight Control Laws*, edited by A. Varga, A. Hansson, and G. Puyou, Lecture Notes in Control and Information Sciences, Vol. 416, Springer Berlin, Heidelberg, 2012, pp. 319–357.

[15] Guo, S.-M., Tsai, J. S.-H., Yang, C.-C., and Hsu, P.-H., "A Self-Optimization Approach for L-SHADE Incorporated with Eigenvector-Based Crossover and Successful-Parent-Selecting Framework on CEC 2015 Benchmark Set," *International Conference on Image and Vision Computing*, Institute of Electrical and Electronics Engineers, Christchurch, New Zealand, 2015, pp. 1003–1010.

[16] Rodriguez Robles, R., Sabaris Boullosa, M., and Asensio Nieto, F., "Flight Control Laws Carefree Handling Clearance of a Highly Manoeuvrable Aircraft Using Multi-Strategy Adaptive Global Optimization," *7th European Conference for Aeronautics and Space Sciences*, Milan, Italy, 2017.

[17] Piprek, P., Dias, P. M., and Schwalb, D., "A Generalised Multi-Objective Island Model for Flight Control System Clearances," *33rd Congress of the International Council of the Aeronautical Sciences*, Stockholm, Sweden, 2022.

[18] Dias, P. M., Piprek, P., and Schwalb, D., "Optimization-Based Flight Control System Clearance Philosophy for Fighter Aircraft," *AIAA Scitech 2023 Forum*, National Harbor, MD, 2023.

[19] Herrmann, A. A., and Ben-Asher, J. Z., "Flight Control Law Clearance Using Optimal Control Theory," *Journal of Aircraft*, Vol. 53, No. 2, 2016, pp. 515–529.

[20] Betts, J. T., *Practical Methods for Optimal Control and Estimation Using Nonlinear Programming*, SIAM Advances in Design and Control, Philadelphia, PA, 2010.

[21] Gerdts, M., *Optimal Control of ODEs and DAEs*, Walter de Gruyter, Berlin/Boston, 2012.

[22] Nocedal, J., "Updating Quasi-Newton Matrices with Limited Storage," *Mathematics of Computation*, Vol. 35, No. 151, 1980, pp. 773–782.

[23] Nocedal, J., and Wright, S. J., *Numerical Optimization*, Springer Series in Operations Research and Financial Engineering, Springer New York, 2006.

[24] Wächter, A., and Biegler, L. T., "On the Implementation of an Interior-Point Filter Line-Search Algorithm for Large-Scale Nonlinear Programming," *Mathematical Programming*, Vol. 106, 2006, pp. 25–57.

[25] Wächter, A., "Short Tutorial: Getting Started with Ipopt in 90 Minutes," *Combinatorial Scientific Computing, Dagstuhl Seminar Proceedings*, Vol. 9061, Schloss Dagstuhl, Germany, 2009, pp. 1–17.

[26] Griewank, A., "On Automatic Differentiation," *Mathematical Programming: Recent Developments and Applications*, Vol. 6, No. 6, 1989, pp. 83–107.

[27] Walther, A., "Getting Started with ADOL-C," *Combinatorial Scientific Computing, Dagstuhl Seminar Proceedings*, Vol. 9061, Schloss Dagstuhl, Germany, 2009, pp. 1–10.

[28] Andersson, J. A. E., Gillis, J., Horn, G., Rawlings, J. B., and Diehl, M., "CasADi: A Software Framework for Nonlinear Optimization and Optimal Control," *Mathematical Programming Computation*, Vol. 11, No. 1, 2019, pp. 1–36.

[29] Rieck, R. M., "Discrete Controls and Constraints in Optimal Control Problems," Ph.D. Dissertation, Technical Univ. of Munich, Munich, 2017.

[30] Fisch, F., "Development of a Framework for the Solution of High-Fidelity Trajectory Optimization Problems and Bilevel Optimal Control Problems," Ph.D. Dissertation, Technical Univ. of Munich, Munich, 2011.

[31] Becerra, V. M., "Solving Complex Optimal Control Problems at No Cost with PSOPT," *IEEE International Symposium on Computer-Aided Control System Design*, Yokohama, Japan, 2010, pp. 1391–1396.

[32] Weinstein, M. J., and Rao, A. V., "Algorithm 984: ADiGator, a Toolbox for the Algorithmic Differentiation of Mathematical Functions in MATLAB Using Source Transformation via Operator Overloading," *ACM Transactions on Mathematical Software*, Vol. 44, No. 2, 2017, Article 21.

[33] Patterson, M. A., and Rao, A. V., "GPOPS-II: A MATLAB Software for Solving Multiple-Phase Optimal Control Problems Using Hp-Adaptive Gaussian Quadrature Collocation Methods and Sparse Nonlinear Programming," *ACM Transactions on Mathematical Software*, Vol. 41, No. 1, 2014.

[34] Rieck, M., Bittner, M., Grüter, B., Diepolder, J., Piprek, P., Göttlicher, C., Schwaiger, F., Hosseini, B., Schweighofer, F., Akman, T., and Holzapfel, F., *FALCON.m User Guide v1.31*, Institute of Flight System Dynamics, Technical Univ. of Munich, Munich, April 24, 2024, www.fsd.ed.tum.de/software/falcon-m.

[35] Bittner, M., "Utilization of Problem and Dynamic Characteristics for Solving Large Scale Optimal Control Problems," Ph.D. Dissertation, Technical Univ. of Munich, Munich, 2017.

[36] Ghazaei Ardakani, M. M., and Magnusson, F., "Ball-and-Finger System: Modeling and Optimal Trajectories," *Multibody System Dynamics*, Vol. 44, No. 2, 2018, pp. 163–189.

[37] Bertolazzi, E., "mexIPOPT," GitHub, January 28, 2025, www.github.com/ebertolazzi/mexIPOPT.

[38] Gill, P. E., Murray, W., and Saunders, M. A., "SNOPT: An SQP Algorithm for Large-Scale Constrained Optimization," *SIAM Journal on Optimization*, Vol. 12, No. 4, 2002, pp. 979–1006.

[39] Bueskens, C., and Wassel, D., "The ESA NLP Solver WORHP," *Modeling and Optimization in Space Engineering*, Vol. 73, Springer New York, 2013, pp. 85–110.

[40] Lai, K.-L., and Crassidis, J. L., "Extensions of the First and Second Complex-Step Derivative Approximations," *Journal of Computational and Applied Mathematics*, Vol. 219, No. 1, 2008, pp. 276–293.

[41] Martins, J. R. R. A., Sturdza, P., and Alonso, J. J., "The Connection Between the Complex-Step Derivative Approximation and Algorithmic Differentiation," *39th Aerospace Sciences Meeting and Exhibit*, AIAA, Reno, NV, 2001, AIAA-2001-0921, pp. 1–11.

[42] Gebremedhin, A. H., Manne, F., and Pothen, A., "What Color Is Your Jacobian? Graph Coloring for Computing Derivatives," *SIAM Review*, Vol. 47, No. 4, 2005, pp. 629–705.

[43] Butcher, J. C., *Numerical Methods for Ordinary Differential Equations*, John Wiley and Sons, Hoboken, NJ, 2016.

[44] Hairer, E., and Wanner, G., *Solving Ordinary Differential Equations I: Nonstiff Problems*, Springer Series in Computational Mathematics, Vol. 8, Springer Berlin, Heidelberg, 1993, p. 212.

[45] Darby, C. L., Hager, W. W., and Rao, A. V., "An Hp-Adaptive Pseudospectral Method for Solving Optimal Control Problems," *Optimal Control Applications and Methods*, Vol. 32, No. 4, 2011, pp. 476–502.

[46] Piegl, L., and Tiller, W., *The NURBS Book*, Monographs in Visual Communications, Springer Berlin, Heidelberg, 1995.

[47] Pontryagin, L. S., Boltyanskii, V. G., Gamkrelidze, R. V., and Mishchenko, E. F., *Mathematical Theory of Optimal Processes*, Wiley-Interscience, New York, 1967.

[48] Bell, D. J., "Lie Brackets and Singular Optimal Control," *IMA Journal of Mathematical Control and Information*, Vol. 1, No. 1, 1984, pp. 83–94.

[49] Hartl, R. F., Sethi, S. P., and Vickson, R. G., "A Survey of the Maximum Principles for Optimal Control Problems with State Constraints," *SIAM Review*, Vol. 37, No. 2, 1995, pp. 181–218.

[50] Bryson, A. E., and Ho, Y.-C., *Applied Optimal Control*, Hemisphere, New York, 1975.

[51] Chang, S. S. L., "Optimal Control in Bounded Phase Space," *Automatica*, Vol. 1, No. 1, 1963, pp. 55–67.

[52] Varga, A., "Clearance Criteria for Civil Aircraft," TR AST5-CT-2006-030768, COFCLUO (Clearance of Flight Control Laws Using Optimization), 2010.

[53] McRuer, D. T., "Pilot-Induced Oscillations and Human Dynamic Behavior," NASA Contractor Rept. 4683, 1995.

[54] Dantzig, G., *Linear Programming and Extensions*, RAND Corp., Santa Monica, CA, 1963.

[55] Ben-Asher, J. Z., *Optimal Control Theory with Aerospace Applications*, AIAA Education Series, Reston, VA, 2010.

[56] Steinert, A. C., "Test-Based Controller Parameter Retuning Using Non-Parametric Frequency Domain Methods," Ph.D. Dissertation, Technical Univ. of Munich, Munich, 2022.

[57] Gal, T., "Linear Parametric Programming: A Brief Survey," *Sensitivity, Stability and Parametric Analysis*, Mathematical Programming Studies, 1984, pp. 43–68.

[58] Adler, I., and Monteiro, R. D. C., "A Geometric View of Parametric Linear Programming," *Algorithmica*, Vol. 8, No. 1, 1992, pp. 161–176.

[59] Greenberg, H. J., "The Use of the Optimal Partition in a Linear Programming Solution for Postoptimal Analysis," *Operations Research Letters*, Vol. 15, No. 4, 1994, pp. 179–185.

[60] Monteiro, R. D. C., and Mehrotra, S., "A General Parametric Analysis Approach and Its Implication to Sensitivity Analysis in Interior Point Methods," *Mathematical Programming*, Vol. 72, No. 1, 1996, pp. 65–82.

[61] Castillo, E., Conejo, A. J., Mínguez, R., and Castillo, C., "A Closed Formula for Local Sensitivity Analysis in Mathematical Programming," *Engineering Optimization*, Vol. 38, No. 1, 2006, pp. 93–112.

[62] Diepolder, J., Ben-Asher, J. Z., and Holzapfel, F., "Flight Control Law Clearance Using Worst-Case Inputs Under Parameter Uncertainty," *AIAA Journal of Guidance, Control, and Dynamics*, Vol. 43, No. 10, 2020, pp. 1967–1974.

[63] Jones, D. R., Perttunen, C. D., and Stuckman, B. E., "Lipschitzian Optimization Without the Lipschitz Constant," *Journal of Optimization Theory and Applications*, Vol. 79, No. 1, 1993, pp. 157–181.

[64] Bjorkman, M., and Holmström, K., "Global Optimization Using the DIRECT Algorithm in MATLAB," *Advanced Modeling and Optimization*, Vol. 1, No. 2, 1999, pp. 17–37.

[65] Slotani, M., "Tolerance Regions for a Multivariate Normal Population," *Annals of the Institute of Statistical Mathematics*, Vol. 16, No. 1, 1964, pp. 135–153.

[66] Varga, A., "Optimisation-Based Clearance," *Advanced Techniques for Clearance of Flight Control Laws*, edited by C. Fielding, A. Varga, S. Bennani, and M. Selier, Lecture Notes in Control and Information Sciences, Vol. 283, Springer Berlin, Heidelberg, 2002, pp. 107–117.

[67] Forssell, L., "Flight Clearance Analysis Using Global Nonlinear Optimisation-Based Search Algorithms," *AIAA Guidance, Navigation and Control Conference and Exhibit*, Austin, TX, 2003, AIAA 2003-5481, pp. 1–8.

[68] De Oliveira, R. F., and Puyou, G., "On the Use of Optimization for Flight Control Laws Clearance: A Practical Approach," *IFAC Proceedings Volumes*, Vol. 44, No. 1, 2011, pp. 9881–9886.

[69] Nelder, J. A., and Mead, R., "A Simplex Method for Function Minimization," *Computer Journal*, Vol. 7, No. 4, 1965, pp. 308–313.

[70] Lagarias, J. C., Reeds, J. A., Wright, M. H., and Wright, P. E., "Convergence Properties of the Nelder-Mead Simplex Method in Low Dimensions," *SIAM Journal on Optimization*, Vol. 9, No. 1, 1998, pp. 112–147.

[71] D'Errico, J., Matlab Central: fminsearchbnd, fminsearchcon, MATLAB Central File Exchange, February 7, 2021, www.mathworks.com/matlabcentral/fileexchange/8277-fminsearchbnd-fminsearchcon.

[72] Kvasnica, M., Grieder, P., Baotic, M., and Morari, M., "Multi-Parametric Toolbox (MPT)," *Hybrid Systems: Computation and Control*, Lecture Notes in Computer Science, Vol. 2993, Springer Berlin, Heidelberg, 2004, pp. 448–462.

[73] Kurzhanski, A. B., and Varaiya, P., "Ellipsoidal Techniques for Reachability Analysis," *Hybrid Systems*, Lecture Notes in Computer Science, Vol. 1790, Springer Berlin, Heidelberg, 2000, pp. 202–214.

[74] Kurzhanskiy, A. A., and Varaiya, P., "Ellipsoidal Toolbox (ET)," *45th IEEE Conference on Decision and Control*, San Diego, CA, 2006, pp. 1498–1503.

[75] Girard, A., "Reachability of Uncertain Linear Systems Using Zonotopes," *Hybrid Systems: Computation and Control*, Lecture Notes in Computer Science, Vol. 3414, Springer Berlin, Heidelberg, 2005, pp. 291–305.

[76] Althoff, D., Althoff, M., and Scherer, S., "Online Safety Verification of Trajectories for Unmanned Flight with Offline Computed Robust Invariant Sets," *IEEE/RSJ International Conference on Intelligent Robots and Systems*, Hamburg, Germany, 2015, pp. 3470–3477.

[77] Kostousova, E. K., "Control Synthesis via Parallelotopes: Optimization and Parallel Computations," *Optimization Methods and Software*, Vol. 14, No. 4, 2001, pp. 267–310.

[78] Kostousova, E. K., "On Tight Polyhedral Estimates for Reachable Sets of Linear Differential Systems," *AIP Conference Proceedings*, Sivasundaram, S. (ed.), Vol. 1493, 2012, pp. 579–586.

[79] Varaiya, P., "Reach Set Computation Using Optimal Control," *Verification of Digital and Hybrid Systems*, NATO ASI Subseries F, Vol. 170, Springer Berlin, Heidelberg, 2012, pp. 323–331.

[80] Baier, R., Büskens, C., Chahma, I. A., and Gerdts, M., "Approximation of Reachable Sets by Direct Solution Methods for Optimal Control Problems," *Optimization Methods and Software*, Vol. 22, No. 3, 2007, pp. 433–452.

[81] Bokanowski, O., Forcadel, N., and Zidani, H., "Reachability and Minimal Times for State Constrained Nonlinear Problems Without Any Controllability Assumption," *SIAM Journal on Control and Optimization*, Vol. 48, No. 7, 2010, pp. 4292–4316.

[82] Mitchell, I. M., "The Flexible, Extensible and Efficient Toolbox of Level Set Methods," *Journal of Scientific Computing*, Vol. 35, 2008, pp. 300–329.

[83] Bayen, A. M., Mitchell, I. M., Oishi, M. M. K., and Tomlin, C. J., "Aircraft Autolander Safety Analysis Through Optimal Control-Based Reach Set Computation," *Journal of Guidance, Control, and Dynamics*, Vol. 30, No. 1, 2007, pp. 68–77.

[84] Lombaerts, T., Schuet, S., Wheeler, K., Acosta, D. M., and Kaneshige, J., "Safe Maneuvering Envelope Estimation Based on a Physical Approach," *Guidance, Navigation, and Control and Co-located Conferences*, AIAA, 2013, pp. 1–20.

[85] Baier, R., Gerdts, M., and Xausa, I., "Approximation of Reachable Sets Using Optimal Control Algorithms," *Numerical Algebra, Control and Optimization*, Vol. 3, No. 3, 2013, pp. 519–548.

[86] Diepolder, J., Heise, C., Bittner, M., Rieck, R. M., Grueter, B., Holzapfel, F., and Ben-Asher, J. Z., "Optimal Control Based Tracking Error Estimation for Model Reference Adaptive Control," *57th Israel Annual Conference on Aerospace Sciences*, Haifa, Israel, 2017, pp. 1–15.

[87] Heise, C. D., "Survivable Flight Control with Guaranteed Stability and Performance Characteristics," Ph.D. Dissertation, Technical Univ. of Munich, Munich, 2017.

[88] Luenberger, D. G., *Optimization by Vector Space Methods*, John Wiley and Sons, Hoboken, NJ, 1997.

[89] Braun, D., Schwaiger, F., Holzapfel, F., Diepolder, J., and Ben-Asher, J. Z., "Continuous Integration of Optimal Control Based Flight Control Law Clearance," *AIAA Scitech 2022 Forum*, San Diego, CA, 2022, pp. 1–12.

[90] Braun, D., *COLIBRY Manual v1.0.1*, Institute of Flight System Dynamics, Technical Univ. of Munich, February 28, 2023, www.fsd.ed.tum.de/software/colibry.

[91] Braun, D., Schwaiger, F., Holzapfel, F., Diepolder, J., and Ben-Asher, J. Z., "COLIBRY - A Counter Optimization Library for MATLAB," *AIAA Scitech 2023 Forum*, National Harbor, MD, 2023, pp. 1–13.

[92] Blumenson, L. E., "A Derivation of N-Dimensional Spherical Coordinates," *American Mathematical Monthly*, Vol. 67, No. 1, 1960, pp. 63–66.

INDEX

Note: Page numbers followed by f indicate figure; those followed by t indicate table.

P

R

S

T

www.ingramcontent.com/pod-product-compliance
Ingram Content Group UK Ltd.
Pitfield, Milton Keynes, MK11 3LW, UK
UKHW020245230826
474451UK00003B/9

9 781624 107344